Astronomy and Cosmology: Historical Perspectives

Astronomy and Cosmology: Historical Perspectives

Edited by
Kane Martin

Willford Press
www.willfordpress.com

Published by Willford Press,
118-35 Queens Blvd., Suite 400,
Forest Hills, NY 11375, USA

ISBN: 978-1-64728-449-7

Cataloging-in-publication Data

Astronomy and cosmology : historical perspectives / edited by Kane Martin.
p. cm.
Includes bibliographical references and index.
ISBN 978-1-64728-449-7
1. Cosmology. 2. Astronomy. 3. Astrophysics. I. Martin, Kane.
QB980 .C67 2023
523.1--dc23

For information on all Willford Press publications
visit our website at www.willfordpress.com

Willford Press

Contents

Preface

I am honored to present to you this unique book which encompasses the most up-to-date data in the field. I was extremely pleased to get this opportunity of editing the work of experts from across the globe. I have also written papers in this field and researched the various aspects revolving around the progress of the discipline. I have tried to unify my knowledge along with that of stalwarts from every corner of the world, to produce a text which not only benefits the readers but also facilitates the growth of the field.

Astronomy is a natural science concerned with the study of celestial phenomena and objects. It elucidates their evolution and origin using chemistry, mathematics and physics. Nebulae, comets, planets, stars, moons and galaxies are among the prominent objects of interest within this field. The phenomena studied within astronomy include blazars, cosmic microwave background radiation, supernova explosions, quasars, pulsars and gamma ray bursts. Cosmology is a field of astronomy that studies the nature of the universe. Some of the major branches of cosmology are physical cosmology and mythological cosmology. Physical cosmology focuses on the physical origins and evolution of the universe. Mythological cosmology is a branch of cosmology which incorporates mythological, religious, and esoteric literature and traditions of creation and eschatology. This book includes some of the vital pieces of work being conducted across the world, on astronomy and cosmology. The topics included herein are of utmost significance and bound to provide incredible insights to readers.

Finally, I would like to thank all the contributing authors for their valuable time and contributions. This book would not have been possible without their efforts. I would also like to thank my friends and family for their constant support.

Editor

1

Pedro Sánchez Ciruelo: A Commentary on Sacrobosco's *Tractatus de sphaera* with a Defense of Astrology

Tayra M. C. Lanuza Navarro

Abstract This paper focuses on the figure of the Spanish scholar Pedro Sánchez Ciruelo (ca. 1470–1548), who published a commentary on Johannes de Sacrobosco's *Sphaera* in 1498. The aim of this study is to analyze the defense of astrology that Ciruelo prepared for the preface of his commentary in order to show that this praise of astrology represents a significant part of the scholar's intellectual interests, which remained constant throughout his career. It is possible to get a richer understanding of Ciruelo's defenses of astrology in several of his works by placing them in their own specific context: the first one in the context of the prohibition of Simon de Phares's activities in Paris in 1494, and the second one in Europe in the first decades of the sixteenth century, during which those interested in astrology were mainly concerned with Pico della Mirandola's *Disputationes* and its consequences for the discipline.

1 Introduction[1]

The Spanish scholar Pedro Sánchez Ciruelo (ca. 1470–1548) published his commentary on Johannes de Sacrobosco's (died ca. 1256) *Tractatus de sphaera, Uberrimum Sphere mundi commentum,* in 1498 (Ciruelo 1498). At that time, he was in Paris, teaching mathematics and studying theology at the University of Paris. Ciruelo had arrived in Paris in 1492, after having spent 10 years at the University of Salamanca. There he was trained at the faculty of arts in the quadrivium, particularly in astrology, as well as in moral and natural philosophy and metaphysics, while teaching disciplines of the *trivium* simultaneously.

---

[1]The research for this article was funded by the 2016–1017 Fellowship Programme of the Herzog August Bibliothek (Wolfenbüttel) and by the Mellon Fellowship (2014–2015) at Villa I Tatti, The Harvard University Center for Italian Renaissance Studies.

Many aspects of Ciruelo's commentary on the *Sphaera* are of interest. First, his insertion of the fourteen *Questions on the Sphere* of Pierre d'Ailly (1351–1420) to complement his own comments and his additions to some of the subjects treated in the text, such as the definitions of the sphere, the number of celestial spheres, or the rise and setting of the signs. Also, the dialogue at the end of the work, where he discussed the role of authorities in a discipline, and his decisions when writing the commentary. And his defense of the certainty of mathematics, and of astrology within the mathematical sciences.

In this paper, I am going to focus particularly on this last aspect, Ciruelo's defense of astrology in the preface of his commentary on Sacrobosco. However, I will also refer to his inclusion of Pierre d'Ailly's *Questions,* to some of the issues he discussed in the comments in addition to what was originally in the *Tractatus de sphaera*, and to some of his explanations in the dialogue, which are related to his interest in astrology. One of the objectives of this paper is to show that Ciruelo defended this discipline from the beginning of his career, while in Paris. Another aim is showing that the preface of the *Sphaera* is not an isolated piece. The text of the preface in praise of astrology represents a significant part of this scholar's intellectual interests, which remained constant throughout his career. The arguments he introduced in the *Sphaera* are present in several of his later works. In 1521, Ciruelo published a complete treatise titled *Apotelesmata astrologiae Christianae,* in which he reflected on his previous assertions and extended his arguments in defense of astrology (Ciruelo 1521).

Despite several historians' emphasis on Ciruelo's relevance as a scholar, the studies considering his figure and works have passed over his astrological ideas—mentioning this part of his intellectual endeavors as an afterthought, if not completely dismissing or ignoring it (Albares 1996; Ayala 1993; Rodríguez Vidal 1981; Lorente Pérez 1921; Flórez et al. 1989, 1990).[2] Theology and becoming a respected theologian were the main focuses of his career.[3] However, astrology and the mathematical sciences in general were indeed some of Ciruelo's scholarly interests.

This paper concentrates on his astrological interests for two reasons. On the one hand, ignoring the influence that the mathematical sciences and astrology had on his career—as is the trend to date—misrepresents his intellectual ventures. On the other hand, I argue, his theological interests were not separated from his astrological ideas. His aim of establishing a clear definition of what constituted a Christian astrology and of differentiating it from non–orthodox practices and superstition, following the Thomistic tradition, was strongly linked to his theological ideals. Both aspects need to be taken into account to understand Ciruelo's writings, and specifically the preface to Sacrobosco as he wrote it in the Paris edition of 1498.

[2]An exception to the attitude of passing over Ciruelo's astrological ideas is Lou Ann Homza's study, which is focused on Ciruelo's religious attitudes and theological works, and therefore does not delve into his astrological interests. See (Homza 1992, 2000).

[3](Ciruelo 1521, f. A2r): "divinam theologorum facultatem..., ad quam totis praecordiis aspirabat."

2 Social and Intellectual Background

Pedro Sánchez Ciruelo was born in the town of Daroca, in the kingdom of Aragon, between 1460 and 1470. Daroca was not a small village in a remote area, as it has been represented in previous studies related to this scholar. It was a city of political, military, and administrative significance for the kingdom, albeit one of small size (Rodrigo Esteban 1999, 11). Daroca was one of the ten biggest towns in Aragon, which ruled a community of villages around it, and by 1495 the census listed 482 *fuegos*, or ca. 2.410 inhabitants (Serrano Montalvo 1995–1997). It had a weekly market and three prosperous annual trade fairs (Rodrigo Esteban 1999, 14, 27; Mateos Royo 1997). It had a collegial cathedral and an arts study particularly relevant to Ciruelo's career.

Both his family and the Darocense Study of Arts played a significant role in Ciruelo's later ambitions. In the biographical information he provided in the prologue to the astrological work *Apotelesmata*, Ciruelo stated that he was an early orphan, and that he was poor and alone in the world.[4] This was hardly the case, not only because a good number of relatives were alive and living in the city of Daroca or the neighboring town of Molina during the 1460s and 1470s (Homza 1992, 24), but also because of the cost of education.[5] Each student had to pay an annual tax to be admitted to the Study (which was twenty-eight *sueldos* in 1525), and although students from Daroca were exempt from paying the five *sueldos* due to the director, and probably also from paying the monthly payment of thirteen more for food and lodging if they did not use it, Ciruelo would hardly have been able to go through the years of schooling without family help or patronage.[6]

Daroca's *Studium Artibus* had been assigned a master to teach "Arts, Logic, and Nature" from 1381 on, as well as a teacher of theology from the local Franciscan convent (Rodrigo Esteban 2004, 144). Even if it was referred to as *Estudio General* in the royal approval of 1381, it is difficult to know to what extent education there went beyond that of a regular grammar school. It was, in any case, the place where the members of the city oligarchy received their education. Sons of merchants and liberal professionals made up most of the student body, but members of more modest social groups were not unheard of. Ciruelo's family does not seem to have been wealthy, as the professions mentioned in the Inquisitorial files are the likes of

[4] (Ciruelo 1521, f. A2r): "mihi humili solo nato, omni etiam humano auxilio parentum et patriae destituto."

[5] Even if Ciruelo's paternal grandparents (Francisco Sánchez Ciruelo and his wife) and his parents (Hernán Ciruelo and his wife) had died when he was a child, he could count with his two paternal uncles (Bartolomé and Alonso Ciruelo) and also an aunt (Francisca) to take him in, as well as his brothers Bartolomé and Jorge, and maybe a sister. He had also, or would later have, several cousins and nephews.

[6] On the cost of education in Daroca, see (Mateos Royo 2002).

shoe–maker (his uncle Bartolomé) and "meat cutter" (his cousin Francisco).[7] So, it would seem he was not really alone in the world, but was probably more or less poor. It is true, however, that he was not the only member of the family to attend university. Whether his studies and those of his brother Bartolomé, who ended up as "maestre" in Alcalá, were paid by his relatives, by the ecclesiastical authorities, or by the city council, which had an effective influence over the Arts Study, is unknown.

Ciruelo's preface to the *Apotelesmata* was of course an exercise in self-promotion, in self-fashioning, as were many dedicatory letters, introductions, prefaces, and prologues in the period.[8] He was creating an image of himself in a piece addressed to the professors of his *alma mater*, the University of Salamanca. In that image, he presented himself as a successful mathematician and theologian who had overcome hard beginnings for the sake of knowledge.[9] He also made a point of explaining that he had funded his university studies by teaching a discipline while studying another one: he used his knowledge of the *trivium* to teach dialectics in Salamanca while he took the courses to become a Master of Arts, and then used his knowledge of mathematics and astrology to teach in Paris while he studied theology.

In promoting himself as a self–made scholar, as an expert in orthodox doctrine, and as the right person to consult when it came to deciding about astrological practices, Ciruelo was also avoiding mentioning his relatives, maybe because of their *converso* origins.[10] He was a descendent of Jews and of judaizers, according to the Inquisitorial files. His grandfather, Francisco Sánchez Ciruelo, had been condemned by the Inquisition for apostasy. His uncle Bartolomé was a *confeso,* that is, he was a Catholic who had confessed reverting to Jewish practices or beliefs.[11] The sons of Bartolomé, Ciruelo's four cousins Benito, Cristóbal, Francisco and Pero, had been forbidden to work in public positions for this reason, making it clear that the family had Jewish ancestors. In 1553, after Ciruelo's death, the Inquisition prosecuted his cousin Benito for heretical blasphemy, and also his nephew Juan Ciruelo, son of his cousin Cristóbal (*via* his other uncle Alonso), for insulting the labor of the Inquisitors.[12]

---

[7] The genealogy of the family has been preserved because of the trials of the Inquisition of his nephew Juan and his cousin Benito in 1553, both of whom mentioned Ciruelo during the lawsuit. Albares presented Ciruelo's family as wealthy, but the Inquisitorial files do not seem to support such an assessment (Albares 1996, 181).

[8] For recent considerations of self-fashioning in the Early Modern University, see the articles in (Kirwan 2016).

[9] The whole sentence was written to present his 'avid search' for knowledge and early struggles: "ut litteras a me quasi fugientes toto pene orbe avidissime persequerer, mihi humili solo nato, omni etiam humano auxilio parentum et patriae destituto foret nimium difficile" (Ciruelo 1521, f. A2r).

[10] "He seems to have tried deliberately to mislead his audience about his family," as put by Lou Ann Homza (1992, 24).

[11] The files of the trials of the grandfather, Francisco Sánchez Ciruelo, and of his wife, whose name does not appear but is also mentioned as a judaizer, as well as the file of the trial of Bartolomé Ciruelo, *confeso*, have not been found to date.

[12] The files are at the Archivo Diocesano de Cuenca, Inquisición, leg. 193, exp. 2175 and 2181. The records of these two trials were summarily described by (Homza 1992, 24–27).

There is no way to know if this family history was one of the motivations behind Ciruelo's insistence in Catholic orthodoxy, or his determination to become a respected theologian and to obtain ecclesiastical positions, which was not unusual among members of *converso* families (Homza 1992, 101–02). What is certain, however, is that he kept ties with his family and with Daroca throughout most of his life.

After his years in Paris, he returned to Daroca on several occasions to stay for months or years. The significance of the Darocense Study of Arts for him is attested to by his acceptance of the study's director position, as well as the way he managed to coordinate this position with his chair at the University of Alcalá for nearly 20 years.

The first time he accepted the position seems to have been at the end of 1499, when he was still supposed to be in Paris, but after only a few months he renounced and left Daroca (Rodrigo Esteban 2004, 146). He might have gone back to Paris, as it is unknown where he was up until his appointment as professor of philosophy at the College of San Antonio de Portaceli—which also made him canon at the cathedral of Sigüenza—from 1502 to 1505 (Lorente Pérez 1921, 276). Then the city council of Daroca, in agreement with the archbishop of Zaragoza, in whose power lay the appointment of the director of the Study of Arts, offered him the position again. Ciruelo accepted it, as well as the later offer to become the teacher of theology, which improved his salary. His presence as a teacher in Daroca is supposed to have caused the huge increase in the number of students in 1505, which spiked to 150 from around 30 in 1496 and 1499 (Mateos Royo 2002, 131).

He remained as *maestro mayor* of the Study for four academic courses, during which he worked to obtain from the city council funding to make reforms and improvements to the buildings of the school, and wrote new statutes for the institution (Rodrigo Esteban 2004; Mateos Royo 2002). After his appointment as professor of Thomistic theology at the University of Alcalá de Henares in 1510, he did not quit his teaching position in Daroca.

Ciruelo managed to remain linked to the Study of Arts of his hometown for many more years. He continued to teach at least part-time in Daroca between 1508 and 1516. By dividing his salary with two assistants and often asking the city council for permissions and leaves, he combined these positions with the chair in Alcalá, with the administration of the vicariate in one of the villages of Daroca (Cuencabuena), and with preaching occasionally in Zaragoza (Rodrigo Esteban 2004, 152). He was also involved in the church of Santo Domingo in Daroca, where his brother Bartolomé was the vicar, after they both had been *racioneros* (canons with right to portion) there for some years.

He definitely left his positions at the Study of Daroca, from which he had been mostly absent for some years already, in 1523, but his links with Daroca and the Study were never forgotten. He went back there in 1524 when his brother died, and in his own testament he donated to the city 400 *sueldos* for works and maintenance of the Study of Arts (Rodrigo Esteban 2004, 153).

3 University Training and Works

Ciruelo's studies at the University of Salamanca, where he arrived around 1482, seem to have inclined him towards logic and the mathematical sciences, and initiated his interest in astrology. He found in Salamanca an intellectual environment with a concentration on astronomy and astrology, strongly influenced by the figure of Abraham Zacut (1452–ca. 1515) and his astronomical tables, even after Zacut left the city in 1485 (and the country in 1492, forced by the expulsion of the Jews). Zacut wrote two of his astrological works, the *Tratado breve de las influencias del cielo* (*Brief treatise on the influences of heavens*) and *De los eclipses del sol y la luna* (*On the eclipses of the sun and the moon*), in 1486, not long after leaving Salamanca. These were also well known among professors and students at Salamanca, as Zacut always stayed in contact with his colleagues, who were professors at the university. The first translator of Zacut's *Great Composition,* later known as the *Almanach perpetuum,* including the tables and the canons to use them, was Juan the Salaya (ca. 1450–ca. 1524). Salaya was teaching logic while Ciruelo studied at Salamanca, meaning he taught Ciruelo one of the disciplines on which he later focused. Before taking the chair of logic, which he obtained in 1469, Salaya had been the professor of astrology since 1464.

The professors of the chair of mathematics and astrology who taught in the last two decades of the century followed Zacut's and Salaya's path. Besides Salaya's lectures on logic, Ciruelo attended the lectures on astrology by one of them, Diego de Torres (ca. 1435–1496), who was the professor from 1482 to 1496. He also met and admired the subsequent professor, Rodrigo Vasurto (ca. 1470–1507), who had the chair of astrology from 1496 to 1504, and who might have previously been the professor of natural philosophy. He wrote a work on the calendar, and also a *Praxis pronosticandi* in 1497 (Flórez et al. 1989, 233–38; Thorndike and Kibre 1963, 504). Diego de Torres wrote an astrological commentary, which remained manuscript, and he seems to have been one of the translators of Sacrobosco's *Sphaera,* writing a version in Spanish around 1484 (Flórez et al. 1988, 22, 1989, 66–75; Burgueño 2009, 8; Gómez Martínez 2006).[13] Ciruelo's appreciation of Zacut's and Vasurto's knowledge is mentioned in several of his texts. In the prefatory letter of his 1521 work *Apotelesmata,* he referred to Zacut among those who had studied the sphere and created astronomical tables, praising the *Almanach,* which gave easy access to the information on the course, place, and aspects of the fixed stars and the planets.[14]

[13] The Spanish translation by Torres of the *Sphaera* is found in manuscript 3385 at the Biblioteca Nacional de Madrid, ff. 189r–199v, and the commentary as well. For other translations of Sacrobosco's *Sphaera,* see (Crowther et al. 2015). For translations into Spanish, and also into Portuguese (Chap. 7), and for the Latin version of André de Avelar and the genre of the *Reportorios de los tiempos* in Spain and Portugal, see (Chap. 10).

[14] (Ciruelo 1521, a2v): "Si eruditioni sphericae a nobis accepte Almanach cum perpetua tum et temporalia superaddant, quae omnes cursus, situs, aspectus, siderum sive fixorum sive errantium brevissime diurneque ostendunt sine ambagibus ac laboribus excessivis. Haec eum Zacutus vester Salmanticensis, Jonannes Nurembergensis et quidam alii recentiores divino quodam ingenio elaborata posteris reliquerunt."

In the *Dialogue* at the end of the commentary to the *Sphaera*, he expressed his admiration for Vasurto, "an Alphonse of our own times."[15]

With this mathematical and astrological background, Ciruelo moved to Paris in 1492 with the aim of studying theology. The intellectual circle he joined there has been studied with an emphasis on those known as the Spanish *calculatores,* a group of mathematicians who later tried to integrate mathematical knowledge into physical science in Spanish universities. They were also students of theology in Paris. This group included Juan Martínez Silíceo (1477–1557) and Fernando de Encinas (died 1523), and later Gaspar Lax (1487–1560) and Fernán Pérez de Oliva (ca. 1494–1531).[16]

Ciruelo also found in Paris a scholarly milieu marked by Jacques Lefèvre d'Étaples's (ca. 1455–1536) vindication of mathematics in his fight against Scotism at the University of Paris, and by his influence on the studies of mathematics and on the printing of mathematical works (Pantin 2013, 23).[17] Ciruelo was following Lefèvre d'Étaples's example when he printed his commentary to the *Tractatus de sphaera* only 3 years after the Parisian professor had published his own in 1495. Ciruelo seems to confirm that this was the case in his own words in the new version of the preface to the *Sphaera* of Sacrobosco that he prepared for publication in Alcalá de Henares in 1526 (Ciruelo 1526). It has been pointed out that Ciruelo seems to have shared some of the ideas expressed by Lefèvre concerning, for instance, the reality of the celestial orbs (in this case, an idea also held by Pierre d'Ailly) (Barker 2011, 15–16) (Chap. 6). The extent to which both shared mathematical notions and natural philosophical ideas on the heavens is a subject still waiting to be explored in depth.

During this period in Paris, Ciruelo published most of his mathematical works: an *Arithmeticae practicae* in 1495, the *Arithmetica Speculativa* in the same year, and the *Geometria Speculativa* by Thomas Bradwardine (ca. 1300–1349) (Ciruelo 1495a, b, c). In 1498, Ciruelo published the commentary on Sacrobosco, which was reprinted in Paris in 1508 and 1515, and in Alcalá in 1526.[18] His interest in preparing materials for teaching the mathematical liberal arts did not end after he left Paris. His course on them (*Cursus quattuor mathematicarum artium liberalium*) was published in 1516, when he was back in Spain (Ciruelo 1516).

[15] (Ciruelo, 1498, n7r.): "Placeret utique si prius hec Rodericus noster Vasurtus rescivisset, qui ita profunde ut nosti mathematica hec callet ut alter nostris temporibus Alphonsus appareat, adde etiam quod nostri amantissimus est et nisi quod verum cognovisse consuluisset nichil."

[16] On the Spanish *calculatores* in Paris, see (Wallace 1969, 1981, 79–90; Rey Pastor 1926, 72–81; Villoslada 1938; Calderón 1991, 247–48; Navarro Brotons 2014, 26–28, 58–72). See also (Elie 1950–1951). On a few of Ciruelo's mathematical notions, see (Cajori 1928; Lorente Pérez 1921). On the interest of these Spaniards in logic, see also (Noreña 1975).

[17] On the mathematical context in Paris at the moment, and on Lefèvre's plans for mathematical reform, see (Chap. 2).

[18] Different editions are sometimes mentioned but it does not seem possible to find them. For the editions of the commentaries on Sacrobosco's *Sphaera*, see the database accessible through the website of the project *The Sphere. Knowledge System Evolution and the Shared Scientific Identity of Europe*.

Later in his life, after his ten years of theological studies in Paris, Ciruelo went back to Spain and in 1510 became professor of theology, in the chair called "of Saint Thomas," at the newly created University of Alcalá de Henares. There, while teaching Thomas Aquinas's (1225–1274) theology and philosophy, Ciruelo published most of his works on logic. These included an edition of Aristotle's (384–322 BCE) logic in 1519 and two commentaries on the *Posteriora Analytica* in 1528 and 1529 (Ciruelo 1519, 1528). While in Alcalá, he also published his major work on astrology, the above mentioned *Apotelesmata*, in 1521, as well as an *Introductio astrologica* in 1523.

Ciruelo would leave the university in 1533 to hold ecclesiastical positions, first in Segovia, and then in Salamanca, where he became the canon preacher of the cathedral. There are no data about him teaching at the university there; however, he certainly continued publishing. This period was focused on religious texts, with sermons, reflections on theological mysticism, and translations of the Bible (Flórez et al. 1990, 18–19; Homza 1992, 175–208).[19] It was in 1537, while living in Salamanca, that he also published what has been labeled his most mature work on logic, his *Summule Petri Hispani* (Muñoz Delgado 1967, 197).

The *Cursus mathematicarum* of 1516 included Bradwardine's *Arithmetic* and *Geometry* with some comments extracted from Jordanus Nemorarius (active ca. 1222) and Campanus of Novara (ca. 1220–1296) (Lorente Pérez 1921, 314), as well as John Peckham's (ca. 1230–1292) and Alhacen's (Ibn–al–Haytam, ca. 965–ca.1040) *Perspective*, and the music of Lefèvre d'Étaples, with Boethius's (active ca. 525) theory.[20] In the prologue of this *Cursus*, which he dedicated to the rector and scholars of the University of Alcalá de Henares, Ciruelo explained that his intention was to serve the university by providing it with an introduction to the four mathematical doctrines mentioned. To these he seemed to intend to add a fifth piece, an astrological introduction. He referred to this work as already written, explaining that it was a small work on the *Sphaera* based on Ptolemy's (ca. 100–ca. 160) *Almagest*, on Zacut's *Almanach*, "and on our own judicial *Quadripartitum* and *Centiloquium*."[21] Ciruelo's "own *Quadripartitum* and *Centiloquium*" are the components of the *Apotelesmata Astrologiae Christianae*, which he printed in 1521. This seems to suggest that the work, or a shorter earlier version of it for students, was already written in 1516. It could also mean that the 1523 *Introductio astrologica* listed among his works—of which no copies are left—was this fifth part of the mathematical course, and that it could have been a summary of the *Apotelesmata*.[22]

[19] On Ciruelo's translations, see also (Valle 1999).

[20] On how the *Cursus* was based on Lefèvre d'Étaples's works, see (Oosterhoff 2018, 108–11).

[21] (Ciruelo 1516, a3r.): "Quinta quoque erit introductio astrologica, quam in sphericum opusculum olim condideram correspondentem *Almagesto* Claudii Ptholomei clarissimi, et *Almanach perpetuum* Zacuti Salmanticensis, cum nostro judiciario *Quadripartito* ac *Centilegio*."

[22] The work was recorded by (Picatoste Rodríguez 1891, 46–52; Latassa Ortiz 1798, 182–91). See the catalogues of Ciruelo's works in (Navarro Brotons et al. 1999, 94–96; Wilkinson 2010, 128–30).

If this was his intention—to create a fifth book with the basics of astronomy and astrology—it would explain why he did not include such knowledge in the *Cursus*.[23]

The kind of defense of astrology which Ciruelo decided to include in his commentary on Sacrobosco's *Sphaera* is different from that of the *Apotelesmata*. This, I argue, is due to the context in which he wrote each text, not in the sense implied by Thorndike, who asserted that the prohibitions of judicial astrology by the Paris faculty would have prevented him from openly standing in defense of the discipline in 1498 (Thorndike 1941, 275–79), but precisely the opposite. It gave him a reason to place his arguments for astrology in the *Sphaera*, and these arguments were later the base for the expanded defense constituted by the *Apotelesmata*. Before discussing this issue, I will describe the actual contents of the preface to the *Sphaera*, and their relationship to the other elements of the commentary mentioned above.

4 The Commentary on Sacrobosco's *Tractatus de sphaera*, the Preface, and It's Context

Ciruelo's edition of Sacrobosco was composed of three parts. It started with the preface containing the defense of astrology, and it ended with a dialogue discussing what would be a crucial theme in sixteenth century works on natural philosophy: the authority of ancient and highly-esteemed medieval scholars versus new ideas and information.

The central part of the work is composed, as expected, by Sacrobosco's text and Ciruelo's glosses. This is where Ciruelo inserted the fourteen *Questions on the Sphere* of Pierre d'Ailly. The text of the *Questions* is placed after Ciruelo's own comments to Sacrobosco. Thus he could introduce issues which were not found originally in the *Sphaera,* or add later opinions on a subject. Many of them were issues which had been regularly discussed by natural philosophers between the thirteenth and the fifteenth centuries, such as the existence of void, the movement of the *primum mobile*, the physical existence of the planetary spheres, the necessity of epicycles and eccentrics, the habitability of all zones of Earth, etc. (Grant 1996). Concerning this last issue, Ciruelo introduced in the commentary news on the travels of Columbus, pointing out that the experience of Spanish and Portuguese sailors had shown that there were territories near the equinox which were inhabited.[24] From the first trip "they brought back with them men of that region," he explained. Ciruelo's description here of the inhabitants of those tropical regions, men who were "not tall but...agreeable...smiling...friendly...with a dark color" (*ceruleo colore*) might

[23] On the teaching of mathematics in Spanish universities during the sixteenth century, see (Navarro Brotons 1995, 1998, 2006).

[24] (Ciruelo 1498, h2r–h2v): "Cum enim anno Christi domini 1491 illustrissimus hyspaniarum rex Fernandus expertissimos nautas versus occidentem equinoctialem ad insulae quererendas miserit; tandem post quattuor fere menses idem nautae reversi insulas multas sub equinoctiali vel prope dicunt se reperisse. In cuius rei testimonium multa genera avium exquisitissima multasque species aromaticas preciosissimas aurumque et homines illius regionis secum advexerunt."

Fig. 3.1 Illustration of a black man. The printer, Guy Marchand, had previously used this image in the widespread almanac *Calendrier des bergers* and other works, so its association with the contents of Ciruelo's *Sphera* is not clear. From (Ciruelo 1498). Biblioteca de Catalunya, Barcelona

have been the inspiration for the inclusion of the illustration of a black man, placed at the beginning of the work, before his preface (Fig. 3.1).[25] The decision must have been taken at the printer's workshop. Guy Marchand, who was printing it on behalf of Jean Petit, had previously used the same image in his printed version of the *Calendrier des bergers* in 1493, and in his edition of Savonarola (1452–1498)'s *Revelatio de tribulationibus* of 1496.[26] Whether Marchand decided to reuse the image because he associated Ciruelo's work with the astrological contents of the *Calendrier des bergers,* or as a call for repentance—as the image seems to be interpreted in Savonarola's work where the image includes the legend "Le maure de sales"—or in relation to the mentioned description, it is difficult to say. In Ciruelo's *Sphera*, the image was present only in the 1498 edition, not to be found in those of

[25] (Ciruelo1498, h2v): "Homines quidem illi non magne stature, sed periocundi, saepius ridentes et bone indolis facile omnibus credentes et aquiescentes, satis ingeniosi, ceruleo colore et capite quadrangulari hyspanis mirabiles apparverunt." The illustration is in fol. a2v.

[26] (Savonarola 1496, a1v). On the role of publishers in the iconographical program of the different editions of the *Sphaera* in the sixteenth century, see (Chap. 9). On the *Calendrier des Bergers*, see (Engammare 2008).

1508 and 1515 which were printed by Jean Petit (fl. 1493–1530) in Paris, and it was absent also from the 1526 edition of Miguel de Eguía (1495–1544) in Alcalá.

Initiating his *Sphaera* with a praise of mathematics was following the tradition of the Boethian claims about their nature and value, and certainly working in the mold of Lefèvre d'Étaples. Actually, Ciruelo started the preface by using the same biblical quotation used sometimes by Lefèvre d'Étaples: Romans, 1:20.[27] All human sciences provided men with knowledge of God, and a science had to be considered worthier if it did so with more certitude than others, he wrote.[28] Thus, he explained, mathematical doctrines had to be preferred to all other sciences after theology, because they fulfilled this role, and therefore had superior dignity (Fig. 3.2).[29] In this first part of the preface, Ciruelo was careful to quote only the Bible and Aristotle, mainly the *Metaphysics*, through Averroes's (Abū l-Walīd'Aḥmad ibn Muḥammad ibn Rušd, 1126–1198) words in some cases. Later natural philosophers and mathematicians, some also editing Sacrobosco's work, would include this same idea—that the study of mathematics and specifically of astronomy led to knowledge of God—among them Oronce Finé (1494–1555) and most famously Philipp Melanchton (1497–1560), who also used this idea to support the legitimacy of astrology (Almási 2014, 10–11).[30]

Other authors, like Nicholas Cusanus (1401–1464), had preceded Ciruelo in insisting on the Aristotelian (Averroistic) idea of the certitude of mathematics and the value of mathematical demonstration (Albertson 2014). These, combined with the also oft-repeated argument that mathematics was superior to other sciences in that it allowed no diversity of opinions, constituted the main ways to praise mathematics in this period, that is, before the 1540s and the burst of debate on the *Quaestio de certitudine mathematicarum.*[31] Ciruelo's work illustrate this previous stage of the question, as all three reprints of his commentary to Sacrobosco were published before the onset of that debate. Ciruelo chose this path to later substantiate the nobility and excellence of astrology not only as one of the mathematical sciences, but as superior to all others.

[27] Ciruelo's version was: "Invisibilia inquit dei per ea quae facta sunt a creatura mundi intellecta conspiciuntur sempiterna quoque virtus eius et divinitas." (Ciruelo 1498, a3v). On the quote and its meaning in Lefèvre d'Étaples, see (Oosterhoff 2018, 127).

[28] (Ciruelo 1498, a3v): "Ideo sciencia omnis a sapientibus huius seculi humanis parta laboribus viribusque omnipotentis et veri dei aliqualem agnitionem hominibus affert...et tanto scienciam aliquam digniorem existimare oportet quanto altius certiusque deum nobis insinuat."

[29] (Ciruelo 1498, a3v): "Cum igitur in mathematicis doctrinis horum utrumque manifeste appareat: merito post primam philosophiam ceteris humanis scienciis preferende sunt easque dignitate antecedent."

[30] On Finé's edition of the *Sphaera* see (Chap. 8). On Finé, see (Pantin 2013), and the articles in (Marr 2009). Specifically on the status of mathematics in Finé, see (Axworthy 2009, 2016), and on Melanchton's letter, see (Pantin 1987). On the theological significance of nature in the sixteenth century, see (Crowther-Heyck 2003).

[31] Ciruelo phrased it thus: "In aliis item scienciis propter earum minorem certitudinem doctorum diversitas et opinionum controversia locum habet; in mathematicis vero nequaquam." (Ciruelo 1498, a3v). On the *Questio*, debated later in the sixteenth century, see (Jardine 1984, 1988, 1998; De Pace 1993; Mancosu 1992, 1996, 8–24; Dear 1995, 34–42; Cozzoli 2007).

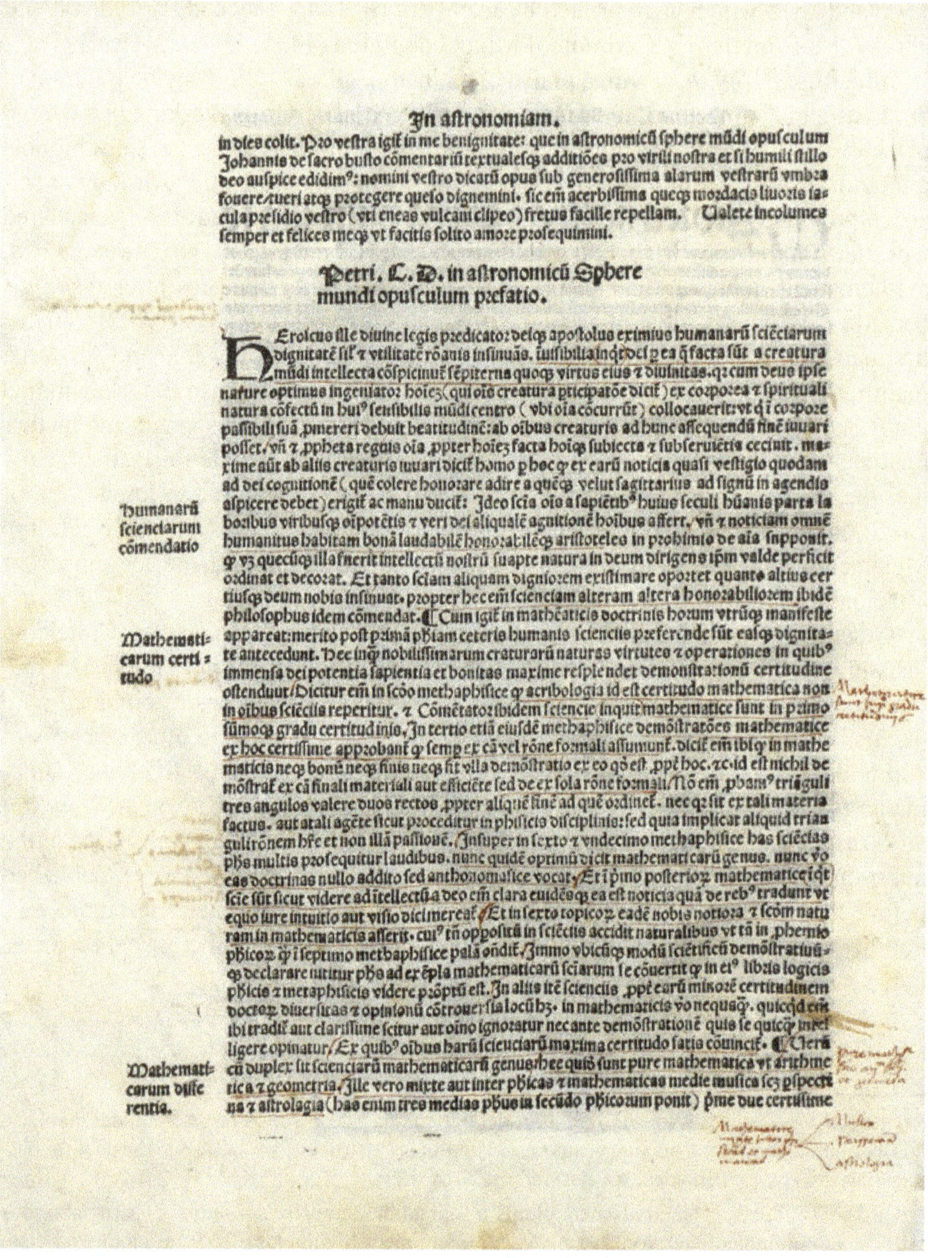

In astronomiam.

Petri. C. D. in astronomicū Sphere mundi opusculum prefatio.

Fig. 3.2 Preface. This first page contains Ciruelo's praise of mathematics as a superior discipline and his discussion on the issue of certitude regarding mathematical doctrines. From (Ciruelo 1498). Biblioteca de Catalunya, Barcelona

This part of the preface closed with Aristotle's categorization of the mathematical disciplines. Those dealing with abstract magnitude (number and quantity) were *pure* mathematical sciences: arithmetic and geometry. Those concerned with magnitude together with some specific natural matter were subordinate to them: mainly astronomy, astrology, and music, but also perspective and optics in general. These did not follow the Aristotelian rule that demonstrations could not jump from one kind of subject matter to another, belonging to what in the medieval period were often called middle (*medie*), intermediate, or, later on, mixed mathematical disciplines.[32] Aristotle's classification of the sciences and his discussion about the scientific status of the mathematical disciplines in the *Posteriora Analytica* and the *Methaphysics*, both of which Ciruelo cites, were the guides for the differentiation developed by medieval authors.[33]

The distinction between purely mathematical sciences and mathematical disciplines which applied mathematics to physical phenomena was common in scholastic philosophy (Galle 2003, 126–27) and provoked later debates on demonstration within the latter sort (Dear 1995; Laird 2016). Thomas Aquinas had explained that there were purely natural sciences, purely mathematical sciences, and *scientiae mediae*. The last received abstract principles from purely mathematical sciences and applied them to sensible matter.[34] On this basis, the Spanish scholar could insist on one of the ideas that appears crucial to his consideration of astrology: that it was demonstrative, like other mathematical sciences, but it also provided information on physical reality. He would reemphasize the importance of this Aristotelian classification again in the dialogue that closed the *Sphaera,* and in some of his other works. To further support this part of the preface, Ciruelo had included Pierre d'Ailly's *Questio Prima* at the beginning of the commentary. Ailly's commentary on the first part of Sacrobosco's first chapter (on the shape of the world) was partly focused on the nature of mathematics and how astrology was part of the mathematical sciences. The bishop's explanation of the two kinds of mathematical disciplines and the nature of perspective, music, and astronomy was summed up in the margin with the words "*mathematice medie vel mixte*" (Ciruelo 1498, b3v). In this part d'Ailly also explained the different branches of astrology, and what *astrologia iudiciaria* was.[35]

Because astrology and the other two mixed mathematical sciences were also physical disciplines, they were nobler, as they transmitted knowledge about natural things with the certitude of mathematical demonstration.[36] In the dialogue at the end

---

[32] Astronomy, astrology, music, perspective, and optics would be the disciplines which during the seventeenth century debates, at least from the publication of Francis Bacon's *Advancement of learning* (if not earlier), were called "mixed mathematics." See (Brown 1991; Dear 1995, 38–39).

[33] Aristotle, *Posterior analytics* I and II, particularly I.7; Aristotle, *Metaphysics* XI and XIII, particularly XIII.3. Also, Aristotle, *Physics*, I.1 and II.2. See (McKirahan 1978, Mandosio 1994).

[34] In the commentary on the *Physics*. See (Gagné 1969; Ribeiro do Nascimento 1974; Laird 2016).

[35] (Ciruelo 1498, b3v): "Quinta parts considerat de effectibus consequentibus coniunctiones vel oppositiones vel aspectus corporum celestium. Et ista vocatur astrologia iudiciaria, et etiam potest proprie vocari astrologia...."

[36] (Ciruelo 1498, a4r): "Hec igitur tres omnium particularium scienciarum nobilissime sunt, quia in eis simul phisice et mathematice pericie pulchritudo reperitur, qua enim phisice sunt obiecti pollent nobilitate, qua vero mathematice demonstrationis gaudent certitudine. Unde rerum naturalium non inductive et debiliter (ut alie sciencie permulte) sed demonstrative et scientifice cognitionem servata materie qualitate tradunt." The idea that among the branches of natural philosophy, astronomy was superior because of the perfection of its subject matter and because of the certainty of its mathematical nature, was a conclusion authors extracted from Aristotle's *Metaphysics,* XII, 8 (Jardine 1988, 145).

of the work, he repeated that *astrologia*, in this case meaning astronomy, was "physical more than mathematical," as it studied the movements of the celestial bodies.[37] The question of categorizing astronomy and astrology this way appeared in other prefaces to Sacrobosco's *Sphaera* (Almási 2014, 11). It was this mixed nature between physics and mathematics that made the science of the stars superior to other disciplines, a notion not unfamiliar to the context of sixteenth century astronomy, nor to later debates about the status of the discipline and the identity of the astronomer (Biagioli 1989, 1993, 211–44; Westman 1980, 2011, 202–9, 353–55).

The core of the preface focused on four arguments to support the assertion of the perfection of astrology among mathematical disciplines. These arguments would be reused by astrologers during the sixteenth and seventeenth centuries in defense of their art. The same questions about the dignity and nobility of astronomy and the value of astrology, although with different views concerning the last, were discussed later, for instance, by Francesco Capuano da Manfredonia (active fifteenth century) (Chap. 4).

Ciruelo's piece proceeded from the Aristotelian idea that the dignity of astrology/astronomy was elevated because its object of study, the celestial bodies, was superior.[38] This was the first place Ciruelo cited Ptolemy, and he was careful not to quote only the *Almagest*, but to mention the *Quadripartitum* as well. This delivered the message that his praise was not only for mathematical astronomy and the study of the movements of the heavens, but for the whole *science of the stars*, including the study of the influences of the heavens, i.e., astrology. Ciruelo had started using the term *astrologia* for the science of the stars in general. But he made sure to keep readers from interpreting his defense as strictly applying to astronomy, the study of the movements, and not necessarily to astrology, the discipline focused on influences. He used a good number of quotes from Aristotle's works on celestial bodies as causes of natural events on earth, on the hierarchical classification of bodies in nature, and on the inferior world as being ruled by the superior bodies of the heavens. Astrology, he insisted—and he had made clear he meant the study of influences as much as any other aspect of the science of the stars—exceeded all other disciplines, because the other disciplines studied only corruptible bodies; they were not focused on the study of the heavens and the whole universe.[39]

Ciruelo continued by stressing that astrology also had to be praised because of the method it used, the way it established knowledge. This was where he discussed the key issue of the study of causes. According to the Aristotelian principles, a discipline only qualified as *scientia* if it concerned the study of causes. The debate of

---

[37] (Ciruelo 1498, n5v): "Astrologiam, o amice, Aristoteles phisicam magis quam mathematicam ostendit, quod scilicet supremorum corporum motus potissimum perscrutatur."

[38] (Ciruelo 1498, a4r): "Quod enim rei digeste dignitate astrologia omnes alias particulares sciencias excedat hec res indicio erit, quia ut Ptholemeus in primo et Almagesti et Quadripartiti habet hec sciencia celorum et ornatissimorum astrorum naturas virtues motus et effectus speculator."

[39] (Ciruelo 1498, a4r): "Cum igitur alie omnes discipline aut nullam aut non ita propriam celorum et totius universi faciant theoriam sed earum precipua intentio circa corruptibilia versetur corpora palam est astrologiam eas omnes hac ex parte antecellere."

the *quaestio de certitudine mathematicarum*, very much focused on this issue as it pertained to mathematics, had not started yet—it developed later, in the sixteenth century. However, scholars were aware that a defense of a discipline had to deal with this question, particularly in the case of a mathematical discipline. Here, the assertion that it was a mixed mathematical discipline played the central role. As a partly physical science, it departed from the effects to get to know the causes. Ciruelo linked this to the previous argument by stressing that astrology investigated the highest causes.[40] The argument was rounded off with the assertion that astrology comprehended the infallible demonstrations of the other four mathematical disciplines. It used geometry to understand the dimensions of the heavens and arithmetic to know the ratio and rules of the celestial movements. The courses of the heavens, Ciruelo stated, were connected with a musical harmonic proportion, and astrology also made use of perspective to understand the influential rays and the aspects of the stars (which had effects on earth and on men).[41] Thus "the procedure of this science is more certain than the ways of all the other [sciences]," he concluded.[42] The fact that the first part of the preface was a praise of the disciplines of the quadrivium gave strength to this part of Ciruelo's defense of astrology.

The next argument was a traditional one: there had been great scholars who studied astrology in the past, and their intellectual relevance and their piety was proof of astrology's nobility. Using a history of mathematics, astronomy, or any scientific discipline was a strategy developed to justify its study and give legitimacy to it, which both humanist and scholastic writers used (Grafton 1997; Goulding 2006). The authors used such narratives to distinguish the acceptable from the unacceptable parts of their discipline, the licit from the illicit (Popper 2006), which was particularly fitting for a defense of astrology designed in the context of a prohibition such as existed in Paris after the Simon de Phares (1444–ca. 1499) affair (see below). Highlighting ancient authorities for mathematics—and astrology—the authors tried to contribute to the process of legitimizing such knowledge (Popper 2006, 106). This was actually the path chosen by Simon de Phares to defend astrology when some of his practices and books on astrology were condemned by the Parliament and by the University of Paris's faculty of theology. He had intended to radically distinguish magical, divinatory, and superstitious arts from astrology, an objective he shared with Ciruelo. He decided to describe the historical precedents,

[40] (Ciruelo 1498, a4r): "Ex suo etiam procedendi modo certissimo astrologia super omnes humanas sciencias laudatur, siquidem enim vigoris modus habet procedendi phisicus ex effectibus scilicet ad causas perveniendo hoc in astrologia que partim phisica est tanto excellentiori modo reperitur quanto altiores causas suo processu investigat."

[41] On musical harmony in nature, and the spheres, see (Voss 1998).

[42] (Ciruelo 1498, a4r–a4v): "Infallibilem etiam demonstrandi modum aliarum quattuor mathematicarum hec sola comprehendit. Ibi enim geometricis theorematibus...celorum et stellarum...comprehenduntur et mesurantur, motuum quoque celestium norme arithmeticis numerorum rationibus cognoscuntur. Tot etiam tamque differentes celorum cursus armonica musicave proportionalitate connectuntur...Non desunt etiam perspective demonstrationes han scienciam condecorantes, irradiationum enim et aspectum stellarum...perspective sunt....Huius igtur sciencie processus omnium aliarum modis certior est."

with an inventory of previous practitioners, which would prove the utility of the discipline and its respectability by way of those previous authorities (Boudet 1994, 10–11). Simon the Phares's *Élucidaire,* which became the *Recueil des plus célebres astrologues*, was still in the making when Ciruelo's *Sphera* went to print. Ciruelo used Josephus's (37–ca. 100) *Antiquities* to mention the tradition that attributed knowledge of astrology to Noah and Abraham, proceeding then to Hermes and Ptolemy, then medieval authors, citing among others Tebit (Thābit ibn Qurra, 826–901) and Alfarganus (Ahmad ibn Muhammad ibn Kathir al–Farghani, ca. 805–870) before getting to Sacrobosco himself and king Alphonse X of Castile, followed by Pierre d'Ailly, whom he couldn't fail to cite, and ending with the two most relevant scholars of fifteenth century astronomy, Peuerbach (1426–1461) and Regiomontanus (1536–1476).[43] "Enjoy the regal and imperial astrology, which is commended by the piety, dignity and sublimity of its authors," Ciruelo concluded.[44]

His last argument appealed to the kind of piety he had just mentioned in association with those authors. Astrology was a discipline which aimed at raising men to the knowledge of the work of God, he said.[45] Establishing the general notion of nature as a divine epiphany allowed for a justification of the sciences in general, and in this case of astrology in particular, as it fit coherently into such a view of nature. This was one of the most important ways to defend astrology when it was confronted with the principles of Christianity. It was obviously necessary to hold up astrology's *pietas,* meaning both its morality and its respect for divinity (Ernst 1991a, 251–53, 1991b). This argument was connected in the preface to the one insisting on the utility of astrological predictions, which was also a traditional defense of astrological knowledge.

Ciruelo had constructed a solid defense of astrology in this preface to the *Sphaera,* with some of the same arguments that would be repeatedly used by astrologers of the sixteenth and seventeenth centuries. His intention was not, however, to turn Sacrobosco's treatise into an introduction to astrology. The work was an introduction to astronomy, and, unlike Cecco d'Ascoli (1257–1327) and Robertus Anglicus (active ca. 1326), Ciruelo did not add the basics of astrology to his commentary.[46] There was Alcabitius for that (Abu al–Saqr Abd al–Aziz al–Qabisi, active ca. 950), namely his *Introduction to the Art of Judgments of the Stars*. In deciding to place Pierre d'Ailly's questions on the *Sphaera* after the chapters to which they applied, Ciruelo gave himself the chance to introduce some astrological content, such as a piece on the nature of the planets and their influences on men (Chapter one of the *Sphaera*, *Questio quinta* by d'Ailly). But Ciruelo's aim when choosing to include d'Ailly's ideas could also be related to the defense of astrology as a discipline. D'Ailly's writings demonstrated a fundamental interplay and essential

[43] On the literary sources of the *Sphaera,* see (Martins 2003).

[44] (Ciruelo 1498, a5r): "Gaudeat igitur regia et imperialis astrologia que tanta inventorum auctorumque suorum santitate, maiestate et sublimitate commendatur."

[45] (Ciruelo 1498, a5r): "Finis quoque huius sciencie pius strenuus et divinus est per creaturas scilicet sensibiles ad dei cognitionem homines elevare."

[46] On Cecco d'Ascoli's and Robertus Anglicus's commentaries, see (Thorndike 1949, 48–54).

harmony between his astrological and his theological beliefs (Smoller 1994, 22). His work *Concordantia astronomie cum theologia* (1414), which had been printed by Erhard Ratdolt (1442–1528) only a few years before in 1490, was one of the main references when authors attempted to resolve possible conflicts between the practice of astrology and the Christian religion (Smoller 1994, 122–30). The constant presence of d'Ailly's ideas in the commentary could be taken as a subtle reminder of the "concordance" of astrology and theology. Inserting d'Ailly's questions into the core of the text gave Ciruelo the chance to insist, after his first gloss on Sacrobosco, on the initial subject he had introduced in the preface, the place of astrology among the mathematical sciences (Chapter one, *Questio prima*).

The last part of the preface contained not only further arguments in defense of astrology, but specifically those which were needed due to what happened in Paris in 1494. While Ciruelo was there teaching the mathematical disciplines and studying theology, he witnessed the condemnation of astrological prognostications of a judicial nature by the University of Paris in 1494. The theology faculty was acting against the activity of Simon de Phares, condemning his astrological activities and part of his astrological library (Boudet 1994, 1999; Thorndike 1934, 544–61).

Simon de Phares was an astrologer who also claimed to be a physician.[47] He had established an astrological practice in Lyon, where he consulted clients and seems to have mainly exercised the branch of astrology known as interrogations. After a royal visit in 1490 and the subsequent fame, he was accused of using a "familiar demon" for his divinatory practice and arrested. This charge was later cancelled, but he was ordered to stop his practice of astrology, and his library was confiscated. He appealed to the Parliament of Paris in 1491. The Parliament asked the faculty of theology at the University of Paris to examine de Phares's library, at least his works on astrology. In 1494, the theologians of the university censored the content of eleven of his astrological works and published a notice condemning judicial astrology. This document was quite a wide condemnation of the discipline, or more specifically of nativities, interrogations, and elections. Knowledge of all the secrets of the future, said the Parisian theologians, as the Papal Bull of 1586 would say a century later, was reserved to God.[48] The theologians were actually focusing on the condemnation of interrogations and elections, and in that, Ciruelo agreed with them. He pointed out the same idea, that those two branches of astrology were to be forbidden, both in the commentary to Sacrobosco and in the *Apotelesmata*. However, despite how radical the notice of the faculty of theology seemed to early modern

[47] Boudet pointed out that there is no evidence of Phares receiving a medical education. For all the details on Simon de Phares's biography, activity, arrest, and condemnation, and for a transcription of the sources related to them, see (Boudet 1988, 1994, 1999). The document of the Parliament was published by (Plessis d'Argentré 1724, I: 324–30).

[48] (Plessis d'Argentré 1724, 325): "...eam nos mathematicam artem seu astrologiam divinatoriam appellamus, quae ex horarum natalium consideratione futuros hominum mores et qua item ex quarumcumque rerum vel operationum horis initialibus futuros eorum successus et casus, fortunas determinative et particulariter praedicere quae etiam cuiusvis interrogationis hora notata et veritatem respondere et omne prorsus cum praesens, tum praeteritum ac futurum secretum revelare ac manifestare se posse, quod divinitas proprium est."

Europeans in 1494, it specified that it was not condemning "true astronomy," which besides the movements of the celestial bodies "conjectured about some of their natural effects."[49] Again, Ciruelo agreed that there was a part of astrology which studied natural influences, and that it was a licit and true liberal art—this he established at length in the preface to Sacrobosco. He did not seem to agree completely with the wide prohibition the document implied, as he defended certain predictions using genethlialogy in his later work, as well as revolutions. Jean-Patrice Boudet explained that the trial of Simon de Phares developed in a repressive context in which any deviation from orthodox faith was addressed with severity, and that divination, including astrology, was associated with blasphemy, as can be seen in royal orders published during the period (Boudet 1999, 108–10). The University—and following its instructions, the Parliament—of Paris prohibited divinatory astrology (*astrologia divinatoria*), but Ciruelo thought that it was necessary to discuss which aspects of astrology were divinatory and which were natural. And those were the issues he addressed in the preface to Sacrobosco, after he made clear that not all astrology was to be considered judicial or divinatory, and that it was a valid, licit, and worthy discipline.

Pedro Ciruelo's main work in defense of astrology, *Apotelesmata astrologiae Christianae*, was not published until much later, in 1521, while he was back in Spain as professor of Thomistic philosophy in the very nominalist environment of the University of Alcalá de Henares. Thorndike assumed that Ciruelo published his work in defense of astrology after he had returned to Spain because he then felt on safer ground, due to the theological opposition found in Paris during the last years of the fifteenth century.[50] This idea is not consistent with the analysis of Ciruelo's strong defense of the discipline in the preface of his commentary to the *Sphaera* given above. Actually, I argue, it was precisely because of the situation in Paris that he decided to defend astrology so vehemently in his preface to Sacrobosco. He was not hiding his opinion on astrology, as his defense was very explicit, nor did he pretend that his words were unrelated to the condemnation of astrology resulting from the De Phares affair. He was addressing the problems created for the discipline because of the decision of the University of Paris about that case. He stated it both in the same preface and later again in the *Apotelesmata*. He explained that the same discipline he was praising had been the object of a recent decision by the theology faculty. The watchers of the Catholic truth wanted to keep all disciplines clean of falsity, so they had to condemn "certain writers, liars and corruptors of truth, who

[49] (Plessis d'Argentré 1724, 325): "Ne quis autem veram astronomiam, quae corporum coelestium magnitudinem, oppositiones, motusque considerat, quae Solis et lunae aliorumque planetarum conjunctiones vel oppositiones caeteras vel habitudines praedicit, quae item effectus quosdam eorum naturales probabiliter ac prudener in universali conjiciit, quam ut liberalem, nobilem, utilemque artem veneramur." On the distinction in this case, see (North 2009, 292).

[50] Thorndike expressed this theory as follows: "Whether Ciruelo expressed such favorable views towards astrology during his residence at Paris before 1502 may be doubted. At any rate he did not print them until 1521 in Spain, and one suspects that he felt somewhat freer to express himself on the subject there and surer of a sympathetic audience." (Thorndike 1941, 275–79).

had tried to mix the chaste astrology with courtesan sciences."[51] This was one of the main defenses on which astrologers relied: it was the specific practitioner who mixed false ideas and superstitions with his astrology who had to be condemned. Ciruelo made a point of this in the preface when making the reference to the condemnation of Paris of 1494. In the *Epistola proemialis* of the *Apotelesmata,* Ciruelo mentioned again the situation in Paris as he introduced his discussion on who was entitled to decide about the nature of astrology and its legitimacy, and on the training of theologians and their lack of knowledge of astrology.[52]

Further evidence that the notice condemning the books and practices of Simon de Phares in Paris in 1494 had a role in the way Ciruelo defended astrology in the preface of his commentary to Sacrobosco's *Sphaera* is the new version of it he prepared for the 1526 edition (Fig. 3.3). The *prohemium* to this edition was a shortened version of the original one. In it, Ciruelo praised Lefèvre d'Étaples and mentioned his experience in Paris making his living as professor of mathematics.[53] He explained that he wanted to make the new version easier to read for the students of the University of Alcalá, and to eliminate his mistakes and youthful verbosity (Fig. 3.4).[54] The context had changed completely in terms of his own status and that of astrology as a discipline. He would not have worried about the 1494 decision against Simon de Phares anymore, because it was by then a past event without much effect. After the publication of Pico della Mirandola's (1463–1494) work against astrology, hardly anything else would have rivaled its importance. In the decade of the 1520s, Ciruelo addressed this issue with the *Apotelesmata* and the *Reprobación.* The preface to Ciruelo's commentary to the *Sphaera* of 1526 did not mention De Phares. However, Ciruelo had not abandoned his objective of placing a defense of

---

[51] (Ciruelo 1498, a5r): "Nec tamen qui divinam astrologiam tot tantisque laudibus efferre conor determinationi cuidam alme facultatis theologie doctorum parisiensium super quibusdam mathematicis libris nuper edite inficior….Ideo predicti doctores et veritatis catholice zelatores volentes tam dominam quam eius ancillas omni falsitate purgatas inmunesque custodire: falsarios quosdam scriptores et veritatis corruptores (qui castissimam astrologiam cum scienciis meretriculis nixi sunt collocare eamque apud dominam suam de magno superstitionis crimine falsa quedam immiscentes diffamare tentarum) a generosissima astrologorum schola perpetuo eiiciendos et dampnandos iamque dampnatos pronunciarunt."

[52] (Ciruelo 1521, a2v): "Verum de iudiciis astrologicis tunc temporis nihil egimus Parisii, eo quod vulgaribus theologis essent invisa atque ludibrio exposita. Nam tota eorum facultas plures iudiciorum astronomicorum libros anno Christi domini 1494 examinans, quesdam illorum manifeste erroneos infide et moribus alios valde suspectos aut dubios damnavit." On Ciruelo presenting himself as the perfect judge for what was orthodox in astrology, see (Lanuza 2017, 9–14).

[53] (Ciruelo 1526, a2v): "Cum in adolescentia mea Parisiensium theologorum scholas famigeratissimas perlustrarem, atque ex professione matheseos mihi necessaria ad victum et indumenta compararem, accidit tunc Iacobum Stapulensem praeclarum certe nostra tempestate philosophum, egregium quendam edere in opusculum sphaericum Ioannis de Sacro Busto commentarium et mox ipsum excussoribus imprimendum trader."

[54] (Ciruelo 1526, a2v): "…hanc sphaericam meam elucubrationem revidere, corrigere, emmendare breviorem facere, atque in melius permutare, eliminata prorsus ab ea. omni prolixitate, omni barbarie omnique iuvenili loquacitate, sicque renovatam eisdem complutensibus calcographis et quidem artificiosissimis formanda tradere, ad usum nostrorum adolescentum hic et alibi philosophantium."

Fig. 3.3 Cover page. This copy at Seville is one of the many with marginal notes denoting the intense use of this commentary of Sacrobosco's *Sphere*. From (Ciruelo 1526). Courtesy of the Biblioteca Universidad de Sevilla

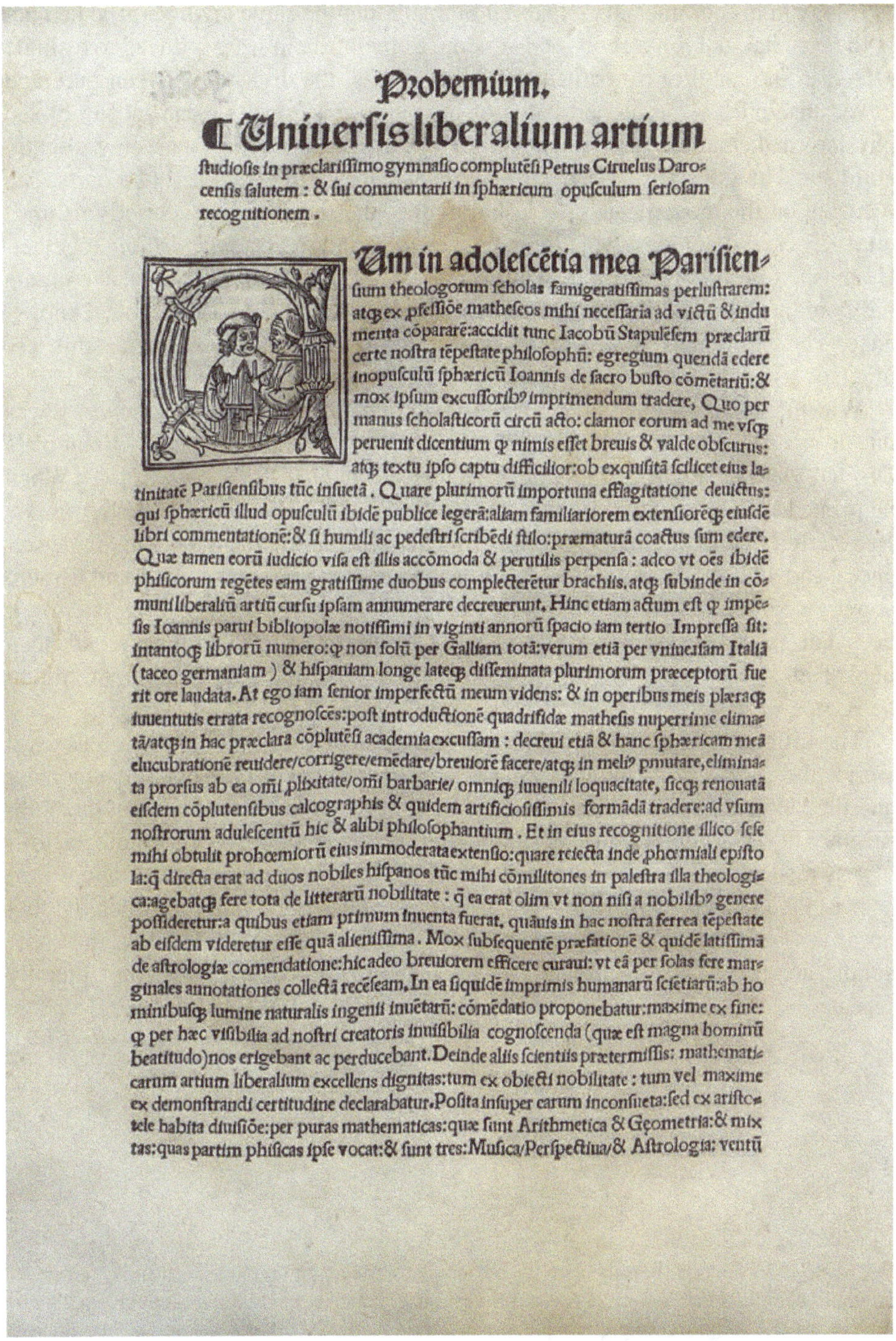

Prohemium.

¶ Uniuerſis liberalium artium ſtudioſis in præclariſſimo gymnaſio complutẽſi Petrus Ciruelus Daroceñſis ſalutem: & ſui commentarii in ſphæricum opuſculum ſerioſam recognitionem.

Um in adoleſcẽtia mea Pariſienſium theologorum ſcholas famigeratiſſimas perluſtrarem: atqꝫ ex pfeſſiõe matheſeos mihi neceſſaria ad victũ & indumenta cõparare: accidit tunc Iacobũ Stapulẽſem præclarũ certe noſtra tẽpeſtate philoſophũ: egregium quendã edere in opuſculũ ſphæricũ Ioannis de ſacro buſto cõmẽtariũ: & mox ipſum excuſſorib⁹ imprimendum tradere. Quo per manus ſcholaſticorũ circũ acto: clamor eorum ad me vſqꝫ peruenit dicentium ꝙ nimis eſſet breuis & valde obſcurus: atqꝫ textu ipſo captu difficilior: ob exquiſitã ſcilicet eius latinitatẽ Pariſienſibus tũc inſuetã. Quare plurimorũ importuna efflagitatione deuictus: qui ſphæricũ illud opuſculũ ibidẽ publice legerã: aliam familiariorem extenſioreqꝫ eiuſdẽ libri commentationẽ: & ſi humili ac pedeſtri ſcribẽdi ſtilo: præmaturã coactus ſum edere. Quæ tamen eorũ iudicio viſa eſt illis accõmoda & perutilis perpenſa: adeo vt oẽs ibidẽ phiſicorum regẽtes eam gratiſſime duobus complecterẽtur brachiis, atqꝫ ſubinde in cõmuni liberaliũ artiũ curſu ipſam annumerare decreuerunt. Hinc etiam actum eſt ꝙ impẽſis Ioannis parui bibliopolæ notiſſimi in viginti annorũ ſpacio iam tertio Impreſſa ſit: intantoqꝫ librorũ numero: ꝙ non ſolũ per Galliam totã: verum etiã per vniuerſam Italiã (taceo germaniam) & hiſpaniam longe lateqꝫ diſſeminata plurimorum præceptorũ fuerit ore laudata. At ego iam ſenior imperfectũ meum videns: & in operibus meis plæraqꝫ iuuentutis errata recognoſcẽs: poſt introductionẽ quadrifidæ matheſis nuperrime elimatã/atqꝫ in hac præclara cõplutẽſi academia excuſſam: decreui etiã & hanc ſphæricam meã elucubrationẽ reuidere/corrigere/emẽdare/breuiorẽ facere/atqꝫ in meli⁹ pmutare, eliminata prorſus ab ea omĩ plixitate/omĩ barbarie/ omniqꝫ iuuenili loquacitate, ſicqꝫ renouatã eiſdem cõplutenſibus calcographis & quidem artificioſiſſimis formãdã tradere: ad vſum noſtrorum adoleſcentũ hic & alibi philoſophantium. Et in eius recognitione illico ſeſe mihi obtulit prohœmiorũ eius immoderata extenſio: quare reiecta inde phœmiali epiſtola: q̃ directa erat ad duos nobiles hiſpanos tũc mihi cõmilitones in paleſtra illa theologica: agebatqꝫ fere tota de litterarũ nobilitate: q̃ ea erat olim vt non niſi a nobilib⁹ genere poſſideretur: a quibus etiam primum inuenta fuerat, quãuis in hac noſtra ferrea tẽpeſtate ab eiſdem videretur eſſe quã alieniſſima. Mox ſubſequentẽ præfationẽ & quidẽ latiſſimã de aſtrologiæ comendatione: hic adeo breuiorem efficere curaui: vt eã per ſolas fere marginales annotationes collectã receſeam. In ea ſiquidẽ imprimis humanarũ ſcĩetiarũ: ab hominibuſqꝫ lumine naturalis ingenii inuẽtarũ: cõmẽdatio proponebatur: maxime ex fine: ꝙ per hæc viſibilia ad noſtri creatoris inuiſibilia cognoſcenda (quæ eſt magna hominũ beatitudo) nos erigebant ac perducebant. Deinde aliis ſcientiis prætermiſſis: mathematicarum artium liberalium excellens dignitas: tum ex obiecti nobilitate: tum vel maxime ex demonſtrandi certitudine declarabatur. Poſita inſuper earum inconſueta: ſed ex ariſtotele habita diuiſiõe: per puras mathematicas: quæ ſunt Arithmetica & Geometria: & mixtas: quas partim phiſicas ipſe vocat: & ſunt tres: Muſica/Perſpectiua/& Aſtrologia: ventũ

Fig. 3.4 Preface. This first page of the 1526 edition can be compared to Fig. 3.2. It contains the same initial arguments and praise of mathematics, discussing the dignity of mathematical sciences, in a briefer and more concise structure. From (Ciruelo 1526). Courtesy of the Biblioteca Universidad de Sevilla

astrology in his commentary. He did it by including the same arguments he had put in the original, only in a shortened version. Some of them were reduced to a single sentence: the nobility of mathematical disciplines, the division between pure and mixed disciplines, the higher status of astrology due to its object and its mixed physico-mathematical nature. The genealogy of the discipline was only slightly abridged. The last part maintained that the science of the stars had two parts, one focusing on the movements of celestial bodies, the other on their (strictly natural) effects.[55] He did not leave out citing the *Almagest* and the *Quadripartitum* together. He finished this new version by insisting on the excellence of astrology, and on the utility of the *Tractatus de sphaera* of Sacrobosco to learn basic astronomical knowledge. The shortened preface of 1526, less prolix indeed, seems to transmit less urgency, which is consistent with its publication in a dissimilar context.

When Pedro Ciruelo published the *Apotelesmata astrologiae Christianae*, it was due to a different situation. He felt the need to answer Pico della Mirandola's *Disputationes adversus astrologiam divinatricem* and his attack on astrology. When he printed the commentary to the *Sphaera* he already knew some of Pico's works. He referred to him as a man who had arrived at the top of all human and divine sciences, and who shone as a new star. "What he ignores, nobody knows," said Ciruelo about Pico.[56] But when he made a first defense of the discipline in the *Sphaera*, it does not seem to have been due to the *Disputationes*, although they had been published already in 1496. He did it because of the condemnation of Simon de Phares.

The differences between the kind of defense that Ciruelo created in these two works (the commentary and the *Apotelesmata*) are due to the context of the moment and the kind of attack against astrology to which he was responding. But the commentary to the *Sphaera*, with its solid combination of arguments defending astrology by associating it with mathematics, and the *Apotelesmata,* where the aim was to answer Pico's work while renovating astrology, constitute a unity in which the first one functions like an introduction to the second. In the *Apotelesmata,* the same simple arguments are mentioned—developed in different ways, but not literally repeated.

---

[55] (Ciruelo 1526, a3r): "...pars theorica seu contemplativa de magnitudinibus, mensuris, numeris et motibus coelorum et stellarum.... Altera vero est quasi practica, quia iudiativa effectuum naturalium ex actione corporum coelestium in terra et mari in hominibus et brutibus animalibus atque in terrae vegetabilibus contingentium, voacturque ab eodem proprie astrologia."

[56] (Ciruelo 1498, a3r): "...qui...et virtutes et scientias licito amore prosequuntur unde et Johannes Picus italorum principum gloria nitidissimumque ornamentum diebus hiis velut novum sidus emicuit. Qui omnium tam humanarum quam divinarum scientiarum fastigium adeo celeriter optinuit ut omne genus scibile antequam adolescentiam excederet calluerit et quod is ignoraverit sapiat nemo."

5 The Structure of the *Apotelesmata* and the Renovation of Astrology

Between the first printing of the *Disputationes* in Bologna in 1496 and the publication of the response to it by the Pedro Sánchez Ciruelo in 1521, several other answers to Pico della Mirandola's work had been already published, mainly in Italy. Before Ciruelo wrote his *Apotelesmata astrologiae Christianae*, Lucio Bellanti (died ca. 1499) had already printed his *Defensio astrologiae* twice (in 1498 and 1502) and Giovanni Pontano (1426–1503) had published his answer in *De rebus coelestibus* in 1512 (Bellanti, 1498; Pontano, 1512). Agostino Nifo (ca. 1473–ca. 1540) had printed his commentary on the *Tetrabiblos* in 1513, and according to the studies of Paola Zambelli, Tiberio Russiliano (1490–ca. 1570) had also mentioned his intention to "destroy the disputations" around 1519 (Zambelli 1986, 1994; Nifo 1513).

The *Apotelesmata* was in a certain way a combination of Bellanti's choice to respond to Pico's work with that of Pontano—if we see Bellanti as a model in which each and every argument in the *Disputationes* is summed up and responded to in the same order—and Pontano's *De rebus coelestibus*, written with the objective of answering Pico by constructing a complete treatise on astrology. In his work, Ciruelo did both.

When he decided to structure his work in four books after Ptolemy's *Tetrabiblos,* it was an exercise inspired by Renaissance *imitatio* and by the general principle of going back to the authority of antiquity for the *renovatio* of the discipline. In the context of the debates about astrology in sixteenth century Europe, this meant a clear position, that of those who based their defense of astrology on the idea that the valid discipline, "pristine" as it was contained in Ptolemy's work, had been corrupted by later medieval—particularly Arab—additions, from which it had to be cleaned to restore a "truer" astrology. Ciruelo was one of the scholars with an interest in astrology who proposed such a view. The condemnation in Paris and Pico's work made him aware of the need for a renovation of the discipline. At that time, other scholars were also defending a return to Ptolemaic astrology. In this respect Ciruelo was not different from Pontano, from Nifo, from Albert Pigghe (1490–1542), and others, before Girolamo Cardano (1501–1576) famously advocated for such a way to recreate the discipline in his commentary to the *Tetrabiblos* in 1547.[57] The argument of the "return to Ptolemy" was also to be found in Melanchton (Pantin 1987). Cardano's role in this debate was crucial during the second half of the sixteenth century. His commentary on Ptolemy's astrological work is the best-known example of the larger theme—found in many authors of the period—of the need to reform astrology by a systematic program of study. Ciruelo would have concurred with this idea. Cardano's opinion, like Ciruelo's, was that the best way to reestablish his art was through the study of Ptolemy (Grafton 1999, 134). He claimed that he

[57] On Bellanti and Pontano, see (Faracovi 2008a, b, a, b; Akopyan 2017). On Pontano, see also (Desantis 1986; Rinaldi 2002). On Pigghe, see (Vanden Broecke 2003, 85–99). On Cardano, see (Grafton 1999, 134–53; Ernst 2001). See also (Rabin 1997).

had clarified the validity of astrological principles to be found in Ptolemy's *Tetrabiblos,* but he also highlighted the limits of the discipline, its imperfection, and he criticized other practitioners in the frame of distinguishing good from bad astrology (Grafton 1999, 91–108, 127–55).

This was one of the ways in which Ciruelo connected the arguments he had introduced in the commentary on the *Sphaera* with those he developed in the *Apotelesmata.* He had already stated in the preface to Sacrobosco that Ptolemy's astrology was the "pure" and legitimate one, and that later—medieval—additions had mixed with it false ideas and practices. He particularly cited as erroneous the astrological ideas of Albumasar (Abū Maʿshar Jaʿfar ibn Muḥammad ibn ʿUmar al-Balkhī, 787–886), Abenragel (Abū l-Ḥasan ʿAlī ibn Abī l-Rijāl, active late tenth–early eleventh century), and Guido Bonatti (1210–1296). To reform astrology, it was necessary to return to Ptolemaic astrology, creating a new discipline by eliminating certain theories from the intervening period.[58] What had constituted a few paragraphs in the prefaces to the *Sphaera* became the basis for Ciruelo's main work in defense of astrology. The idea of the renovation of astrology was often structured upon the distinction between a true and a false astrology. The first would be learned and legitimate; the second would be superstitious and incompatible with Christian doctrine (Walker 1958; Ernst 2001, 39–68). Medieval authors had already attempted to establish criteria to guide the efforts to distinguish between natural and judicial astrology. Natural astrology would interpret the effects that the properties and movements of celestial bodies caused naturally on earthly bodies and events, neither requiring supernatural intervention nor denying human free will. Judicial astrology predicted human activities which depended on the decisions of men, but its deterministic interpretation presupposed the annulment of free will, of the possibility of unpredictable events, and of divine intervention. These theoretical descriptions of the two kinds of astrology were the subject of many works during the early modern period. They insisted on the existence of a good astrology which was true and licit from the point of view of the Church, and a false, superstitious, illicit one. But opinions on which theories and practices must be classified as one or the other were varied and contradictory. Medieval antecedents were to be found in the works of Roger Bacon and Albertus Magnus (Crombie and North 1970, 377–85; Molland 1997, 68–71, Zambelli 1986, 1992). Ciruelo discussed the issue in the *Apotelesmata* for a learned audience, and later, in his *Reprobacion de supersticiones y hechicerias* (*Reprobation of superstitions and sorceries*), first published in 1537, he tried to explain the difference between the two kinds of astrology in the vernacular to reach a wider audience. True astrology, he explained, was based on the natural effects of the stars on air and water, thus affecting the weather and human health, complexions, and inclinations. False astrology predicted things related to human free will

[58] (Ciruelo 1498, a5v–a6r): "Hanc tamen astrologiam multi ut dictum est plura falsa immiscentes late ampliarunt, nam Albumazar, Hali Abenragel, Guydo Bonati, Abraham Avenezre et multi alii…."

and chance.[59] While brief in form, this was an argument that had also been present in Ciruelo's commentary on Sacrobosco. When discussing the contents of the third chapter on the risings and settings of the signs, and on climates and the habitability of the diverse zones of the earth, he repeated the differences among the mathematical disciplines explained in the preface. He then explained that several of them were "*artes mathematice prohibite ab ecclesia que sunt divinatorie*" (Ciruelo 1498, h4v). The list of the prohibited arts is reminiscent of those he criticized in the *Reprobación*. His efforts to establish a distinction between licit and illicit astrology, and to distinguish the former from any divinatory art, was quite constant throughout his career. He was successful in spreading these distinctions. The three reprints of the *Sphaera* and the *Apotelesmata* were received only by scholarly audiences, but his vernacular *Reprobación,* containing the same distinction, had a huge impact. It was printed at least ten times during the sixteenth century since its first edition in 1537 (Mateo Ripoll 2002).[60] He was repeatedly cited as an authority on the subject during the seventeenth century debates on astrology at the core of the Inquisition (Lanuza Navarro 2017).

The determination of which astrological doctrines had to be considered unacceptable from the point of view of Catholic orthodoxy was associated in the *Apotelesmata* with the explanation of all aspects that constituted a 'true' Ptolemaic astrology. Both were directly linked, which was also the reason why Ciruelo presented himself as a perfect judge for astrological issues. His knowledge of astrology, which other theologians lacked, was the reason he considered himself better qualified (Lanuza Navarro 2017).

Ciruelo devoted a great portion of the *Apotelesmata* to forming an answer to each and every argument contained in Pico's attack.[61] His response to the *Disputationes* was organized in two parts. The first part responded to Pico's books I and II, and was placed before the astrological treatise. Then Ciruelo's work became a manual

---

[59] (Ciruelo 1551, 16r–17r): "La verdadera astrología habla de cosas que se causan por las virtudes de los cielos, que con sus movimientos y luces alteran el ayre y la mar, y la tierra, y assí causan diversos effetos de tiempos..., y porque los cielos y las estrellas alterando el ayre y la tierra también alteran a los hombres..., assí los cielos causan a nuestros cuerpos diversas calidades, complexiones, passiones y enfermedades, diversas inclinaciones y abilidades.... Y en estos juicios no hay vanidad ni superstición alguna, porque aplica a los effectos sus causas que tienen virtud natural para los hazer..., y esta astrología es lícita y verdadera sciencia como filosofía natural, o la medicina.... La falsa astrología no es arte ni sciencia verdadera, antes es una superstición, porque de los cielos y estrellas presumen de juzgar de cosas que ellas no pueden ser causas dellas..., las cosas de acaescimientos por diversos casos de fortuna...y los secretos del coraçón y voluntad del hombre... [que es] libre."

[60] On the *Reprobación*, see (Ebersole 1962; Zamora Calvo 2008; Tausiet Carles 1992).

[61] (Ciruelo 1521, a4r): "In secundo prologo ponuntur argumenta et responsiones circa primos duos libros operis Mirandulani Contra astrólogos;" (Ciruelo 1521, C4r): "Restat etiam pro nostri operis complemento ad argumenta que contra veritates astrologicas fieri consueverunt respondere. Et que nobilis quidam Joannes Picus Mirandula insignis orator et sumus in arte dicendi, putavit se recollegisse omnia argumenta que contra haec facultatem apparentiam aliquam habent in suo opere dicto contra astrologos...."

explaining astrological theories, followed by his glosses on the pseudo-Ptolemaic *Centiloquium.* At the end of the work he placed the second part of his answer to Pico. In this, he responded more briefly to the rest of the books (III to XII) of the *Disputationes*.

The internal structure of the texts of these 'Responsiones' is, like Bellanti's work, that of the scholastic tradition. Ciruelo summed up the content of each of the chapters of the *Disputationes*, following its order, and then answered Pico's arguments with his own opinions.

The arguments with which he rebuked Pico's criticisms in the second book were not innovative, which is unsurprising as Pico also drew his attacks from the critical tradition. The re–elaboration of medieval and classical arguments which were previously pointed out in the *Disputationes* happens thus in both the condemnation and the responses, sometimes lending a kind of circularity to the two in tandem, as Pico had considered in advance the usual answers of astrologers to his objections and created his work by responding to them preemptively.[62]

Claims for and against astrology had classical origins and passed through medieval recreations. According to Bouché–Leclerc, after the first attacks had been issued, nothing new followed; they had already presented every argument and the ensuing criticisms only repeated themselves without innovations (Bouché-Leclercq 1899, 571). This is not as straightforward as he claims. Even if the participants in the early modern polemics against astrology made use of the same arguments, this does not mean they were mere repetitions. In the first place, Pico's *Disputationes* is much more than a list of the same arguments, and, as put by Cesare Vasoli, it is not a work reducible to the traditional argumentations of the anti–astrological polemics of an Augustinian nature and origin.[63] In the second place, more generally, even if the arguments are the same, the context—socially speaking, politically speaking, scholarly speaking—was seriously different in each moment, and that altered not just the way they were presented, but probably also their meaning.[64] It would be wise to follow the arguments through time and see how they were reproduced in different contexts to understand how they changed and the roles they played at different points in this long-lasting debate.

6 Conclusion

Pedro Sánchez Ciruelo was a scholar of scholastic training, educated in the nominalist tradition. He has repeatedly been described as a humanist (Flórez et al. 1990, 50–53; Albares 1996). It is difficult to agree with applying such a label to him, due as much to the nature of his writings and activities as to the problematic definition

[62] On the re–elaboration of arguments, see (Caroti 2008, 72–78).

[63] Arguments which according to Boll and Bouché–Leclercq were traceable to Carneades through Cicero (Bouché-Leclercq 1899, 571, citing Boll 1894, 182). See, against that idea, (Vasoli 2008, 5).

[64] On the difference that social and political context made in arguments concerning astrology, see (Vanden Broecke 2016).

of the term "humanism." Created as a historian's category in the nineteenth century after the Renaissance term "humanist," humanism has been the subject of much controversy and of varied and often contradictory views. Even after leaving behind nineteenth and twentieth century debates over the meaning of humanism and over its main traits, scholars currently working on several aspects of Renaissance culture and society still disagree about how best to categorize it.[65] Assuming certain points of agreement, I will try to briefly expose why applying them to Ciruelo's works is problematic. Some characteristics described as typical of humanism can be found in his writings; however, the weight of medieval scholasticism in them cannot be denied either.

According to the traditional definition of humanism, humanists promoted the disciplines concerning language—grammar, rhetoric, and poetics—the *studia humanitatis,* as opposed to the predominant scholastic Aristotelian philosophy. One of the points of discussion about it is, of course, to what extent it really opposed scholastic philosophy, as the borders between them cannot be taken as so neat and clear as the historical actors, and many historians, affirmed.[66] The not-so-straightforward definition of what medieval Scholasticism was has been another point of discussion, as well as the role of many medieval legacies in humanist attitudes (Fubini 2006). It is reasonable to agree with Paul Oskar Kristeller that reducing fifteenth century culture to a confrontation between these two cultural movements is a misrepresentation of it, that the culture of this period was more varied, and that humanism and Scholasticism coexisted during the fifteenth and sixteenth centuries (Mazzoco 2006a, 17). Seeing Ciruelo's life and works in a context of coexistence seems quite reasonable as well. If his writings can be said to share some of the traits of humanist works, it is also certain that he cultivated literary genres usually associated with scholastic activity. He wrote commentaries on Aristotle's *Posteriora analytica,* he used pedagogical dialogues, disputations, and disputations in the form of *quaestio*.[67] His dedication to logic and his role as the professor of Thomistic theology for three decades can hardly be ignored either.[68]

The values of the humanists pervaded scholarship as a whole, including scholastic commentaries of the sixteenth century (Dear 2001, 32). Humanists put emphasis on the value of education to create better citizens, on the relevance of discarding medieval Latin in favor of classical Latin because they saw it as inseparably related to the wisdom of the content, and on the idea of a rebirth of classical culture,

---

[65] For a summary of the nineteenth and twentieth century debates over humanism, and the relevant roles played by different scholars, including the significant works of Burckhardt and Haskins, and of Baron, Garin, and Kristeller, see (Mazzoco 2006a). For several current views of humanism and the most relevant aspects of it, some of them contradictory, see the varied essays in this edited volume. See also (Grendler 2002; Fubini 2001, 25–27; Celenza 2004, 16–57).

[66] Kristeller insisted that humanism evolved out of the grammatical and rhetorical medieval traditions. Discussion on the real meaning of Kristeller's thesis and influence, in (Mazzoco 2006b, 176–77).

[67] See an analysis of the *Dialogue* at the end of the *Sphaera*, and of the genres in the works of Ciruelo, in (Gómez Sierra 1999, 2000).

[68] On Ciruelo's logic, see (Ayala 1993; Muñoz Delgado 1964, 1966, 1967).

of renewing culture by a return to the language and knowledge of antiquity, which in their opinion were linked. Language or the recovery of ancient texts for better translations seems not to have been one of Ciruelo's concerns, at least not concerning scientific works. He did, however, embark on a translation from Hebrew of parts of the Old Testament, though late in his life. This must be analyzed in the context of another debate concerning humanism, this one from the point of view of religious studies and focused specifically on Spain. Marcel Bataillon's immeasurably influential *Erasme et l'Espagne* defended the idea that humanism in Spain arose exclusively from the supporters of Erasmus and his doctrines, spiritual ideals, and philological criticism (Bataillon 1966).[69] Bataillon presented Ciruelo as the epitome of conventional Catholicism and an enemy of Erasmus, and therefore as the absolute opposite of a humanist.[70] Lou Ann Homza demonstrated how biased this view of humanism, and of this scholar, was, by analyzing Ciruelo's theology and his religious works. She concluded that although Ciruelo's early works displayed no respect for the integrity of the text or interest in exegetical issues, he evolved, became a Hebrew scholar and showed philological interests, embraced the humanist historicist approach and developed a new exegetical awareness (Homza 1992, 60–101, 175–209). His "altered approach to biblical hermeneutics qualifies him as a Renaissance humanist" (Homza 1992, 11, 284).

From the point of view of the historian of science there is also a long debate on the relationship of humanism and science to be taken into account. Scholarship discarded long ago the traditional view of an opposition between Renaissance humanism and scientific thought, once defended by Thorndike (Blair and Grafton 1992, 535). In the last decades, several scholars have made a 'reassessment' of humanism and its interaction with science, pointing out, among other issues, the impact of ancient texts of a scientific nature made available by humanist scholarship and the press, and the spread of scientific ideas through humanist works (Blair and Grafton 1992; Grafton 1991).

Regiomontanus has been held up as an example of a humanist specialist in Latin literature who was simultaneously a mathematician and astronomer—an example of how the ideal of renewing culture by a return to antiquity was applied to the mathematical sciences. In Peter Dear's words, he transferred the language of the humanists regarding decline and renewal to the specific arena of mathematics, and his humanist rhetoric of 'restoration' in the *Epitome* found a receptive audience (Dear 2001, 33–34). It is not difficult to relate Regiomontanus's defense of mathematics as well as his regard of astronomy as the highest of the mathematical sciences to Ciruelo's words in the preface to Sacrobosco. Ciruelo's insistence on returning to Ptolemaic astrology could be read as participation in the humanist revival of ancient sources and practices. In this he would be following Regiomontanus's, as well as

[69] For a description of the scholarship concerning humanism in Spain as presented by Bataillon, and a criticism of what it erroneously entailed, see (Homza 1992).

[70] A transcription of Ciruelo's opinions on Erasmus in the Assembly of Valladolid in 1527, can be found in (Beltran de Heredia 1972, 50–53). See an analysis of it in (Homza 1992, 209–81, 1997, 2000, 49–76).

Lefèvre d'Étaples's, humanist way of addressing the issue. The same could be said about basing his argument for the value and nobility of astrology on an appeal to ancient authorities: the strategy of creating a genealogy of a discipline that extends as far back into antiquity as possible and ties in the most famous individuals to legitimize it and justify its practice was also a humanist move (Grafton 1992). In this sense, it must be pointed out that in the dialogue at the end of the commentary Ciruelo also discusses the function of *authoritas.* The use of historical narrative to work through problems of the legitimacy and nature of disciplines, as used by Regiomontanus and others, has been termed "the most humanist of activities" (Goulding 2006; Popper 2006). Still, one should not lose sight of how Ciruelo, like Lefèvre d'Étaples, also built his defense on Boethius's claims for the value of mathematics, and that this medieval tradition dovetailed with the humanist interest in mathematics (Moyer 2014). Nor should one lose sight of the fact that as much as a mathematician and astrologer, Ciruelo wrote as a theologian, and that his main concern was to establish the parts of astrology acceptable from the point of view of orthodox Catholic doctrine.

Ciruelo's preface is strikingly similar to Regiomontanus's inaugural oration on the history and utility of the mathematical arts at the University of Padua in 1464.[71] The main features were shared: praise of the mathematical arts and its practitioners, a history of mathematics, the utility of the discipline and its supremacy based on certainty, and among the mathematical sciences, the superiority of astrology because of the knowledge it provided of the secrets of God. Another idea used by Ciruelo in the commentary on Sacrobosco, that astrology encompassed all other branches of mathematics (arithmetic, geometry, etc.), could also be found in the letter contained in some copies of Regiomontanus' *Epitome,* as Robert Westman has recently pointed out. This letter, written by Giovanni Battista Abiosi (active in the last decades of the fifteenth century), was also an encomium to astrology (Westman 2016, 17).

Regiomontanus's *oratio* is considered an example of a humanist rhetorical text on the revival of ancient mathematics. James Byrne has shown, however, that even if it is stylistically consistent with humanist norms, its vision is grounded in the medieval university's mathematical curriculum and does not reject medieval authorities (Byrne 2006). Stylistically speaking, Ciruelo's preface would probably not be considered humanist, as he shares the vision of the scholastic mathematical university training and the respect for medieval authorities.

In this paper I did not intend to place Ciruelo into one or another category or to discuss his humanist versus his scholastic traits. The aim was to analyze his defense of astrology, and to show that we get a richer understanding of Ciruelo's works involved in that defense, by placing them in their own specific context.

In the case of the preface to Sacrobosco, he wrote it while trying to establish his relevance as a professor of the mathematical disciplines, including astrology, in Paris. When Simon de Phares and his astrology of elections and interrogations, two

[71] For a study and description of Regiomontanus's *oratio* at the University of Padua in 1464, see (Byrne 2006).

astrological practices very much in conflict with Christian doctrine, were condemned by the faculty of theology in 1494 in a decree that was critical of all branches of astrology, Ciruelo had to defend the legitimacy and importance of the discipline he was teaching. He placed this defense in his commentary on the *Sphaera* of Sacrobosco, the traditional work for teaching astronomy, and used the arguments that aimed to praise the virtues of astrology and the mathematical disciplines and to differentiate the "good" from the "bad" astrology. Those were the arguments that would help him vindicate his place at the university and reinforce his status there. Placing the defense in the most widely read of all works on astronomy in academic circles, in a commentary addressed to an audience of students and professors, was a wise strategy.[72]

When he published the *Apotelesmata* two decades later, in the middle of the debates about the discipline renewed at the end of the fifteenth century by the publication of Pico de la Mirandola's *Disputationes*, the arguments Ciruelo highlighted were different. He focused then on the creation of a whole corpus of astrological knowledge not in conflict with Christian doctrine. One of his objectives was to set up guidelines that would make himself seem a valid authority in decisions about astrology. To this end, he emphasized his training as a theologian at the University of Paris, one of the reference colleges for theology. This, combined with his knowledge of astrology resulting from his earlier training in mathematics and astrology at the University of Salamanca—knowledge well exposed in the work—would make him the perfect judge.

The works aiming to defend astrology before and after Pico della Mirandola's attack followed different strategies according to the concrete ideas of their authors. They shared many arguments, as is obvious from the overlap between Ciruelo's arguments and Regiomontanus's and Bellanti's, for instance. Those arguments were inherited from the medieval tradition, but their particular context determined what was highlighted in each case and in what way the arguments were presented. The case of Ciruelo shows that his way of presenting the defense was immersed in the specific context in which he was living, and that it grew in part from the perceived need to vindicate his own status and that of the discipline within two different scholar communities at two points in time.

References

Manuscripts

Torres, Diego de. *Sphaera de Sacrobosco*. In *Tratados de astronomía*. mss/3385, fols. 189r–99v. Biblioteca Nacional de Madrid, Spain.

[72] On the diffusion of diverse commentaries to Sacrobosco's *Tractatus de Sphaera* and the general relevance of it, see (Valleriani 2017; Crowther et al. 2015).

Primary Sources

Bellanti, Lucio. 1498. *Liber de astrologica veritate et In disputationes Ioannis Pici aduersus astrologos responsiones*. Florence: Gherardus de Haerlem.

Ciruelo, Pedro. 1495a. *Arithmetica speculativa Thome Bravardini bene revisa et correcta a Petro Sánchez Ciruelo Aragonensi mathematicas legente*. Paris: Guy Marchand. (reprinted Paris, 1502).

———. 1495b. *Geometría speculativa Thome Bravardini*. Paris: Jean Petit. (reprinted 1502, 1508, 1511, 1530).

———. 1495c. *Tractatus Arithmeticae Practice qui dicitur algorismus*. Paris: Guy Marchand. (reprinted 1502, 1505, 1509, 1513, 1514).

———. 1498. *Uberrimum sphere mundi comentum intersertis etiam questionibus domini Petri de Aliaco*. Paris: Jean Petit. http://hdl.handle.net/21.11103/sphaera.100038 and http://hdl.handle.net/21.11103/sphaera.100274.

———. 1508. *Uberrimum Sphere mundi commentum. Johannis de sacro busto anglici, una cum textualibus optimisque additionibus ac uberrimo commentario Petri Ciruelli; intersertis etiam questionibus domini Petri de Aliaco*. Paris: Jean Petit. http://hdl.handle.net/21.11103/sphaera.100642.

———. 1515. *Habes lector Iohannis de sacro busto sphere textum una cum additionibus non aspernandis Petri Ciruelli. D.* Paris: Jean Petit. http://hdl.handle.net/21.11103/sphaera.100988.

———. 1516. *Cursus quattuor mathematicarum artium liberalium*. Alcalá: Arnaldo Guillén Brocar. This work was reprinted in Alcalá in 1526 and 1528.

———. 1519. *Prima pars logices ad veriores sensus textus Aristotelis*. Alcalá: Arnaldo Guillén Brocar.

———. 1521. *Apotelesmata astrologiae Christianae*. Alcalá de Henares: Juan de Eguía.

———. 1526. *Opusculum de sphera mundi Joannis de sacro busto: cum additionibus: et familiarissimo commentario Petri Ciruelli Darocensis: nunc recenter correctis a suo autore: intersertis etiam egregijs questionibus domini Petri de Aliaco*. Alcalá de Henares: Miguel de Eguía. http://hdl.handle.net/21.11103/sphaera.100884.

———. 1528. *In Posteriora analytica commentarius*. Alcalá: Miguel de Eguía. (two new editions in 1529).

———. 1551. *Reprobación de supersticiones y hechicerías*. Medina del Campo: Guillermo de Millis.

Le Kalendrier des bergers. 1493. Paris: Guy Marchant.

Nifo, Agostino. 1513. *Ad Apotelesmata Ptolomaei eruditiones*. Naples: P.M. De Richis.

Pontano, Giovanni. 1512. *De rebus coelestibus*. Naples: Sigismundus Mayr.

Savonarola, Girolamo. 1496. *Revelatio de tribulationibus nostrorum temporum*. Paris: Guy Marchant.

Secondary Literature

Akopyan, Ovanes. 2017. *Controversies on astrology in Renaissance Italy (late 15th–early 16th centuries)*. PhD Thesis. University of Warwick.

Albares Albares, Roberto. 1996. El humanismo científico de Pedro Ciruelo. In *La Universidad Complutense Cisneriana. Impulso filosófico, científico y literario*, ed. Luis Jiménez Moreno, 177–205. Madrid: Editorial Complutense.

Albertson, David. 2014. *Mathematical theologies: Nicholas of Cusa and the legacy of Thierry of Chartres*. Oxford: Oxford University Press.

Almási, Gábor. (2014). Rethinking Sixteenth-Century 'Lutheran Astronomy'. *Intellectual History Review* 24 (1): 5–20.

Axworthy, Angela. 2009. The epistemological foundations of the propaedeutic status of mathematics according to the epistolary and prefatory writings of Oronce Finé. In *The worlds of Oronce Fine. Mathematics, instruments and print in Renaissance France*, ed. A. Marr, 31–51. Donington: S. Tyas.

———. 2016. *Le Mathématicien renaissant et son savoir. Le statut des mathématiques selon Oronce Fine*. Classiques Garnier.

Ayala, Jorge M. 1993. El Maestro Darocense Pedro Sánchez Ciruelo. *Aragón en la Edad Media* 10–11:85–99. Zaragoza: Universidad de Zaragoza.

Barker, Peter. 2011. The reality of Peurbach's orbs: Cosmological continuity in fifteenth and sixteenth century astronomy. In *Change and continuity in early modern cosmology*, ed. Patrick J. Boner, 7–32. Heidelberg/London/New York: Springer.

Bataillon, Marcel. 1966. *Erasmo y España*. Madrid: Fondo de Cultura Económica. (French original: *Erasme et l'Espagne. Recherches sur l'histoire spirituelle du XVIe siècle*, first ed. 1937).

Beltrán de Heredia, Vicente. 1972. *Cartulario de la Universidad de Salamanca*. Salamanca: Universidad de Salamanca.

Biagioli, Mario. 1989. The social status of Italian mathematicians, 1450–1600. *History of Science* 27: 41–95.

———. 1993. *Galileo courtier. The practice of science in the culture of absolutism*. Chicago: The University of Chicago Press.

Blair, Ann, and Anthony Grafton. 1992. Reassessing humanism and science. *Journal of the History of Ideas* 53 (4): 535–540.

Boll, Franz. 1894. *Studien über Claudius Ptolemaeus. Ein Beitrag zur Geschichte der griechischen Philosophie und Astrologie*. Leipzig: Teubner.

Bouché-Leclercq, Auguste. 1899. *L'astrologie grecque*. Paris: E. Lerous.

Boudet, Jean-Patrice. 1988. Simon de Phares et l'astrologie a Lyon a la fin du XV siècle. In *Lyon, cité de savants*. Paris: Éditions du C.T.H.S.

———. 1994. *Lire dans le ciel: la bibliothèque de Simon de Phares*. Bruxelles: Centre d'Étude des Manuscrits.

———. 1999. *Le Recueil des plus célèbres astrologues de Simon de Phares*. Paris: Champion.

Brown, Gary I. 1991. The evolution of the term 'mixed mathematics'. *Journal of the History of ideas* 52: 81–102.

Burgueño, Susana. 2009. Astronomia, filosofia y humanismo cientifico en Diego de Torres (1482–1496). *Duererías, Analecta Philosophiae. Revista de Filosofía* 1: 1–11.

Byrne, James S. 2006. A humanist history of mathematics? Regiomontanus's Padua oration in context. *Journal of the History of Ideas* 67 (1): 41–61.

Cajori, Florian. 1928. Ciruelo on the names arithmetical and geometrical proportions and progressions. *ISIS* 10: 363–366.

Calderon, Calixto. 1991. The 16th century Iberian Calculatores. *Revista de la Unión Matemática Argentina* 35: 245–258.

Caroti, Stefano. 2008. Le fonti medievali delle Disputationes adversus astrologiam divinatricem. In *Nello specchio del cielo. Giovanni Pico della Mirandola e le Disputationes contro l'astrologia divinatoria*, ed. Marco Bertozzi, 67–93. Florence: Leo S. Olschki.

Celenza, Christopher S. 2004. *The lost Italian Renaissance: Humanists, historians, and Latin's legacy*. Baltimore: Johns Hopkins University Press.

Cozzoli, Daniele. 2007. Alessandro Piccolomini and the certitude of mathematics. *History and Philosophy of Logic* 28: 151–171.

Crombie, A.C., and J.D. North. 1970. Roger Bacon. In *Dictionary of scientific biography*, ed. Ch. Gillispie, vol. I, 377–385. New York: Charles Scribner's Sons.

Crowther, Kathleen, Ashley Nicole McCray, Leila McNeill, Amy Rodgers, and Blair Stein. 2015. The book everybody read: Vernacular translations of Sacrobosco's *Sphere* in the sixteenth century. *Journal for the History of Astronomy* 46 (1): 4–28.

Crowther-Heyck, Kathleen. 2003. Wonderful secrets of nature. Natural knowledge and religious piety in reformation Germany. *ISIS* 94: 253–273.

De Pace, Anna. 1993. *Le matematiche e il mondo: ricerche su un dibattito in Italia nella seconda metà del cinquecento*. Milan: Franco Angeli.

Dear, Peter. 1995. *Discipline and experience: The mathematical way in the scientific revolution*. Chicago: The University of Chicago Press.

———. 2001. *Revolutionizing the sciences: European knowledge and its ambitions, 1500–1700*. Basingstoke: Palgrave.

Desantis, G. 1986. Pico, Pontano e la polemica astrologica. *Annali Della Facoltà di Lettere e Filosofia* 29: 155–192.

Ebersole, A.V. 1962. Pedro Ciruelo y su Reprobación de hechicerías. *Nueva Revista de Filología Hispánica* 3/4: 430–437.

Elie, Hubert. 1950–1951. Quelques Maitres de l'université de Paris vers l'an 1500. *Archives d'histoire doctrinale et littéraire du Moyen Âge* 18: 193–243.

Engammare, Max. 2008. *Calendrier des bergers, préface de Max Engammare*. Paris: PUF, collection Sources, Fondation Martin Bodmer.

Ernst, Germana. 1991a. Astrology, religion and politics in counter-reformation Rome. In *Science, culture and popular belief in Renaissance Europe*, ed. S. Pumfrey, P. Rossi, and M. Slawinski, 249–273. Manchester/New York: Manchester University Press.

———. 1991b. *Religione, ragione e natura. Ricerche su Tommaso Campanella nel tardo Rinascimento*. Milan: Franco Angeli.

———. 2001. Veritatis amor dulcissimus: Aspects of Cardano's astrology. In *Secrets of nature. Astrology and alchemy in early modern Europe*, ed. W. Newman and A. Grafton, 39–68. Cambridge, MA/London: MIT Press.

Faracovi, Ornella Pompeo. 2008a. In difesa dell'astrologia: risposte a Pico in Bellanti e Pontano. In *Nello specchio del cielo. Giovanni Pico della Mirandola e le Disputationes contro l'astrologia divinatoria*, ed. Marco Bertozzi, 47–66. Firenze: Olschki.

———. 2008b. La riforma dell'astrologia, in *Il Rinascimento italiano e l'Europa*, eds. G.L. Fontana and L. Molà, vol. V, *Le scienze*, eds. A. Clericuzio, G. Ernst, and M. Conforti, 59–71. Treviso-Costabissara.

———, ed. 2012a. *Lo Specchio Alto: astrologia e filosofia fra Medioevo e prima età moderna*. Pisa: F. Serra.

———. 2012b. La polemica antiastrologica di Giovanni Pico della Mirandola. In *Il linguaggio dei cieli. Astri e simboli nel Rinascimento*, ed. Germana Ernst and Guido Giglioni, 91–107. Rome: Carocci.

Flórez Miguel, Cirilo, Pablo García Castillo, and Roberto Albares Albares. 1988. *El humanismo científico*. Salamanca: Caja de Ahorros.

———. 1989. *La ciencia del cielo: astrología y filosofía natural en la Universidad de Salamanca*. Salamanca: Caja de Ahorros.

———. 1990. *Pedro S. Ciruelo: Una enciclopedia humanista del saber*. Salamanca: Caja de Ahorros.

Fubini, Ricardo. 2001. *L'umanesimo italiano e i suoi storici: Origini rinascimentali-critica moderna*. Milan: Franco Angeli.

———. 2006. Humanism and scholasticism: toward an historical definition. In *Interpretations of renaissance humanism*, ed. Angelo Mazzoco, 127–136. Leiden/Boston: Brill.

Gagné, J. 1969. Du quadrivium aux *Scientiae Mediae*. In *Arts libéraux et philosophie au moyen âge*, 975–986. Paris: Vrin.

Galle, Griet. 2003. *Peter of Auvergne, questions on Aristotle's De Caelo—A critical edition with an interpretative essay*. Leuven: Leuven University Press.

Gómez Martínez, Marta. 2006. *Sacrobosco en castellano*. Salamanca, Ediciones Universidad de Salamanca.

Gómez Sierra, Esther. 1999. Home and away in Paris: Pedro Sánchez Ciruelo and his *Disputatorius dyalogus*. In *Latin and vernacular in Renaissance Spain*, ed. Barry Taylor and Alejandro Coroleu, 83–104. Manchester: Manchester Spanish and Portuguese Studies, Cañada Blanch Centre for Advanced Hispanic Studies.

———. 2000. Pedagogia y amicitia. Los dialogos latinos de Pedro Ciruelo. In *Actas del VIII Congreso de la Asociación Hispánica de Literatura Medieval*, ed. M. Freixas and S. Iriso, 843–856. Santander: Asociación Hispánica de Cultura Medieval.

Goulding, R. 2006. Histories of science in early modern Europe. Introduction. *Journal of the History of Ideas* 67 (1): 33–40.

Grafton, Anthony. 1991. *Defenders of the text: The traditions of scholarship in an age of science*. Cambridge, MA: Harvard University Press.

———. 1992. Kepler as a reader. *Journal of the History of Ideas* 53 (4): 561–572.

———. 1997. From Apotheosis to Analysis: Some Late Renaissance Histories of Classical Astronomy. In *History and the Disciplines: The Reclassification of Knowledge in Early Modern Europe*, ed. Donald R. Kelley. Rochester, 261–76. New York: University of Rochester Press.

———. 1999. *Cardano's cosmos. The worlds and works of a Renaissance astrologer*. Cambridge, MA: Harvard University Press.

Grant, Edward. 1996. *Planets, stars, and orbs: The Medieval cosmos, 1200–1687*. Cambridge: Cambridge University Press.

Grendler, Paul. 2002. The Italian Renaissance in the past seventy years: Humanism, social history, and early modern in Anglo-American and Italian scholarship. In *The Italian Renaissance in the twentieth century*, ed. Allen J. Grieco, Michael Rocke, and Fiorella Gioffredi Superbi, 3–24. Florence: Olschki.

Homza, Lou Ann. 1992. *Religious humanism, pastoral reform and the* Pentateuch *in early modern Spain. Pedro Ciruelo's journey from grace to law*. PhD Thesis, University of Chicago.

———. 1997. Erasmus Hero or heretic Spanish Humanism and the Valladolid assembly of 1527. *Renaissance Quarterly* 50: 78–118.

———. 2000. *Religious authority in the Spanish Renaissance*. Baltimore/London: The Johns Hopkins University Press.

Jardine, Nicholas. 1984. *The birth of history and philosophy of science: Kepler's 'a defence of Tycho against Ursus' with essays on its provenance and significance*. Cambridge: Cambridge University Press.

———. 1988. Epistemology of the sciences. In *The Cambridge history of Renaissance philosophy*, ed. C.B. Schmitt, Q. Skinner, E. Kessler, and J. Kraye, 685–711. Cambridge: Cambridge University Press.

———. 1998. The places of astronomy in early modern culture. *Journal for the History of Astronomy* 29: 49–62.

Kirwan, Richard, ed. 2016. *Scholarly self-fashioning and community in the early modern university*. London/New York: Routledge.

Laird, W.R. 2016. Galileo and the mixed sciences. In *Method and order in Renaissance philosophy of nature: The Aristotle commentary tradition*, ed. Daniel A. di Liscia and Eckhard Kessler, 253–270. London: Routledge.

Lanuza Navarro, Tayra M.C. 2017. Astrology in court: The Spanish inquisition, authority, and expertise. *History of Science* 55 (2): 187–209.

Latassa Ortiz, Félix. 1798. *Biblioteca nueva de los escritores aragoneses que florecieron desde el año de 1500 hasta 1599*. Pamplona: J. Domingo.

Lorente Pérez, J.M. 1921. *Biografía y análisis de las obras de matemática pura de Pedro Sánchez Ciruelo*. Madrid: Junta para la Ampliación de Estudios.

Mancosu, P. 1992. Aristotelian logic and Euclidean mathematics: Seventeenth-century developments of the *Quaestio de Certitudine Mathematicarum*. *Studies in History and Philosophy of Science* 23 (2): 241–265.

———. 1996. *Philosophy of mathematics and mathematical practice in the seventeenth century*. New York/Oxford: Oxford University Press.

Mandosio, Jean-Marc. 1994. Entre mathématiques et physique: Note sur les 'sciences intermédiaires' à la Renaissance. In *Comprendre et maitriser la nature au Moyen Age: Mélanges d'histoire des sciences offerts à Guy Beaujouan*, 115–138. Geneve: Droz.

Marr, Alexander, ed. 2009. *The worlds of Oronce Fine. Mathematics, instruments and print in Renaissance France*. Donington: Shaun Tyas.

Martins, Roberto de Andrade. 2003. Las Fuentes Literarias del Tratado de la esfera de Sacrobosco. In *Epistemología e Historia de la Ciencia. Selección de Trabajos de las XIII Jornadas*, ed. Víctor Rodríguez and Luis Salvatico, 307–314. Córdoba: Universidad Nacional de Córdoba.

Mateo Ripoll, Verónica. 2002. Sobre una edición ignota de la *Reprobación de supersticiones* del maestro Ciruelo. *Dynamis* 22: 437–459.

Mateos Royo, J.A. 1997. Sobre tasas y monedas, ferias y usuras: municipio y mercado en Daroca bajo Juan II y Fernando el Católico (1459–1516). *Aragón en la Edad Media* 13: 195–216.

———. 2002. Municipio y enseñanza pública en Aragón: la escuela de Gramática de Daroca (ss. XVI-XVII). *Revista de Historia Moderna* 20: 117–148.

Mazzoco, Angelo, ed. 2006a. *Interpretations of Renaissance humanism*. Leiden/Boston: Brill.

———. 2006b. Kristeller and the vernacular. In *Kristeller reconsidered: Essays on his life and scholarship*, ed. John Monfasani, 176–177. New York: Italica Press.

McKirahan, Richard D. 1978. Aristotle's subordinate sciences. *The British Journal for the History of Science* 11: 197–220.

Molland, George. 1997. Roger Bacon's *De laudibus mathematicae*. A preliminary study. In *Texts and contexts in ancient and Medieval science*, ed. Edith Sylla and Michael McVaugh, 68–83. Leiden: Brill.

Moyer, Ann E. 2014. The quadrivium and the decline of Boethian influence. In *A companion to Boethius in the Middle Ages*, ed. Noel H. Kaylor and Philip E. Phillips, 479–517. Leiden: Brill.

Muñoz Delgado, Vicente. 1964. *La lógica nominalista en la Universidad de Salamanca (1510–1530)*. Madrid: Publicaciones del Monasterio de Poyo.

———. 1966. La lógica como *scientia sermocinalis* en la obra de Pedro Sánchez Ciruelo. *Estudios* 22: 23–52.

———. 1967. La lógica en Salamanca durante la primera mitad del siglo XVI. *Salmanticensis* 14: 171–207.

Navarro Brotons, Víctor. 1995. El renacimiento científico (siglo XVI) y la enseñanza de las disciplinas matemáticas en las universidades de Valencia y Salamanca. In *II congreso internacional sobre historia de las universidades hispánicas*, ed. Mariano Peset, 141–159. Valencia: Universitat de València.

———. 1998. El Renacimiento científico (siglo XVI) y la enseñanza de las disciplinas matemáticas en las Universidades de Valencia y Salamanca. In *Doctores y Escolares*, ed. J.L. Peset, vol. I, 141–159. Valencia: Universidad de Valencia.

———. 2006. The teaching of the mathematical disciplines in sixteenth-century Spain. *Science and Education* 15: 209–233.

———. 2014. *Disciplinas, saberes y prácticas: Filosofía natural, matemáticas y astronomía en la sociedad española de la época moderna*. Valencia: Universidad de Valencia.

Navarro Brotons, Víctor, Vicente L. Salavert Fabiani, Victoria Roselló Botey, and Víctor Darás Román. 1999. *Bibliographia Physico-Mathematica Hispanica*. Vol. I, 1475–1600. Valencia: Instituto de Historia de la Ciencia y Documentación 'López Piñero' Universitat de València-CSIC.

Noreña, Carlos G. 1975. *Studies in Spanish renaissance thought*. The Hague: M. Nijhoff.

North, John D. 2009. Types of inconsistency in the astrology of Ficino. In *Christian Humanism*, ed. A. McDonald et al., 281–302. Leiden: Brill.

Oosterhoff, Richard J. 2018. *Making mathematical culture: University and print in the circle of Lefèvre d'Étaples*. Oxford: Oxford University Press.

Pantin, Isabelle. 1987. La lettre de Melanchton a Simon Grynaeus: Avatars d'une défense de l'astrologie. In *Divination et controverse religieuse en France au XVIe siècle*, ed. R. Aulotte, 85–101. Paris: École normale supérieure de jeunes filles.

———. 2013. Oronce Finé mathématicien et homme du livre: la pratique éditoriale comme moteur d'évolution. In *Mise en forme des savoirs à la Renaissance. À la croisée des idées, des techniques et des publics*, ed. Isabelle Pantin and Gérald Péoux, 19–50. Paris: Armand Colin.

Picatoste Rodríguez, F. 1891. *Apuntes para una biblioteca científica española del siglo XVI*. Madrid: Tello.

Plessis d'Argentré, Charles Du. 1724. *Collectio judiciorum de novis erroribus*. Vol. I. Paris: Duchesne.

Popper, Nicholas. 2006. Abraham, Planter of Mathematics: Histories of mathematics and astrology in early modern Europe. *Journal of the History of Ideas* 67 (1): 86–106.

Rabin, Sheila. 1997. Kepler's attitude toward Pico and the anti-astrology polemic. *Renaissance Quarterly* 50 (3): 750–770.

Rey Pastor, Julio. 1926. *Los matemáticos españoles del siglo XVI*. Toledo: A. Medina.

Ribeiro do Nascimento, C.A. 1974. Le statut épistémologique des "sciences intermédiaires" selon S. Thomas d Aquin. *Cahiers d'Etudes d'Études Médiévales* 2: 33–95.

Rinaldi, Michele. 2002. *Sic itur ad astra. Giovanni Pontano e la sua opera astrologica nel quadro della tradizione manoscritta della Mathesis di Giulio Firmico Materno*. Naples: Loffredo.

Rodrigo Esteban, María Luz. 1999. *La ciudad de Daroca a fines de la Edad Media. Selección documental (1328–1526)*. Daroca: Centro de Estudios Darocenses/Institución Fernando el Católico.

———. 2004. Pedro Sánchez Ciruelo y el estudio general de Artes de Daroca. *El Ruejo* 5: 139–156.

Rodríguez Vidal, R. 1981. *Pedro Ciruelo. Darocense (Un intelectual ejemplar)*. Zaragoza: Institución Fernando el Católico/CSIC.

Serrano Montalvo, Antonio. 1995–1997. *La población de Aragón según el Fogaje de 1495*. Zaragoza: Institución Fernando el Católico/ Instituto Aragonés de Estadística.

Smoller, Laura A. 1994. *History, prophecy, and the stars: The Christian astrology of Pierre D'Ailly, 1350–1420*. Princeton, NJ: Princeton University Press.

Tausiet Carles, María. 1992. Religión, ciencia y superstición en Pedro Ciruelo y Martín de Castañega. *Revista de Historia Jerónimo Zurita* 65–66: 139–147.

Thorndike, Lynn. 1934. *A history of magic and experimental science*. Vol. IV. New York: Columbia University Press.

———. 1941. *A history of magic and experimental science*. Vol. V. New York: Columbia University Press.

———. 1949. *The Sphere of Sacrobosco and its commentators*. Chicago: The University of Chicago Press.

Thorndike, Lynn, and Pearl Kibre. 1963. *A Catalogue of Incipits of Mediaeval Scientific Writings in Latin. Rev. and augm*. London: Mediaeval Academy of America.

Valle, Carlos del. 1999. Un poema hebreo de Alfonso de Zamora en alabanza de la versión latina bíblica de Pedro Ciruelo. *Sefarad* 59 (2): 419–437.

Valleriani, Matteo. 2017. The tracts on the *Sphere*: Knowledge restructured over a network. In *The structures of practical knowledge*, ed. Matteo Valleriani, 421–473. Dordrecht: Springer.

Vanden Broecke, Steven. 2003. *The limits of influence. Pico, Louvain and the crisis of Renaissance astrology*. Leiden/Boston: Brill.

———. 2016. From cosmic governance to governmentality: Shaping subulunary order in seventeenth century French critiques of astrology. In *Unifying heavens and earth: Essays in the history of early modern cosmology*, ed. M.A. Granada, P.J. Boner, and D. Tessicini, 317–342. Barcelona: Universitat de Barcelona.

Vasoli, Cesare. 2008. La polemica antiastrologica di Giovanni Pico. In *Nello specchio del cielo. Giovanni Pico della Mirandola e le Disputationes contro l'astrologia divinatoria*, ed. Marco Bertozzi, 1–18. Firenze: Olschki.

Villoslada, R.G. 1938. *La Universidad de Paris durante los estudios de Francisco de Vitoria, O.P. (1507–1522)*. Roma: Università Gregoriana.

Voss, Angela. 1998. The music of the spheres: Marsilio Ficino and Renaissance harmonia. *Culture and Cosmos* 2 (2): 16–38.

Walker, D.P. 1958. *Spiritual and demonic magic. From Ficino to Campanella*. London: Warburg Institute.

Wallace, William A. 1969. The *Calculatores* in early sixteenth-century physics. *British Journal for the History of Science* 4: 221–232.

———. 1981. Chapter. 5: The Calculatores in the sixteenth century. In *Prelude to Galileo. Essays on Medieval and sixteenth century sources of Galileo's thought*, 78–90. Dordrecht: Reidel.

Westman, Robert S. 1980. The astronomer's role in the sixteenth century: A preliminary study. *History of Science* 18: 105–147.

———. 2011. *The Copernican question. Prognostication, skepticism, and celestial order*. Berkeley/Los Angeles: University of California Press.

———. 2016. *Copernicus and the astrologers: Dibner library lecture, December 12, 2013*. Washington, DC: Smithsonian Libraries.

Wilkinson, Alexander S. 2010. *Iberian books: Books published in Spanish or Portuguese or on the Iberian Peninsula before 1601*. Leiden/Boston: Brill.

Zambelli, Paola, ed. 1986. *Astrology Hallucinati: Stars and the end of the world in Luther's time*. Berlin: W. de Gruyter.

———. 1992. *The Speculum astronomiae and its enigma: Astrology, theology, and science in Albertus Magnus and his contemporaries*. Dordrecht/Boston: Kluwer.

———. 1994. *Una reincarnazione di Pico ai tempi di Pomponazzi*. Milan: Il Polifilo.

Zamora Calvo, M.J. 2008. Ciruelo y sus juicios contra la superstición. In *La fractura historiográfica: las investigaciones de Edad Media y Renacimiento desde el Tercer Milenio*, ed. Javier San José Lera et al., 781–797. Salamanca: Seminario de Estudios Medievales y Renacentistas.

2

Sacrobosco's *Sphaera* in Spain and Portugal

Kathleen M. Crowther

Abstract This paper analyzes sixteenth-century commentaries on Johannes de Sacrobosco's *Sphaera* in Portuguese and Spanish. Between the fifteenth and seventeenth centuries, there were more translations of the *Sphaera* into Castilian and Portuguese than into all other European vernaculars combined. A major reason for the interest in *Sphaera* translations in Spain and Portugal was that the basic astronomical and geographical knowledge contained in this text could be used for navigation. Because of their enormous interests in overseas exploration and colonization, Iberian monarchs supported the development of cosmography, a subject that combined mathematics, astronomy, and geography. The astronomical information in the *Sphaera* was also valued on the Iberian Peninsula for its applications to astrology. Finally, the *Sphaera* was critical to a Christian education because it taught readers to appreciate the wonders of God's creation. Spanish and Portuguese commentaries on Sacrobosco's *Sphaera* reflect this mix of practical, political, and spiritual concerns.

1 Introduction

In 1792, French philosophe Nicolas Masson de Morvilliers (1740–1789) characterized Spain as "the most ignorant nation in Europe" (Navarro-Brotóns and Eamon 2007, 27). A weak and ineffective government, combined with a powerful and oppressive Catholic clergy and Inquisition, and a lazy and superstitious population meant that Spain was an intellectual backwater. "Spanish science" was an oxymoron. All the progress of the Scientific Revolution and the Enlightenment took place far away from the Iberian Peninsula and had limited impact there. This dismissive and caricatured view of Spain and its intellectual history is, of course, no longer tenable. Several generations of historians of Spanish science have made abundantly clear that early modern Spain and Portugal had a lively scientific culture (Portuondo 2017). This culture was distinctive, characterized by a focus on the practical problems and empirical evidence produced by voyages of exploration and colonization and often conducted in vernacular languages (Leitão and Sánchez 2017). But Spanish and Portuguese scientists were also active participants in European-wide natural philosophical debates and made distinctive contributions to the Scientific Revolution of the sixteenth and seventeenth centuries.

In this paper I examine a group of commentaries on Johannes de Sacrobosco's (died ca. 1256) *Sphaera* produced by Portuguese and Spanish authors. The *Sphaera* was one of the most popular scientific texts in early modern Spain and Portugal. In the sixteenth and seventeenth centuries there were at least ten printed *Sphaeras*, most with commentaries, in Spanish and Portuguese, and many more manuscript versions (Martins 2004, 371–73). The sheer number of *Sphaeras* with commentaries suggests that there was a lively conversation about astronomy and cosmology on the Iberian Peninsula. Close reading of some of these treatises helps to illuminate the scientific culture of early modern Spain and Portugal, and allows us to see the features that both distinguish this scientific culture from that of other parts of Europe as well as what connects it to the larger European intellectual scene. Although historians of science have by and large moved "beyond the Black Legend," there is much about early modern Spanish and Portuguese contributions to and involvement with science that remain unclear, and Iberian science is still poorly integrated into larger histories of the Scientific Revolution. Analysis of *Sphaera* commentaries demonstrates that Spanish and Portuguese authors were engaged with a range of issues, practical, natural philosophical, and theological, that they were engaged with the work of scholars across Europe, and that their work was read far beyond the borders of Iberia.

2 The Iberian Context

Spain and Portugal were engaged in voyages of exploration and colonization earlier than the rest of Europe. A major reason for interest in the *Sphaera* in Spain and Portugal was that the basic astronomical and geographical knowledge contained in this text could be used for navigation (Martins 2004). Because of their enormous interests in overseas exploration, exploitation, and colonization, the Spanish and Portuguese monarchs actively supported the development of cosmography, a subject that combined mathematics, astronomy, and geography. These intellectual disciplines were vital to the success of colonial operations. Astronomical knowledge served vital state interests—and received state support—in a way that it did not in other parts of Europe.

The Portuguese were the first to venture into parts of the Atlantic Ocean not previously traversed by Europeans. The Portuguese voyages began with the conquest of the North African city of Ceuta in 1415. This military victory was followed by decades of exploration of the west coast of Africa sponsored by the Portuguese monarchy, or more precisely, by Prince Henry the Navigator (1394–1460), younger son of King João I (1357–1433). In 1488, a voyage headed by Bartholomeu Dias (ca. 1450–1500) rounded the southern tip of Africa, the Cape of Good Hope, which gave the Portuguese access to the Indian Ocean and thus a maritime route to Asia (Leitão 2003, 229–30; Leitão 2009). The scale and scope of these voyages was unprecedented, as were the logistical, administrative, technological, and navigational challenges they posed. Sometime in the second half of the fifteenth century, the Portuguese Crown established the *Armazéns da Guineé Índia* (Storehouse of Guinea and the Indies) to meet these challenges. The *Armazéns* coordinated the African voyages and managed the flow of information, people, and objects coming from them. One responsibility of the *Armazéns* was the improvement of navigational tools and techniques as well as the training of navigators (Leitão and Sánchez 2017, 202–03). Much of this work involved the basic astronomical knowledge to be found in Sacrobosco's *Sphaera*. For example, between 1455 and 1475, the Portuguese developed a method for determining latitude at sea using measurements of the height of the North Star taken with a mariner's astrolabe (Leitão 2009, 30). The demand for men with training in practical mathematics and astronomy grew as the global reach of the Portuguese extended. In 1547, the Crown established the position of chief cosmographer, a man tasked with training pilots as well as examining nautical charts and instruments (Leitão 2003, 233).

The other major site for the teaching of astronomy and mathematics in early modern Portugal was the Jesuit Colegio de Santo Antão in Lisbon, which was established in 1553 (Leitão 2003, 234; Leitão 2007). At the request of King João III, the Colegio added the *Aula da Esfera* (Course on the *Sphaera*), a set of lectures on mathematical topics, in 1574. This course was initially only for Jesuit students, but became public in 1590. According to Carolino and Leitão, "great importance [was] attached to nautical questions: elements of cosmography, rules of nautical astronomy and navigation, uses of nautical instruments (astrolabe, quadrant, etc.), the design and construction of nautical charts and globes, etc." (Carolino and Leitão 2006, 162). In keeping with its practical teaching agenda, instruction at the *Aula da Esfera* was in Portuguese, not Latin (Leitão 2006, 374).

In Spain, two state-supported institutions were established "to coordinate the colonization and exploitation of the New World: the *Casa de la Contratación* or House of Trade in Seville and the Council of the Indies" (Portuondo 2009, 4). The *Casa* was founded in 1503 to coordinate and regulate commerce and navigation to the Americas and the Indies (Portuondo 2009; Barrera-Osorio 2006; Sandman 2001). The *Casa* needed men who could prepare instruments and charts for navigation, teach methods of navigation, and make maps. The chief cosmographer at the *Casa* was responsible for licensing pilots and administered an exam to make sure they knew how to use navigational instruments. Cosmographers at the *Casa* produced numerous navigation manuals in Spanish for pilots and navigators in the first half of the sixteenth century (Portuondo 2009; Sandman 2001). Most of these manuals contained material either directly translated from or heavily based on Sacrobosco's *Sphaera* (Portuondo 2009; Martins 2004). The Casa brought together university-educated scholars and practically-trained navigators and sought to foster cooperation between those with theoretical expertise and those with practical experience. However, the relations between these two groups were far from smooth. Cosmographers promoted the use of astronomical instruments—most notably the astrolabe, which was used to find latitude at sea. Navigators, on the other hand, preferred to use the compass and dead reckoning (Sandman 2001). Because Sacrobosco's *Sphaera* offered a basic introduction to astronomy, a number of cosmographers saw it as a valuable resource in convincing pilots, as well as royal and aristocratic patrons, of the usefulness of astronomical knowledge.

In Portugal, astronomy and mathematics were neglected at the universities of Evora and Coimbra, although they were taught at the *Armazéns* and the *Aula da Esfera* (Carolino and Leitão 2006, 162). By contrast, in Spain astronomy and mathematics flourished at the universities of Salamanca, Valencia, Alcalá, and Seville (Navarro-Brotóns 2006). At all of these institutions the *Sphaera* was an important teaching text. The University of Salamanca was a key early site for the development of cosmography (Chabás 2006). The curriculum included subjects like cartography, and incorporated information from voyages to the West Indies and the Americas. Beginning in 1529, the university's statutes specified that the mathematics professor teach arithmetic, geometry, astrology, perspective, and cosmography. This statute was instituted because the Spanish Crown was concerned that there be enough men trained in cosmography, because the subject was critical for navigation and for the mapping of Spain's overseas possessions. The other three universities followed suit over the course of the sixteenth century. These statutes meant that university professors in Spain who taught mathematical subjects, including the *Sphaera*, taught a more diverse body of students than at other European universities. Their students included traditional university students, but also practical men training for careers in navigation and cartography. To accommodate both these groups, instruction in mathematical subjects was often in Spanish rather than Latin. In the sixteenth century, at the University of Salamanca, basic astronomy, including Sacrobosco's *Sphaera*, was taught in Castilian rather than Latin. In 1529, and again in 1538, the university statutes specified that all instruction should be in Latin, except for "grammar, music, rhetoric and astrology [i.e. astronomy]" (Gómez Martínez 2006, 205).

The astronomical information in the *Sphaera* was also valued on the Iberian Peninsula, as elsewhere in Europe, for its applications to astrology. Astrology was critical to medical practice, as well as to political and personal decision-making. Several professors at the Jesuit Colegio da Santo Antão in Lisbon taught courses on astrology as part of the *Aula da Esfera* (Leitão 2006; Carolino 2017). In Spain, astrology was taught at the universities of Valencia, Salamanca, Alcalá de Henares, and Valladolid from the late fifteenth to the early eighteenth centuries (Lanuza-Navarro 2017a, b, 413–18). It was also taught at the Casa de Contratación and the Academia de Matemáticas in Seville. Members of the Spanish royal family and court regularly consulted astrologers throughout the sixteenth and seventeenth centuries (Lanuza-Navarro 2007, 75–77). And in both Portugal and Spain, astrology was considered integral to medical theory and practice (Lanuza-Navarro 2006) (Chaps. 3 and 10).

Alongside a strong emphasis on the practical uses of astronomy and mathematics, Iberian scholars were also engaged with the same natural philosophical questions about the structure of the cosmos that preoccupied scholars in other parts of Europe. Some of their natural philosophical speculations were incorporated into commentaries on the *Sphaera*. Although, as noted above, the teaching at the *Aula da Esfera* was focused on the practical uses of astronomy and mathematics, it also "served as an exceptional channel of communication with the scientific centers of Europe" (Leitão 2003, 236). Two Italian Jesuits who taught at the *Aula da Esfera*, Cristoforo Borri and Giovan Paolo Lembo, both discussed Galileo's telescopic observations of 1610 and 1611. Lembo carried out his own telescopic observations in Lisbon, and his lecture notes indicate that he "addressed the cosmological consequences of these new observations, discarding the traditional Ptolemaic system and proposing a semi-Tychonic system instead" (Leitão 2003, 236).

In Spain, new cosmological discoveries and ideas were discussed at both the universities and at the *Casa* and *Consejo*. For example, Jerónimo Muñoz (ca. 1520–1591), who taught astronomy and mathematics at the universities of Valencia and Salamanca, was one of the many European scientists to observe and write about the supernova of 1572. For Muñoz, the supernova challenged the Aristotelian notion that change was impossible in the celestial realm. In some of his unpublished work and letters to other European astronomers like Tycho Brahe (1546–1601), he espoused an understanding of the relationship between the celestial and terrestrial realms drawn from Stoic philosophers. He denied the existence of celestial orbs and instead asserted that the planets moved through the heavens like birds through the air or fish through the water. He also discussed Nicolaus Copernicus' (1473–1543) heliocentric system with his students, although he did not endorse it (Navarro-Brotóns 1995, 57). In fact, as Victor Navarro-Brotóns has shown, "the work of Copernicus circulated freely in sixteenth-century Spain, where its technical and empirical aspects were greatly admired and used" (Navarro-Brotóns 1995, 63). In 1561, the statutes of the University of Salamanca specified that in the second year of the astronomy course the professor must teach either "the *Almagest* of Ptolemy, or its *Epitome* by Regiomontanus, or Geber, or Copernicus," and that the students could vote on which text they wanted (Navarro-Brotóns 1995, 55). In 1594, these statutes were amended and the teaching of Copernicus was made mandatory, no longer subject to the vote of the students (Navarro-Brotóns 1995, 59). The 1594, statutes were reproduced with no change in 1625, despite the prohibition of Copernicus' work by the Roman Inquisition in 1616 (Navarro-Brotóns 1995, 60). In fact, *De revolutionibus* was "never placed on any Spanish Inquisitorial index" (Navarro-Brotóns 1995, 63), which does not mean Spanish astronomers were free to adopt heliocentrism but does indicate that it was possible to teach and discuss Copernicus in Spanish universities. As Navarro-Brotóns notes, only one Spanish scholar, Diego de Zúñiga (1536–1597), is known to have actually endorsed the Copernican system. Others used the Prutenic tables, which were calculated using Copernicus' mathematical models, and other parameters drawn from *De revolutionibus*, in much the same way that Copernicus was taught at the University of Wittenberg (Navarro-Brotóns 1995, 59; Westman 1975). Finally, interest in

Copernicus spread outside universities, because the Prutenic tables and other technical aspects of Copernicus' work had applications in navigation. For example, Juan Cedillo Diaz (ca. 1560–1625), who studied at Salamanca and became chief cosmographer at the *Consejo de Indias* and professor at the Mathematical Academy in Seville in 1611, made a free Spanish translation of the first three books of Copernicus' *De revolutionibus* sometime between 1620 and 1625 (Granada and Crespo 2019; Navarro-Brotóns 1995, 63; Esteban Piñeiro and Gómez Crespo 1991).

Beyond these practical and philosophical issues, there was a long tradition, going back to Plato, of regarding the study of astronomy as essential to the cultivation of virtue. For many medieval and early modern people, the *Sphaera* was critical to a Christian education because it taught readers to appreciate the wonders of God's creation. Spanish and Portuguese translations of and commentaries on Sacrobosco's *Sphaera* reflect this mix of practical, philosophical, and spiritual concerns.

2.1 Tractado da Spera *(Lisbon, 1510)*

In around 1510, an anonymous Portuguese translation of the *Sphaera* appeared in Lisbon (Bensaude 1914; Martins 2004, 371). This is the first printed *Sphaera* in any European vernacular, although there were certainly vernacular versions in manuscript both earlier and later. This Portuguese translation, titled *Tractado da Spera* (Treatise on the sphere), does not have a formal commentary in the academic sense, but it does have significant extra material attached to it. This *Sphaera* was appended to a treatise on navigation titled *Regimento do estrolabio & do quadrante pera saber ha declinaçam & ho logar do soll em cada hora in dia & asy pera saber ha estrella do norte* (Rules of the astrolabe and the quadrant to find the declination and the place of the sun every hour of the day and also to know the north [pole] star). This short treatise contained instructions for using a mariner's astrolabe and quadrant. It was one of the very first printed books to contain instructions for two new methods for determining latitude at sea: the first using the height of the North Star and the second using the height of the noon sun. As Leitão points out, both of these methods were devised by the Portuguese sometime in the second half of the fifteenth century to meet the challenges of exploring the west coast of Africa (Leitão 2009, 30–32). The practical information in this translation of the *Sphaera* summarized the results of several decades of collaborative scientific work to solve some of the problems of navigating into previously unknown waters. The book was published in a small handy format that could conceivably have been carried on a sea voyage. This Portuguese *Sphaera* stands at the beginning of a long line of navigational manuals that incorporate the *Sphaera*. Multiple versions of these manuals were produced in Spain and Portugal and subsequently the formula was copied in other countries (Leitão 2009, 36; Portuondo 2008, 63).

In addition to information on navigational instruments, this *Sphaera* includes a Portuguese translation of a letter originally written in Latin by Hieronymus Münzer (ca. 1447–1508) to King João II of Portugal (1455–1495) in the summer of 1493

(Pohle 2017, 63). Münzer was a Nuremberg physician, humanist scholar, traveler, and one of the authors of the famous *Nuremberg Chronicle*. In the letter of 1493, Münzer informs João II that the Holy Roman Emperor Maximilian I (1459–1519) recommended that João pursue a westward route to "Cathay." He further recommended Martin Behaim (1459–1507), a German cosmographer already known to João (Pohle 2017), to lead this voyage. What Maximilian, Münzer, and Behaim had in mind was a westward passage across the Atlantic to China, but at a much more northerly latitude than the route Christopher Columbus (1451–1506) had taken in 1492. There was already doubt that Columbus had actually succeeded in finding a route to the Indies, although Columbus himself vehemently asserted that he had. It would not have been unreasonable for any of the people involved in this scheme to believe that a better route might be found (one that actually ended in the Far East) (Hunter 2011). Münzer visited João in 1494 as an emissary of Maximilian (Horst 2017, 18; dos Santos Lopes 2017, 77), and it seems likely he discussed the plan further with the Portuguese monarch.

The letter from Münzer to João II functions as a kind of commentary on the contents of the *Sphaera*. Münzer urges the king to take both ancient authorities and more recent experience into account in considering the feasibility of finding a new sea route to China. Münzer writes that Aristotle (348–322 BCE), Seneca (ca. 1–65), Pierre d'Ailly (1351–1420), "and many other illustrious men, have said that the beginning of the habitable Orient is very close to the end of the habitable West. The proof is in the elephants, which abound in these two places, and similarly in the reeds which the storm from the east throws on the coast of the Azores islands" (Bensaude 1914, 61). But he is also critical of the errors of ancient and medieval authorities and opposes their errors to recent experience, which must carry greater weight. "Do not let yourself be troubled by Alphraganus and others without experience," he writes, "who say that only a quarter of the land is above the sea level, and that the earth in the other three quarters is drowned under the sea, because in things pertaining to the dwelling of the earth, we must believe rather the experience and probable stories than fantastic imaginations. You certainly know that many authoritative astronomers have denied that there is any habitable land in the tropics and in the equinoctial regions. But you have proved by experience that these things were false and foolish" (Bensaude 1914, 61).

We have evidence that this *Sphaera* was read outside of Portugal and by at least one humanist with interests in overseas exploration. There is only one extant copy of the book, in the Bavarian State Library (Bayerische Staatsbibliothek). This copy originally belonged to the German humanist Conrad Peutinger (1465–1547), who assembled one of the largest private libraries in Europe. At his death he owned around 200 manuscripts and 10,000 printed books. He collected many books and manuscripts having to do with geography, travel, and navigation, and was very interested in the overseas "discoveries" of the Portuguese in Africa and the "New World" (dos Santos Lopes 2017, 78). He had a number of Portuguese books and manuscripts in his collection. Peutinger had multiple connections to Portugal and multiple reasons for being interested in Portuguese voyages of exploration. Born into a wealthy Augsburg family, Peutinger was educated at the universities of Basel,

Padua, and Bologna, and he was acquainted with the leading humanist scholars of his age. He achieved a high-level administrative position in the city of Augsburg, and eventually became one of the Holy Roman Emperor Maximilian I's chief confidants, advisors, and diplomats. In this role, he may certainly have been aware of Maximilian's overture, through Münzer, to the Portuguese King João II, to convince him to support an expedition to China by sailing across the Atlantic Ocean. Peutinger had other reasons to be interested in and familiar with Portuguese voyages. He was married to Margarete Welser (1481–1552), member of a wealthy Augsburg patrician family. The Welsers were merchants with far ranging operations. They were the first Germans who, after the discovery of the Indies, established a sales office in Lisbon. They were also the first Germans who joined their ships with Portuguese fleets heading East (dos Santos Lopes 2017, 78).

Many of the Portuguese books in Peutinger's collection were sent to him by Valentim Fernandes, a German printer who moved to Lisbon and was active there from 1494 to 1518 (dos Santos Lopes 2012, 2017, 78; Hendrich 2017; Blackmore 2009, 25–26). Fernandes is a reasonable candidate for the printer of this Portuguese sphere although he is not identified on the title page. In this period there were only about eleven printers active in Lisbon. Fernandes published other books connected to exploration, including a translation into Portuguese (which he prepared) of the travels of Marco Polo (1254–1324) (*Livro de Marco Paulo*, 1502). Further, he had a close relationship with both the Portuguese court and with Germans resident in or visiting Portugal. He would certainly have been familiar with the representatives of the Welser family. When Hieronymus Münzer visited Lisbon in 1494 on a diplomatic errand for Maximilian I, Fernandes served as his translator. As noted, Fernandes had an ongoing relationship with Conrad Peutinger and sent him a number of Portuguese books. Although this *Sphaera* is small and unprepossessing, it reflects a dense network of connections between the Holy Roman Empire and Portugal in the early sixteenth century, and intense interest in voyages of exploration on the part of a wide variety of people—humanists, diplomats, monarchs, printers, and sailors (Horst et al. 2017).

2.2 *Francisco Faleiro,* Tratado del esphera y del arte del marear *(Seville: Juan Cromberger, 1535)*

Another translation of the *Sphaera* ostensibly aimed at navigators was Francisco Faleiro's *Treatise on the sphere and on the art of navigation*, published in Seville in 1535 (Faleiro 1535). Faleiro was a Portuguese cosmographer who came to Spain in 1518 and worked at the House of Trade from 1519 to 1532. Faleiro's text is, as the title suggests, two separate books: the first a partial translation of Sacrobosco's *Sphaera* and the second a treatise on navigation. The translation of the sphere contains numerous illustrations to facilitate understanding, and some commentary on the original text. In the treatise on navigation, Faleiro proposed a solution to a problem that plagued navigators, the magnetic variation of the compass (Collins 2013).

Magnetic variation causes compass needles to deviate from true north. For navigation in smaller areas like the Mediterranean the deviation is not great enough to cause problems. But on long ocean voyages it is. As Maria Portuondo notes, "Navigation manuals became a forum for advocating different ways of coping with the problem, either by designing instruments that 'corrected' the compass reading or trying to use the needle's deviation as an indicator of longitude" (Portuondo 2008, 64).

Faleiro translated portions of the *Sphaera*, rearranged them, and added new material, reflecting the book's function as a teaching text for men who were going to sea. Indeed, like the Portuguese *Sphaera* discussed above, the text is small enough that it was conceivably taken to sea for use as a reference. However, despite this clearly practical context, Faleiro's *Sphaera* is notable for the repeated emphasis on the importance of cosmographical knowledge in cultivating Christian piety. In the introduction, Faleiro claims he has translated the *Sphaera* "for those like me who do not have polished Latin," so that they might be able to learn about "the admirable works and marvels of God."[1] Contemplation of the divine works of creation will lead the reader to a deeper knowledge and love of God. The Christian who learns about the heavens "will know much more clearly the greatness, power, and wisdom with which such work was made, and with much more understanding, joy and knowledge give praise to the Lord, as the psalmist says: The heavens declare the glory of God."[2]

Faleiro expanded on various sections of the *Sphaera* and cut others entirely. In his discussion of the planets he comments on their influences on human life and health. For example, he informs readers that Saturn is cold and dry, connected to the humor melancholy, and has a generally negative effect on human life. This was all quite standard astrological fare, but it was not in Sacrobosco's original text. Faleiro uses it to emphasize the interconnectedness of the heavens and the earth, and the magnificence of God's plan for the world. Faleiro cut out the last section of Sacrobosco's *Sphaera*, which introduced mathematical models for the motions of the sun and the moon. Perhaps he deemed them too technical and unnecessary for his projected audience.

Faleiro also corrected and updated Sacrobosco's original text by referring to knowledge gained through the recent experiences of Portuguese and Spanish explorers. For example, Sacrobosco had asserted that there were five climactic zones: a torrid zone around the equator which was too hot to sustain life, two frigid zones around the Arctic and Antarctic poles which were too cold to sustain life, and two temperate zones which were the right temperature to sustain life. Faleiro pointed out

[1] (Faleiro 1535, fol. a ii v): "…para que los que como yo no alcancaren la polida latinidad: a esta falta no dexen de saber algo por natural razon delas admirables obras y maravillas de dios."

[2] (Faleiro 1535, fol. a iii): "Mas el christiano que por todo esto passare contemplando y viendo como el esphera y la orden della es la mas excelente y admirabile obra entre todas las obras despues de la que dios a su semejanca hizo: con mucha mas claridad conocera la grandeza / poder / y saber del que tal obra hizo: y con much mas conocimiento / gozo y saber dara loores al señor: y conel psalmist dira. Celi enarrant gloriam dei."

that Portuguese voyages to the Cape of Good Hope and Brazil demonstrated that no portion of the earth was uninhabitable. And far from being arid deserts, the equatorial regions were remarkably fertile. According to travelers's reports, "cucumbers and melons, pears, lettuce, eggplant, and many other fruits"[3] grew year-round in the torrid zone, rather than only in a particular season as in the temperate zone. However, the peoples of this region were sometimes strange—Magellan allegedly found people almost the size of giants on his voyages of 1520–1521. In general, Faleiro informs his readers, "The temperate zone is populated by the best part of people, most endowed with reason and of better understanding and more skill than people who inhabit the other zones."[4]

In this section, Faleiro uses Sacrobosco to represent received and authoritative knowledge, and he opposes this authoritative knowledge drawn from books with experience. But Sacrobosco's statement that the torrid and frigid zones of the earth were uninhabitable was by no means a universally accepted claim in antiquity or the Middle Ages. For example, in the 1260s Campanus of Novara (ca. 1220–1296) composed an astronomical textbook, one that Lynn Thorndike describes as "evidently…indebted" to Sacrobosco's *Sphaera*, in which he argued that the climate at the equator was temperate (Thorndike 1949, 28). Long before European voyages of exploration began in the fifteenth century, commentators on Sacrobosco's *Sphaera* refuted his claim that the torrid and frigid zones were uninhabitable, drawing on reason, experience, and authoritative texts. One of the earliest commentaries on the *Sphaera* is by Robertus Anglicus (13th cent.) and was composed around 1271 (Thorndike 1949, 28). On the subject of the torrid zone, Robert argued "that the equatorial region is habitable and not merely habitable but fine to live in" (Thorndike 1949, 239). He cited Avicenna (980–1037) and Isidore as authors who asserted that the torrid zone was habitable, but he also reasoned that since days and nights are always equal at the equator, the heat of the day would always be balanced out by the cool of the night, creating a temperate climate. Nevertheless, these passages on the climactic zones in Faleiro's *Tratado del esphera* reflect the high value he placed on experiential knowledge. He could have cited various ancient and medieval authorities who argued that the equatorial region was temperate, but he chose to highlight the reports of mariners. In so doing he created a sense of progress, of modern knowledge surpassing ancient wisdom. Many historians of Spanish and Portuguese science identify this kind of valorization of experiential and empirical evidence as a distinctive feature of "Iberian science" and a major contribution to the development of modern science (Leitão and Sánchez 2017; Portuondo 2009; Barrera-Osorio 2006). As Henrique Leitão and Antonio Sánchez write, "All Iberian literature connected to maritime discoveries in this period exhibits a vigorous and self-conscious departure from ancient authorities. The new knowledge was not acquired by logical

---

[3] (Faleiro 1535, fol. ciii): "…pepinos y melones / peras / lechugas / berenjenas: y otras muchas frutas…."

[4] (Faleiro 1535, fol. ciiii): "Esta templada zona es poblada por la mayor parte de gentes mas acogidas a razon y de mejores entendimientos y mas abiles y para mas que las otras gentes de que son abitadas las otras zonas."

syllogisms and deductive reasoning, but by experience—i.e., by direct empirical evidence" (Leitão and Sánchez 2017, 208).

Faleiro is also concerned to explain how this newly acquired geographical knowledge can be incorporated into biblical interpretation. He concludes that Asia and Europe are in the temperate zone, while most of Africa is in the torrid zone. And it is in the temperate zone that all the important historical events have occurred. The first human beings were created and sinned for the first time in the temperate zone. Noah built his ark in the temperate zone. The twelve tribes of Israel were dispersed in the temperate zone. And so on. All the events of biblical and world history occurred in Europe and Asia. Africa was in this account devoid of history and significance (Faleiro 1535, fol. ciiii–ciiii v). It is worth pointing out that by the time Faleiro wrote, the Portuguese slave trade had been going on for almost a century and was one of the most lucrative aspects of Portuguese overseas operations (Saunders 1982). Faleiro's low estimation of the peoples of Africa must be seen in the context of ongoing efforts to provide legal, philosophical, and moral justification for human trafficking.

Significantly, there is very little in Faleiro's translation of the *Sphaera* that refers to the practical uses of astronomy in navigation. That is all reserved for the second half of the book, on the art of navigation. The material in the *Sphaera* was clearly necessary for the reader to make sense of the navigational tools and techniques described in the second part, but Faleiro cast his version of the *Sphaera* as a pious contemplation of the wonders of the cosmos God created and the unfolding of human history according to God's plan.

2.3 Pedro Nuñes, **Tratado da sphera com a Theorica do Sol e da Lua** *(Lisbon: Germão Galharde, 1537)*

In 1537, Pedro Nuñes (1502–1578), one of the most prominent European mathematicians of the sixteenth century, published a Portuguese translation of the *Sphaera* with commentary (Nuñes 1537). Nuñes's book was called *Tratado da sphera com a Theorica do Sol e da Lua* (Treatise on the sphere and *Theorica* of the sun and moon). It also included a translation of the first book of Ptolemy's (died 160) *Geography*, a translation of the chapters on the sun and moon from Georg Peuerbach's (1423–1461) *Nova Theorica*, and two treatises on navigation (the first on "certain doubts about the navigation" and the second on the nautical chart). Nuñes was educated at the University of Salamanca and then the University of Coimbra in Lisbon. Between 1544 and 1562 he held the Chair of Mathematics at Coimbra, a position that required him to teach arithmetic, geometry, perspective, music, and astronomy (Leitão 2009, 81; Carolino and Leitão 2006, 159; Leitão 2002 15–27). Although he was a brilliant mathematician, he seems to have been an indifferent teacher. Carolino and Leitão comment dryly that his "classes in Coimbra seem to have been constantly plagued by lack of attendance, both by students and by Nuñes himself" (Carolino and Leitão 2006, 159). In 1547, he was appointed to the newly created position of chief cosmographer, a position that gave him responsibility over navigators, cartographers, and instrument makers, as well as access to the royal court (Leitão and Sánchez 2017, 204; Leitão 2003, 233). Nuñes's translation and commentary on the *Sphaera* reflects his multiple roles as university professor, royal adviser, and trainer of pilots. His book contained a great deal more theoretical information on astronomy and mathematics than did either of the two *Sphaerae* I have discussed so far. In the first half of the sixteenth century there was considerable disagreement about just how much theoretical knowledge navigators needed (Sandman 2001; Portuondo 2009; Almeida 2008). Some, like Nuñes, felt that navigators needed a very solid grounding in mathematics, astronomy, and cosmology to do their jobs safely and effectively. At the other extreme were those who were satisfied with rote memorization of rules and the use of tables. As Bruno Almeida notes, Nuñes distinguished between "*ars navigandi*" and "*ratio navigandi*." The "art" of navigation was based on known sets of rules, procedures, and instruments." "Rational" navigation, which Nuñes held to be far superior, was "based in the understanding and use of mathematical principles" (Almeida 2012, 462).

Nuñes, like Faleiro, referred to Portuguese voyages of exploration to refute the idea that there were parts of the earth that were uninhabitable: "The navigations of the Portuguese show us that there is no land so distempered by hot or by cold that it is not inhabited."[5] In addition, he tried to introduce greater mathematical precision and rigor into Sacrobosco's *Sphaera*. For example, he adds a reference to the spherical geometry of Theodosius (ca. 160–ca. 100 BCE) to Sacrobosco's discussion of the movement of the sun and the unequal lengths of the day. The passage in Sacrobosco reads,

> In the right sphere the horizon, since it passes through the poles of the world, divides all those circles into equal parts, whence the arcs of days are the same as those of nights for persons living at the equator. Hence it is evident that for persons living at the equator it is always equinox, wherever the sun may be in the firmament. (Thorndike 1949, 133).

Nuñes adds,

> Theodosius, in proposition 19 of his first book, demonstrates that every major circle passing through the poles of another circle cuts it in two equal parts and right angles. And in [proposition] 18, he demonstrates that if you cut a minor [circle] in half, you must pass through the poles.[6]

Nuñes also includes a new mathematical demonstration about the climactic zones (Nuñes 1537, fols. c–ciii v). At the end of book three in the *Sphaera* Sacrobosco describes the seven climes into which the habitable portion of the earth

---

[5] (Nuñes 1537, fol. b): "As navegaçones dos portugueses nos a mostratan que não ha terra tam destemperada per quente ne per fria em que não aja homes."

[6] (Nuñes 1537, fol. Biii.): "Theodosio demonstrou a 19. proposição do seu primeiro livro que todo circulo mayor que passa pellos polos de outro circulo o corta en duas partes yguais e per angulos reytos: e na. 18. demostra que se cortar a algun menor pella metade he necessario que passe pellos polos delle."

was divided. Sacrobosco defines these by the average length of the longest day of the year in each, ranging from 13 h in the southernmost to 16 in the northernmost. The climes were of unequal size as well, ranging from 440 miles across to 185. In other words, the climes closer to the equator were wider, those closer to the north pole narrower. But Sacrobosco provides no rationale for these numbers. Nuñes proposes a mathematical demonstration of the varying width of the climate zones. Matteo Valleriani has traced the way in which this particular section was incorporated into a large number of subsequent *Sphaera* commentaries (Valleriani 2017). The French humanist Elié Vinet (1509–1587) translated this section of Nuñes's commentary into Latin and incorporated it into his own commentary on Sacrobosco's *Sphaera*. Vinet's commentary was reprinted numerous times, first in Paris and then in several major European cities. In this way, Nuñes' mathematical demonstration spread all over Europe.

2.4 *Jerónimo de Chaves,* Tractado de la sphera que compuso el doctor Ioannes de Sacrobusto con muchas additiones *(Seville: Juan de Leon, 1545)*

In 1545, Jerónimo de Chaves (1523–1574) published a translation of Sacrobosco's *Sphaera* with commentary (de Chaves 1545). Jerónimo de Chaves was a mathematician, astronomer, and cosmographer who held important positions at the House of Trades. He followed in the footsteps of his father, Alonso de Chaves (1492–1586), who was also a cosmographer at the House of Trades (Sánchez Martínez 2010; Martins 2004, 376; Sandman 2001). Jerónimo de Chaves's sphere was called *Treatise of the sphere composed by doctor Johannes de Sacrobosco with many additions*. His treatise contains more theoretical material and less practical material than any of the *Sphaeras* I discussed above. Chaves' translation of Sacrobosco's *Sphaera* contains a quite extensive commentary. However, despite his later role as a teacher of navigators, the commentary contains very little directly related to navigation. In the prologue, he claims that he chose to translate the *Sphaera* "for the pleasure and common usefulness of my country and my friends."[7] But he makes clear that he had far more in mind than the usefulness of this text as a basis for navigation. He claims that he wanted to make accessible to those who could not read Latin, "the most sublime science and art of Astronomy,"[8] which reveals the "Majesty, Wisdom and infinite Power"[9] of the creator (de Chaves 1545, fol. iii). Chaves' view of the "utility" (*utilidad*) of astronomy was quite broad (Omodeo 2017). He asserts that the material in Sacrobosco's *Sphaera* was essential for philosophers, physicians, and astrologers. It was also necessary if one wanted to comprehend "certain passages

[7] (de Chaves 1545, fol. iii): "...dela delectation y utilidad commun de mipatria y amigos...."

[8] (de Chaves 1545, fol. iii): "...tan sublimada sciencia y arte de la Astronomia...."

[9] (de Chaves 1545, fol. iii): "...su Magestad, Saber, y infinito Poder...."

and obscure verses of poets."[10] Here he seems to have had in mind primarily Virgil (70–19 BCE), whose *Georgics* is cited throughout the commentary, as well as Lucan (39–65). Navigation only comes in at the end of Chaves's list of reasons for learning astronomy. There are only a couple of references in the entire book to "those who sail the Atlantic."[11] Very little material in the book touches on problems of navigation or cosmography relevant to exploration of the "New World." Chaves includes a discussion of the location of the meridian at the "Fortunate Isles" (or the "Canaries"), a topic of much contention between Spanish and Portuguese cosmographers (Sandman 2001). He includes a table of the positions of fixed stars but does not include any "new" stars or constellations, like the Southern Cross (de Chaves 1545, fol. xlviii v). Similarly, he includes a table of the longitudes and latitudes of various places, but all in Spain, nothing in the "New World" (de Chaves 1545, fol. l v–li).

While Chaves's translation of Sacrobosco's *Sphaera* may well have appealed to people who could not read Latin (or who could not read Latin well), he had a more learned audience in mind as well. He claimed that his version of the *Sphaera* had more extensive commentary, including "many demonstrations, figures and supplementary tables,"[12] than most other editions of Sacrobosco, even those in Latin. And indeed, his explanations and expansions on Sacrobosco's original text are long and detailed, and his text is richly illustrated. Much of his commentary engages with natural philosophical issues that concerned his contemporaries, both in Spain and in other parts of Europe. For example, he discusses celestial orbs in much greater detail than Sacrobosco did. In his commentary on Sacrobosco's first book, Chaves explains the difference between an orb and a sphere and includes a picture of an orb whose concavity has a different center than its outer surface. This material was certainly not necessary for navigation, but it was an important prelude to a discussion of the structure of celestial orbs in the more advanced astronomy text by Georg Peuerbach, the *New Theorica* (Crowther and Barker 2013). Chaves wrote this in a period when European astronomers, including Spanish astronomers like Jerónimo Muñoz, were debating whether celestial orbs existed or not, and if they did, what their internal structure was (Barker 2011) (Chap. 6). Chaves also discusses the phenomenon of trepidation (de Chaves 1545, fols. xiii, xvii) and debates about the number of celestial spheres (de Chaves 1545, fol. xv v). Again, none of this material was necessary for navigation, but it was all part of learned discussions of astronomy in this period.

Chaves updated Sacrobosco's original text with new information gained from the Portuguese and Spanish voyages of exploration. For example, he explains how to determine the longitude of a place using the timing of lunar eclipses (fol. xlviii), a method much discussed by cosmographers interested in mapping Spain's overseas possessions. Another good example is in the section on the five climactic zones, which Chaves, like Faleiro and Nuñes, notes are all both habitable and inhabited.

[10] (de Chaves 1545, fol. iii v): "...algunos lugares y versos obscuros de Poetas...."

[11] (de Chaves 1545, fols. xxvii v, xxx): "Los que navegan el mar Oceano...."

[12] (de Chaves 1545, fol. iii–fol. iii v): "...muchas demonstrationes, figuras y tablas supputatorias...."

Like Faleiro, Chaves comments on the supposed characters of people in the different climactic zones. According to Chaves, the people in the torrid and frigid regions were wilder and more animal-like than the inhabitants of the temperate zone. People in the torrid region were exposed to excessive heat "which burns their bodies, and blackens them, and dries the humors, and bakes their faces, and weakens their limbs."[13] Their hair is "black and frizzy" (*negros, y crespos*) and their social customs are "wild and without temperance" (*silvestres y sin temperamento*). People in the frigid zones are very pale, with red hair and large soft bodies. They too are savage and uncivilized. Chaves claims, "There are in these lands or cold areas, men who eat human flesh, and drink blood: as there are also in the torrid zone."[14] The torrid and frigid zones are full of "evil people, and devils, and beasts who are inimical to human nature."[15] By contrast, the inhabitants of the temperate zone are "better and healthier men."[16] These passages show how *Sphaera* commentators corrected Sacrobosco's text, but also how they fit new information into pre-existing intellectual frameworks, in this case climactic theories of human difference that were derived from Hippocrates (ca. 460–ca. 370 BCE). And as I commented in reference to Faleiro's text, by the time that Chaves wrote this, the Portuguese slave trade had been underway for a century and there were several 1000 men, women, and children of African origins, both enslaved and free, living in Portugal and Spain (Saunders 1982). Chaves did not so much "observe" these people as reproduce racial and ethnic stereotypes that had been prevalent in Europe for centuries. The description of Africans as "black" and of blackness as physically, morally, and spiritually inferior to whiteness long predates voyages of exploration (Heng 2011). The thirteenth-century encyclopedist Bartolomeus Anglicus (1190–1250) also described the people of Africa as black, frizzy-haired, cowardly, and full of guile in his *De proprietatibus rerum* (Heng 2011, 316). As in Faleiro's case, Chaves's comments on racial differences need to be seen as part of a justification for enslavement and colonization.

2.5 *Martin Cortes.* Breve Compendio de la Esfera y de la Arte de Navegar *(Seville, 1551)*

One of the most successful translations and commentaries on Sacrobosco's *Sphaera* produced in Spain was the *Brief Compendium of the Sphaera and of the Art of Navigation* by Martin Cortes (1510–1582), published in 1551 in Seville (Cortes 2003). Martin Cortes taught cosmography and navigation to pilots in Cadiz and was

[13] (de Chaves 1545, fol. liiii v): "…les quema sus cuerpos, y los ennegresce, y desecca los humores, y assa les los rostros, y enflaquece sus miembros."

[14] (de Chaves 1545, fol. liiii v): "Y hallanse enestas terras ò zonas frias, hombres que comen carne humana, y beven la sangre: segun que tambien se hallan enla torrida zona."

[15] (de Chaves 1545, fol. liiii v): "…malignos, y diablos: y crianse bestias que son empecientes a la naturaleza humana…."

[16] (de Chaves 1545, fol. liiii v): "…los hombres mejor y mas sanos…."

at some point royal cosmographer (Barrera-Osorio 2008). Cortes's book is divided into three parts. The first is heavily based on Sacrobosco's *Sphaera*, the second contains material from Sacrobosco's *Sphaera* along with navigational material, and the third is instructions on the construction and use of navigational instruments (Martins 2004, 378). The translation of the *Sphaera* is indebted to Jerónimo de Chaves's earlier translation, and parts of the sections on navigation are copied from Pedro Medina's (1493–1567) *Art of Navigation* of 1545. Cortes, however, claimed great originality for his book. In his dedicatory letter to the Spanish king Charles V, Cortes claims to have been "the first person to reduce navigation to a brief compendium, setting out infallible principles and clear demonstrations, writing the practice and theory of it, giving the true rules to sailors, showing the way to pilots, making them instruments to know how to take the height of the sun, to know the ebb and flow of the sea, ordering charts and compasses for navigation, advising them of the course of the sun, the movement of the moon, and a clock for the day so true that in all the lands it indicates the hours without any defect, another clock infallible for the nights, discovering the secret property of the stone magnet, explaining the [deviation] of their needles."[17]

Although the claims of originality are greatly overstated, the third part of the book does indeed contain instructions for making and using sea charts (fols. lxi v–lxviii), measuring the height of both the North Star and the sun in order to determine one's longitude at sea (fol. lxxiii), as well as the construction and use of the mariner's astrolabe (fol. lxxv v–lxxviii) and other nautical instruments.

Although written in Spanish for the ostensibly practical purpose of instructing navigators, the *Breve compendio* bristles with references to ancient, medieval, and modern writers, including Aristotle (passim), Lucretius (fol. xiii), Pliny (fol. xxi), Albertus Magnus (fol. xii), Avicenna (fol. xii), Averroes (fol. xi), Alfonso X (fol. xv), Arzachel (fol. xv), Regiomontanus (fol. xxvii), Faber Stapulensis (fol. xx), and Franciscus Capuanus de Manfredonia (fols. xvii, xxxi).[18] And Cortes addresses a number of natural philosophical questions that would not have had practical implications for navigators, including debates about the number of celestial spheres (fol. xii v) and the phenomenon of trepidation (fol. xv v). He also raises the "Pythagorean" argument for a moving earth, if only to reject this idea. Cortes offers both physical and scriptural evidence that the earth is stationary (fols. xiii–xiii v).

Like many other Iberian authors, Cortes incorporates information gleaned from voyages of exploration into his commentary on Sacrobosco. While his citations to a

[17] (Cortes 2003, fol. ii v): "…mas digo aver sido yo el primero que reduxo la navegacion a breve compendio / poniendo principios infalibles y demonstraciones evidentes / escriviendo pratica y theorica della / dando regla verdadera a los marineros / mostrando camino a los pilotos / haziendoles instrumentos para saber tomar el altura del sol / para conoscer el fluxo yre fluxo del mar / ordenarles cartas y bruxolas para la navegacion / avisandoles del curso del sol / movimiento de la luna / relox para el dia y tan cierto que en todas las terras señala las horas sin defecto alguno / otro si relox infalible para las noches / descubriendo la propriedad secreta dela piedra yman / aclarando el nordestear y norvestear / delas agujas."

[18] On Francesco Capuano da Manfredonia's commentary on *De sphaera* see (Chap. 4).

variety of authors demonstrate his erudition, he consistently valorizes experience over texts and describes the knowledge of his contemporaries as superior to that of earlier generations. "Navigation, as well as the other arts…, has been increasing little by little until it has come to perfection,"[19] he writes in his dedication letter to Charles V. He uses the evidence of recent voyages to refute Sacrobosco's statement that the torrid and frigid zones are uninhabitable (fols. xxi–xxii), and also notes that it was the moderns, not the ancients, that discovered the West Indies (fol. iiii).

Finally, Cortes sets astronomical and navigational knowledge in an explicitly Christian framework. He begins the book with a chapter on the creation of the world, which starts, "The great God is the beginning and cause of the entire universe,"[20] and goes on to detail the different aspects of creation and the special place of human beings. And in his discussion of the solstices and equinoxes, Cortes states that Jesus Christ was conceived on the vernal equinox and born on the winter solstice, while John the Baptist (first cent. BCE–ca. 28) was conceived on the autumnal equinox and born on the summer solstice (fols. Xxxi–xxxi v). In this way, the solar year is connected to salvation history. Cortes's *Breve compendio* was even more popular in England than it was in Spain. It was translated into English and went through nine English editions between 1551 and 1630 (Martins 2004, 377). This text was one of the means by which a great deal of new information and technology relevant to navigation was disseminated from the Iberian Peninsula to other parts of Europe.

2.6 *Rodrigo Saenz de Santayana y Spinosa,* La Sphera de Iuan de Sacrobosco *(Valladolid: Adrian Ghemart, 1567)*

One of the largest and most extensive *Sphaera* commentaries in Spanish is the edition published by Santayana y Spinosa 1567 in Valladolid (Santayana y Spinosa 1567). Very little is known about this author, other than that he was born in about 1540 and was a member of the nobility (Martins 2007, 331). Santayana y Spinosa asserts that knowledge of God's creation, and especially knowledge of "the composition and order of the celestial machine, the movements and courses of the circular spheres, the virtues and influences of the stars and celestial constellations"[21] will bring the pious Christian to a deeper knowledge and love of the God. On the frontispiece, there is a line from Psalm 19: "the heavens declare the glory of God; and the firmament sheweth his handywork." Throughout the text the emphasis is on contemplating the wonders of God's creation. However, Santayana y Spinosa does

[19] (Cortes 2003, fol. iiii): "La navegacion asi como las otras artes de cada dia se ha ydo augmentando y poco a poco ha venido a tener su perfection…."

[20] (Cortes 2003, fol. ix): "El immenso dios principio y causa de todo el universe…."

[21] (Santayana y Spinosa 1567, fol. 1): "…la compustura y orden de la machina celestial, los movimientos y cursos de sus circulares Spheras, la virtudes e influencias de las Estrellas y constellaciones celestials…."

assert that understanding the heavens is crucial to understanding life on earth, because the heavenly bodies influence earthly bodies, including the human body. As he puts it, "He who knows [astrology] can avoid many influences of the stars, because he knows his own nature, and he can prepare beforehand."[22] Only a man who actually understood how he was influenced by the stars could take charge of his destiny. Santayana y Spinosa includes a significant amount of astrological material in his commentary, including a spirited defense of the utility and validity of astrology, complete with references to Thomas Aquinas (1225–1274), Pedro Ciruelo (1470–1554) (Chap. 3), Albumasar (787–886), and Cecco d'Ascoli (1269–1327) (Martins 2007, 332).

Santayana y Spinosa includes natural philosophical speculations about the cosmos in his commentary on Sacrobosco's text, including consideration of the heliocentric system of Nicolaus Copernicus. In one section of the *Sphaera*, Sacrobosco explains that the earth is at rest in the center of the cosmos, and he briefly gives some Aristotelian arguments about why this must be the case. Santayana y Spinosa, like a number of sixteenth-century commentators on Sacrobosco, expanded on this, clearly in response to the newly proposed heliocentric system. Although Santayana y Spinosa does not refer to Copernicus by name, he refers to "Pythagoreans" (Santayana y Spinosa 1567, fol. 20), a common designation of Copernicans in the period. Of course, when Sacrobosco wrote the *Sphaera* in the first half of the thirteenth century, there was little reason to question this view. But after the 1543 publication of Copernicus' *De revolutionibus orbium coelestium*, some *Sphaera* commentators felt the need to address the question of the motion or rest of the earth in more detail. Most, including Santayana y Spinosa, did not accept Copernicus' admittedly rather weak arguments for a moving earth, but it is nonetheless interesting that they were clearly aware of Copernicus' book and expected that at least some of their readers were as well. Santayana y Spinosa explains that the earth must be immobile because "every heavy body naturally desires the lowest and deepest place, and this is the center and middle of the entire Firmament."[23] And because it is "the natural place of the [element] earth, [thus once it reaches this place] it remains in firmness and stillness."[24] Any movement of the earth away from the center "would be by violence and not naturally."[25] This reiterates an argument found in Aristotle's *De caelo*, and was one of the standard objections to the physical reality of the Copernican system.

Santayana y Spinosa was only in his mid-twenties when he published his translation and commentary on the *Sphaera*. Unlike the other authors I have discussed in this paper, his motivations were unclear—he was not affiliated with a university or

[22] (Santayana y Spinosa 1567, fol 1 v): "Porque el que es sabio…puede evitar muchas influencias delas Estrellas, conosciendo su naturaleza, y preparar se antes de su venida."

[23] (Santayana y Spinosa 1567, fol 19 v): "…que todo cuerpo pesado naturalmente desea el lugar mas baxo y mas hondo, y este es el Centro y medio de todo el firmament…."

[24] (Santayana y Spinosa 1567, fol 19 v): "Y por ser aquel el natural lugar de la tierra, permanesce en firmeza y quietud…."

[25] (Santayana y Spinosa 1567, fol 19 v): "…lo qual seria por violencia y no naturalmente…."

with the Casa de Contratacion. Further, as Roberto de Andrade Martins has pointed out in his analysis of Santayana y Spinosa's *Sphera*, the man's grasp of astronomy appears rather weak at points. He makes a number of basic errors, including confusing the phases of the moon with eclipses, misstating Aristotelian arguments for the rotundity of the earth, and misunderstanding the concepts of the horizon and the major circles (Martins 2007, 333–34). Further, parts of the text appear to have been copied from Chaves' translation (Martins 2007, 335). While it is impossible to draw any firm inferences from Santayana y Spinosa's translation and commentary on the *Sphaera* since so little is known of his life and career, his enthusiastic if naïve text seems to indicate interest in astronomical and cosmological knowledge beyond the circles of university professors, cosmographers, and navigators.

2.7 Diego Pérez de Mesa, Comentarios de Sphera *(1596)*

The last *Sphaera* commentary that I will discuss is one by Diego Pérez de Mesa (1563–ca. 1632), composed sometime around 1596. According to Víctor Navarro-Brotóns, Pérez de Mesa studied at the University of Salamanca under Jerónimo Muñoz, arguably the most famous Spanish astronomer of the period. Muñoz was then chair of astronomy and mathematics. He wrote a widely read work on the supernova of 1572. As I noted before, several Spanish universities of the period specified that Copernicus either could or should be taught by the professor of astronomy. Muñoz discussed Copernicus' work in writing and with students, although he did not endorse heliocentrism. After studying in Salamanca, Pérez de Mesa took up the chair in astronomy at Alcalá de Henares. In 1595, at the request of the King Philip II, Pérez de Mesa moved to Seville. Here he took up a position newly created to train navigators and cosmographers. He taught both university students and men training to be pilots, giving classes on arithmetic, algebra, geometry, astrology, and navigation. All his classes were in Spanish. Although he wrote several books, most, including his *Comentarios de Sphera*, remained in manuscript (Navarro-Brotóns 2006, 92–93).

In his preface, he describes cosmography as "a science of the heavens and the world" (*una siencia del cielo y mundo*). It concerns practical subjects like geography and navigation, but it also engages "many beautiful questions of philosophy, such as whether the fourth element fire is in the concavity of the moon, and if it is possible that the earth and the celestial bodies move together, and if the stars move by themselves or together with their orbs."[26] Pérez de Mesa's commentary ranges over all of these natural philosophical questions and more, as well as practical applications of astronomy. Pérez de Mesa also wrote separate treatises on navigation and

[26] (Pérez de Mesa 2009, fol. 1): "…muchas questiones hermossimas de la misma filosofia como es hay fuego quarto Elementi en el concavo de la luna puesto si es posible que se muevan juntamente la tierra y los cuerpos celestiales y si se mueven las estrellas por si solas o juntamente con las orbes…."

astrology. Both of these, like the *Comentarios*, remained in manuscript. His work reflects familiarity with authors both ancient and modern. Among the moderns, he cites Nicolaus Copernicus, Girolamo Cardano (1501–1576), Francisco Valles (1524–1592), and Jerónimo Muñoz. He includes discussion of Copernicus' heliocentric system, devoting a section to arguments for and against a moving earth. He also incorporates new information gleaned from voyages of exploration into his philosophical discussions. A particularly striking example of this is in his discussion of the four elements. He begins, conventionally enough, by stating that "earth has the lowest place of all the elements."[27] Earth is the heaviest element and it naturally moves down to the center of the cosmos. Here he notes that this is an objection to the Pythagoreans, Aristarchus (310–230 BCE), and Copernicus, all of who would "put the earth among the stars."[28] This is an argument that comes straight out of Aristotle and would have surprised no one. His discussion of the element water, however, is considerably less conventional. Water is the second heaviest element, according to Pérez de Mesa, and it covers the earth. But unlike earth, which moves in straight lines, water has a circular motion: "water moves in circles over the surface of the earth."[29] His evidence for this position, which he acknowledges is a controversial one, is "the experience and certain reports of all navigators."[30] Sailors, he writes, know that ocean currents move in circles. He gives multiple examples of this phenomenon, ranging from the Pacific Ocean, the Indian Ocean, the Atlantic Ocean, the North Sea, and the Mediterranean.

In addition to his discussion of natural philosophical questions, Pérez de Mesa includes practical uses of astronomy. For example, when he discusses the meridian, which Sacrobosco defined as "a circle passing through the poles of the world and through our zenith," he explains how to calculate one's longitude using a table of the height of the sun at midday. His commentary, like Nuñes work, is full of mathematical demonstrations that lend greater rigor and precision to Sacrobosco's text.

3 Conclusions

My goal has been to demonstrate the rich variety of Iberian treatises of *De sphaera* in the sixteenth century, and to examine what they can tell us about scientific culture on the Iberian Peninsula in this period. The seven treatises that I have described are different in tone, format, and content. Some were aimed at practical men—pilots and navigators who would need to use cosmographical knowledge at sea. But most also included considerable information relevant to astrology, history, and religious

[27] (Pérez de Mesa 2009, fol. 4): "La tierra tiene el lugar mas baxo de todos los Elementos…."

[28] (Pérez de Mesa 2009, fol. 4): "…las quales la ponía entre las estrellas…."

[29] (Pérez de Mesa 2009, fol. 4): "…el agua se mueve sircularmente sobre la superficie de la tierra…."

[30] (Pérez de Mesa 2009, fol. 4): "…la espiriencia y siertas rrelaciones de todos los navegantes…."

piety. Some commentators, like Chaves and Santayana y Spinosa, situated astronomical knowledge in a religious and moral context rather than a practical context, although they too included some information about astrology and navigation. These texts were aimed at more elite, highly educated readers who might be more interested in natural philosophical questions. But what is distinctive and fascinating about Spanish and Portuguese science in this period is that the boundaries between these groups—practical and elite—were much more porous than they were in other parts of Europe. It is also the case that many Spanish and Portuguese treatises of *De sphaera* were read and used all around Europe.

References

Primary Sources

Bensaude, Joaquim. 1914. Histoire de la science nautique portugaise à l'époque des grandes découvertes. In *Regimento do estrolabio e do quadrante. Tractado da spera do mundo. Reproduction fac-similé du seul exemplaire connu appurtenant à la Bibliothèque Royale de Munich*, vol. 1. Munich: Carl Kuhn. http://hdl.handle.net/21.11103/sphaera.100916.

Cortes, Martin. 2003. *Breve Compendio de la Sphera y de la Arte de Navegar (Facsimile)*. Valladolid: Maxtor.

de Chaves, Jerónimo. 1545. *Tractado de la sphera que compuso el doctor Ioannes de Sacrobusto con muchas additiones*. Seville: Juan de Leon. http://hdl.handle.net/21.11103/sphaera.101033.

Faleiro, Francisco. 1535. *Tratado del esphera y del arte del marear: con el regimie[n]to de las alturas: co[n] algu[n]as reglas nueuame[n]te escritas muy necessarias*. Seville: Juan Cromberger. http://hdl.handle.net/21.11103/sphaera.101182.

Nuñes, Pedro. 1537. *Tratado da sphera com a Theorica do Sol e da Lua. E ho primeiro liuro da Geographia de Claudio Ptolomeo Alexa[n]drino*. Lisbon: Germão Galharde. http://hdl.handle.net/21.11103/sphaera.101010.

Pérez de Mesa, Diego. 2009. *Comentarios de sphera. Reproducción digital del manuscrito original*. Localización: Biblioteca Nacional (España). Sig. Mss. 8882. http://www.cervantesvirtual.com/nd/ark:/59851/bmchx1t7. Accessed June 2019.

Santayana y Spinosa, Rodrigo de Saenz. 1567. *La Sphera de Iuan de Sacrobosco. Nueva y fielmente traduzida de Latin en Romance, por Rodrigo Saenz de Santayana y Spinosa*. Valladolid: Adrian Ghemart. http://hdl.handle.net/21.11103/sphaera.101077.

Secondary Literature

Almeida, Onésimo T. 2008. Science during the Portuguese maritime discoveries: A telling case of interaction between experimenters and theoreticians. In *Science in the Spanish and Portuguese empires, 1500–1800*, ed. Daniela Bleichmar, Vos De Paula, Kristin Huffine, and Kevin Sheehan, 78–92. Stanford: Stanford University Press.

Almeida, Bruno. 2012. On the origins of Dee's mathematical programme: The John Dee-Pedro Nuñes connection. *Studies in History and Philosophy of Science* 43: 460–469.

Barker, Peter. 2011. The reality of Peurbach's orbs. In *Change and continuity in early modern cosmology*, ed. Patrick J. Boner, 7–32. New York: Springer.

Barrera-Osorio, Antonio. 2006. *Experiencing nature: The Spanish American empire and the early scientific revolution*. Austin: University of Texas Press.

———. 2008. Empiricism in the Atlantic world. In *Science and empire in the Atlantic world*, ed. James Delbourgo and Nicholas Dew, 178–193. New York/London: Routledge.

Blackmore, Josiah. 2009. *Moorings: Portuguese expansion and the writing of Africa*. Minneapolis: University of Minnesota Press.

Carolino, Luís Miguel. 2017. The Jesuit paradox: Intellectual authority, political power, and the marginalization of astrology in early modern Portugal. *Early Science and Medicine* 22: 438–463.

Carolino, Luís Miguel, and Henrique Leitão. 2006. Natural philosophy and mathematics in Portuguese universities, 1550–1650. In *Universities and science in the early modern period*, ed. Mordechai Feingold and Victor Navarro-Brotons, 153–168. Dordrecht: Springer.

Chabás, José. 2006. The university of Salamanca and the renaissance of astronomy during the second half of the 15th century. In *Universities and science in the early modern period*, ed. Mordechai Feingold and Victor Navarro-Brotons, 29–36. Dordrecht: Springer.

Collins, Edward. 2013. Francisco Faleiro and scientific methodology at the *Casa de la Contratación* in the sixteenth century. *Imago Mundi: The International Journal for the History of Cartography* 65: 25–36.

Crowther, Kathleen, and Peter Barker. 2013. Training the intelligent eye: Understanding illustrations in early modern astronomy texts. *Isis* 104: 429–470.

dos Santos Lopes, Marília. 2012. From discovery to knowledge: Portuguese maritime navigation and German humanism. In *Portuguese humanism and the republic of letters*, ed. Maria Berbara and Karl A.E. Enenkel, 425–445. Leiden: Brill.

———. 2017. Importing knowledge: Portugal and the scientific culture in fifteenth and sixteenth century's Germany. In *Renaissance craftsmen and humanistic scholars: Circulation of knowledge between Portugal and Germany*, ed. Thomas Horst, Marília dos Santos Lopes, and Henrique de Sousa Leitão, 73–89. Frankfurt am Main: Peter Lang Edition.

Esteban Piñeiro, Mariano, and Félix Gómez Crespo. 1991. La primera version castellana de *de revolutionibus orbium caelestium*: Juan Cedillo Diaz (1620–1625). *Asclepio* 1: 131–191.

Gómez Martínez, Marta. 2006. *Sacrobosco en castellano*. Salamanca: Ediciones Universidad de Salamanca.

Granada, Miguel Ángel, and Félix Gómez Crespo, eds. 2019. *Ydea astronomica de la fabrica del mundo y movimiento de los cuerpos celestiales. Traducción de De revolutionibus I-III, de Nicolás Copérnico [by] Juan Cedillo Díaz*. Barcelona: Universitat de Barcelona.

Hendrich, Yvonne. 2017. *De insulis et peregrinatione lusitanorum*–Valentim Fernandes als Vermittler von Information zwischen Portugal und Oberdeutschland zu Beginn des 16. Jahrhunderts. In *Renaissance craftsmen and humanistic scholars: Circulation of knowledge between Portugal and Germany*, ed. Thomas Horst, Marília dos Santos Lopes, and Henrique de Sousa Leitão, 103–120. Frankfurt am Main: Peter Lang Edition.

Heng, Geraldine. 2011. The invention of race in the European middle ages I: Race studies, modernity, and the middle ages. *Literature Compass* 8/5: 315–331.

Horst, Thomas. 2017. The relationship between Portugal and the holy Roman Empire at the beginning of the early modern period: A brief introduction. In *Renaissance craftsmen and humanistic scholars: Circulation of knowledge between Portugal and Germany*, ed. Thomas Horst, Marília dos Santos Lopes, and Henrique de Sousa Leitão, 9–35. Frankfurt am Main: Peter Lang Edition.

Horst, Thomas, Marília dos Santos Lopes, and Henrique de Sousa Leitão, eds. 2017. *Renaissance craftsmen and humanistic scholars: Circulation of knowledge between Portugal and Germany*. Frankfurt am Main: Peter Lang Edition.

Hunter, Douglas. 2011. *The race to the new world: Christopher Columbus, John Cabot, and a lost history of discovery*. London: St. Martin's Press.

Lanuza-Navarro, Tayra M.C. 2006. Medical astrology in Spain during the seventeenth century. *Cronos* 9: 59–84.

———. 2007. Astrological prognostications in seventeenth-century in Spain. In *Más allá de la Leyenda Negra: Espana y la Revolución Científica/beyond the black legend: Spain and the scientific revolution*, ed. Victor Navarro-Brotóns and William Eamon, 73–86. Valencia: Instituto de Historia de la Ciencia y Documentación López Piñero.

———. 2017a. Astrology in court: The Spanish inquisition, authority, and expertise. *History of Science* 55: 187–209.

———. 2017b. From intense teaching to neglect: The decline of astrology at the university of Valencia and the role of the Spanish Novatores. *Early Science and Medicine* 22: 410–437.

Leitão, Henrique, ed. 2002. *Pedro Nuñes 1502–1578 novas terras, novos mares e o que mays he: novo ceo e novas estrellas*. Lisbon: Biblioteca Nacional.

———. 2003. Jesuit mathematical practice in Portugal, 1540–1759. In *The new science and Jesuit science: Seventeenth century perspectives*, ed. Mordechai Feingold, 229–247. Dordrecht: Kluwer Academic Publishers.

———. 2006. Entering dangerous ground: Jesuits teaching astrology and chiromancy in Lisbon. In *The Jesuits II: Cultures, sciences and the arts 1540–1773*, ed. John O'Malley et al., 371–389. Toronto: University of Toronto Press.

———. 2007. *A Ciencia na* Aula da Esfera *no Colégia de Santo Antão 1590–1759*. Lisbon: Comissariado Geral das Comemorações do V Centenário do Nascimento de São Francisco Xavier.

———. 2009. *The Portuguese discoveries and the science in Europe*. Lisbon: Alêtheia Editores.

Leitão, Henrique, and Antonio Sánchez. 2017. Zilsel's thesis, maritime culture, and Iberian science in early modern Europe. *Journal of the History of Ideas* 78: 191–210.

Martins, Roberto de Andrade. 2004. La herencia de Sacrobosco en España: intercambio entre estudios universitarios y la práctica de navegación durante el siglo XVI. In *Epistemología e Historia de la ciencia. Selección de trabajos de las XIV jornadas*, ed. Pío García and Patricia Morey, 371–379. Córdoba: Universidad Nacional de Córdoba.

———. 2007. El comentario de Santayana y Spinosa (1568) al Tratado de la esfera de Sacrobosco. In *Epistemología e historia de la ciencia. Selección de trabajos de las XVII jornadas*, ed. Luis Salvatico and Pío García, 331–338. Córdoba: Universidad Nacional de Córdoba, Facultad de Filosofía y Humanidades.

Navarro-Brotóns, Victor. 1995. The reception of Copernicus in sixteenth-century Spain: The case of Diego de Zuniga. *Isis* 86: 52–78.

Navarro-Brotóns, Víctor. 2006. The cultivation of astronomy in Spanish universities in the latter half of the 16th century. In *Universities and science in the early modern period*, ed. Mordechai Feingold and Victor Navarro-Brotóns, 83–98. Dordrecht: Springer.

Navarro-Brotóns, Víctor, and William Eamon. 2007. Spain and the scientific revolution: Historiographical questions and conjectures. In *Más allá de la leyenda Negra: Espana y la revolución científica/Beyond the black legend: Spain and the scientific revolution*, ed. Victor Navarro-Brotóns and William Eamon, 27–38. Valencia: Instituto de Historia de la Ciencia y Documentación López Piñero.

Omodeo, Pietro D. 2017. *Utilitas astronomiae* in the renaissance: The rhetoric and epistemology of astronomy. In *The structures of practical knowledge*, ed. Matteo Valleriani, 307–332. Dordrecht: Springer.

Pohle, Jürgen. 2017. Kaiser Maximilian I. und die Rezeption der portugiesischen Entdeckungen in Nürnberger Kaufmanns- und Gelehrtenkreis am Ende des 15. Jahrhunderts. In *Renaissance craftsmen and humanistic scholars: Circulation of knowledge between Portugal and Germany*, ed. Thomas Horst, Marília dos Santos Lopes, and Henrique de Sousa Leitão, 57–71. Frankfurt am Main: Peter Lang Edition.

Portuondo, María M. 2008. Cosmography at the *Casa*, *Consejo*, and *Corte* during the age of discovery. In *Science in the Spanish and Portuguese Empires, 1500–1800*, ed. Daniela Bleichmar, Paula De Vos, Kristin Huffine, and Kevin Sheehan, 57–77. Stanford: Stanford University Press.

———. 2009. *Secret science: Spanish cosmography and the new world*. Chicago: University of Chicago Press.

———. 2017. Iberian science: Reflections and studies. *History of Science* 55: 123–132.

Sánchez Martínez, Antonio. 2010. Los métodos pedagógicos de la Corona para disciplinar la experiencia de los navegantes en el siglo XVI. *Anuario de Estudios Americanos* 67: 133–156.

Sandman, Alison Deborah. 2001. *Cosmographers vs. pilots: Navigation, cosmography, and the state in early modern Spain.* PhD dissertation, University of Wisconsin-Madison, Chicago.

Saunders, A. 1982. *A social history of black slaves and freedmen in Portugal, 1441–1555.* Cambridge: Cambridge University Press.

Thorndike, Lynn. 1949. *The* sphere *of Sacrobosco and its commentators*. Chicago: University of Chicago Press.

Valleriani, Matteo. 2017. The tracts on the *Sphere*: Knowledge restructured over a network. In *The structures of practical knowledge*, ed. Matteo Valleriani, 421–470. Dordrecht: Springer.

Westman, Robert. 1975. The Melanchthon circle, Rheticus, and the Wittenberg interpretation of the Copernican theory. *Isis* 66: 165–193.

3

André do Avelar and the Teaching of Sacrobosco's *Sphaera* at the University of Coimbra

Roberto de Andrade Martins

Abstract André do Avelar was the mathematics professor of University of Coimbra from 1592 to 1616. During his lifetime he published two astronomical works: a popular vernacular book called *Reportorio dos tempos* (1585) and *Sphaerae vtrivsque tabella* (1593). Avelar's *Sphaera*, composed for use at the university, was the only Latin version of Sacrobosco's work ever published in Portugal. Avelar's book was inspired by Sacrobosco's work and followed it closely, but it might be regarded as a new work, because he changed the wording and the order of presentation of the topics, added new information and several tables. After his retirement, Avelar was accused of Jewish beliefs and practices, and was condemned by the Inquisition in 1620, dying in prison. This paper analyzes Avelar's life, his work at the University of Coimbra, the content and role of his *Sphaera*, his career, and his trial by the Inquisition.

1 Introduction[1]

All over Europe, the *Tractatus de sphaera* by Johannes de Sacrobosco (died ca. 1256) remained the most widely known astronomical textbook for centuries. Four versions of Sacrobosco's *Sphaera* were published in Portugal in the sixteenth century. Three of them were in Portuguese, and only one in Latin (do Avelar 1593) (Fig. 10.1).

The title page of the Latin version contains some biographical information about the author, André do Avelar (1546–ca. 1623): born in Lisbon (*Olysipo*, in Latin), he was a Master of Arts and Philosophy (to be exact, in theology), and at the time of

[1]The author is grateful for the support provided by the Brazilian National Council for Scientific and Technological Development (CNPq). The author is also grateful to Charlie Zaharoff for his careful revision of the paper.

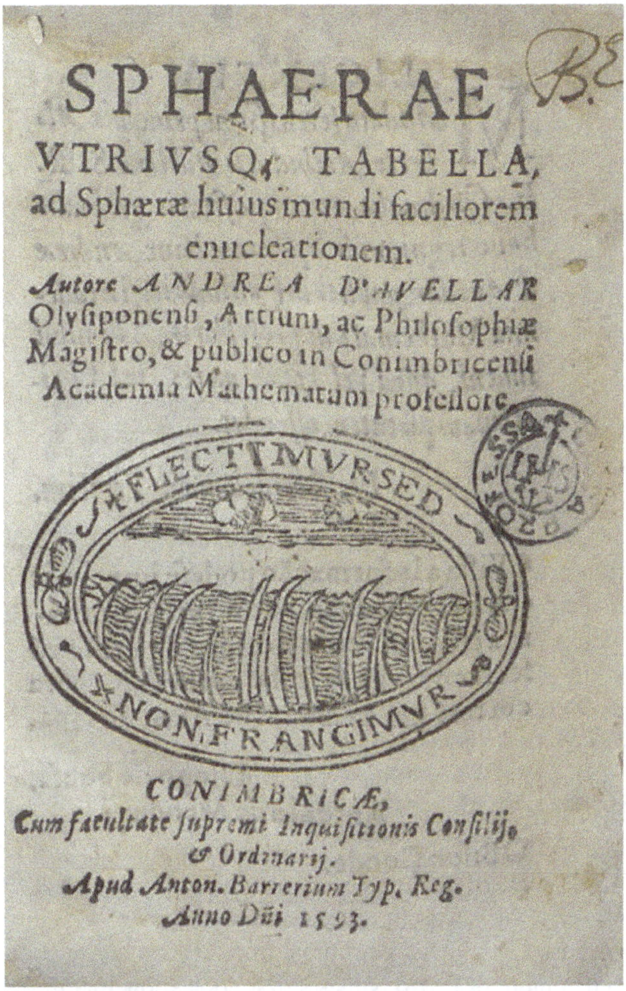
SPHAERAE
VTRIVSQ; TABELLA,
ad Sphæræ huius mundi faciliorem
enucleationem.
Autore ANDREA D'AVELLAR
Olysiponensi, Artium, ac Philosophiæ
Magistro, & publico in Conimbricensi
Academia Mathematum professore.

FLECTIMVR SED NON FRANGIMVR

CONIMBRICÆ,
Cum facultate supremi Inquisitionis Consilij,
& Ordinarij.
Apud Anton. Barrerium Typ. Reg.
Anno Dñi 1593.

Fig. 10.1 Title page of André do Avelar's commentary on the *Sphaera*. From (do Avelar 1593, fol. 1). Courtesy of the Biblioteca Universidad de Sevilla

the book's publication he was a professor of mathematics at the University of Coimbra. As one can see on the title page, the work was published with the permission of the Inquisition. The elliptical emblem contains the expression "Flectimur sed non frangimur" (we are bent, but not broken), meant to exhibit a proud and defiant attitude in the face of adversity. Here, on the very first page of the book, we find the main topics to be addressed in the following paper: a Latin commentary on the *Sphaera*; its author's biography; the teaching of the *Sphaera* at the University of Coimbra; and the involvement of the author in the Inquisition.

2 Pedro Nuñes and the *Sphaera* at the University of Coimbra

The first Portuguese university was created in Lisbon in 1290 (Rodrigues 2006). During the fourteenth century it moved to Coimbra, then back to Lisbon, back again to Coimbra, and once more to Lisbon (1377), where it remained until 1537, when it was permanently relocated to Coimbra by King João III (1502–1557). Only after this final move was the university endowed with its earliest chair of mathematics, which was first occupied by Pedro Nuñes (1502–1578) in 1544 (da Fonseca 2001; da Fonseca 2004).

The Society of Jesus was officially approved by Pope Paul III (1468–1549) in 1540. In the same year, two Jesuits arrived in Portugal to begin their religious and educational work. Their first school in Portugal was the Colégio de Jesus in Coimbra, founded in 1542. The Colégio do Espírito Santo began its activities in Évora in 1551, and it officially became the second Portuguese university in 1559 (Dias et al. 2012, 117). No other university was created in Portugal until the twentieth century.

Pedro Nuñes studied medicine at the universities of Salamanca and Lisbon, obtaining his title in 1525. During his medical studies he became deeply involved with mathematics and astronomy, and in Portugal he obtained the position of "Royal Cosmographer" in 1529 (da Fonseca 2004). From 1531 onwards, he was the tutor of two of the younger brothers of King João III: Luís, Duke of Beja (1506–1555), and Fernando, Duke of Guarda (1507–1534). Later on, he also taught the prince and future king Dom Sebastião (1554–1578). Some noblemen also benefited from Nuñes's teachings during this period, including Dom João de Castro (1500–1548), who became famous for his navigations and was later viceroy of Portuguese India.

In 1537, Nuñes published his first book, containing a Portuguese translation of the *Sphaera* with commentary as well as a partial translation of Georg Peuerbach's (1423–1461) *Theorica planetarum* and other tracts (Nuñes 1537). It was dedicated by his author to Prince Luís. This was not the first Portugese translation of Sacrobosco; another one, by an anonymous translator, had been published twice before 1520 (Bensaude 1912; de Albuquerque 1965) (Chap. 7).

It is clear that Nuñes's translation was not intended to be nor ultimately used as an academic textbook, because the statutes of the university prescribed the exclusive use of Latin as the official idiom for teaching (Leite 1963, 314). The language and content of Pedro Nuñes's first book show that his envisioned audience included people associated with navigation (pilots and cosmographers). That is what the author himself stated in the preface of his book:

> Having seen that the treatise on the *Sphaera* and the theory of the Sun and the Moon, and also the first book on the Geography of Ptolemy, contain those principles that must belong to any person who wants to know something about Cosmography, I translated them into our language so that those who do not know Latin would not be deprived of them. (Nuñes 1537, 5)

Nuñes had taught philosophy at Lisbon University (1531–1533), and became its initial professor of mathematics after it was moved to Coimbra (da Fonseca 2001; da Fonseca 2004). He certainly used Sacrobosco and Peuerbach as the basis of his astronomical teaching. Another of his books was possibly produced for the use of his students: *Astronomici introdvctorii de spaera epitome* (Nuñes 1940, vol. 1, 244–67). It is a very short version (twelve pages) of the first three books of the *Sphaera*. There is only one known copy of this book and it lacks the title page; for that reason, its place of publication, publisher, and date are uncertain. Some historians (including Joaquim Bensaúde) claim that it was written before the translation of the *Sphaera*; others (such as Rodolfo Guimarães) suppose that it was composed later (Nuñes 1940, vol. 1, 329–33). It is doubtful that it could be of any use outside the academic environment. Hence, I think that it was composed for use at the university, probably after 1544. This book was published only once in the sixteenth century and was not reproduced in Nuñes's *Opera* (printed in 1566, 1573, and 1592), although that work contained his comments on Peuerbach's *Theorica planetarum*.

In 1547, Nuñes was appointed as the first "cosmógrafo-mor" (high cosmographer) of the Portuguese kingdom, holding that position until his death in 1578 (da Mota 1976, 54). The duties of the high cosmographer during this period are not well known. The oldest official regulations about the *cosmógrafo-mor* were established in 1559, but they are unknown; only the directives of 1592 survived. Those were established under Spanish rule, and might have been widely different from their predecessors.

Pedro Nuñes taught mathematics only from 1544 to 1557, with frequent calls by the king to Lisbon. In 1557, he took a leave and moved to Lisbon for 4 years, possibly because of the needs of his position as high cosmographer. He retired from the university in 1562, shortly after his return to Coimbra (da Fonseca 2004). It is odd that his chair was not assigned to any other professor for three decades. Mathematics had been a relevant part of the Liberal Arts curriculum in all European universities since the Middle Ages, and it was also required for the study of medicine because of its astrological components. It is known that there were temporary teachers of mathematics in some periods after Nuñes's retirement; but the chair remained vacant until 1592.

3 Mathematics at the *Colégio das Artes* of Coimbra

The extended absence of a mathematics professor at the University of Coimbra can be explained, however, by the fact that this discipline was taught at another institution. In 1542, King Dom João III also founded the *Real Colégio das Artes e Humanidades* (Royal School of Arts and Humanities), a preparatory school in Coimbra, which was attached to the university. Its aim was providing the prerequisite studies for the main university courses. It was created as a general college for grammar, rhetoric, poetry, logic, philosophy, mathematics, Greek, and Hebraic. Its faculty numbered sixteen, one of them a professor of mathematics (Teixeira 1899, 4–5).

Its activities began in 1547 under the direction of André de Gouveia (1497–1548), who had formerly been the dean of the University of Paris. André de Gouveia brought from France several outstanding teachers and the school soon gained prominence (Hopkins 2016, 165). One of the foreign teachers brought by Gouveia to Coimbra was Élie Vinet (1509–1587), who became a friend of Pedro Nuñes and helped to disseminate his works abroad. In 1555, Dom João III decided to hand the direction of the *Colégio das Artes* to the Company of Jesus (Dias et al. 2012, 118).

Several archival documents of the 1550s refer to the professor of mathematics (Teixeira 1899, 282, 305, 403, 407), although there is no information about who occupied the chair. The directive of the *Colégio das Artes* was contained in the 1559 statutes of the University of Coimbra, which ascertained that the students of arts should have lessons on arithmetic, geometry and perspective during the third term of the second year, and during the third term of the third year they should read the *Sphaera* (Leite 1963, 317–18). Perhaps the subjects that were lectured on were not so different from the curriculum taught at other Jesuit schools. Jerónimo Nadal (1507–1580) wrote the statutes of the Jesuit School of Mesine and included the following mathematical topics in the study of philosophy: some books of Euclid (died 285 BCE), practical arithmetic, the *Sphaera* or the cosmography of Oronce Fine (1494–1555), the astrolabe of Johannes Stoeffler (1452–1531), and the astronomy of Georg von Peuerbach (1423–1461). In 1552, Nadal proposed a general plan of study for all Jesuit schools, including a more complete program of mathematics, adding to the previous list the study of the theory of music and perspective; all the students of philosophy should study mathematics for 3 years (Fuentes 2012, 136–37). Baltasar Torres (1518–1561), the first professor of mathematics at the *Collegio Romano*, wrote two poposals for the teaching of mathematics at the Jesuit schools that were similar to those of Nadal (Fuentes 2012, 139). It is unlikely, however, that the teaching of mathematics at the *Colégio das Artes* was so complete. Nadal himself visited the school of Coimbra and complained in his report that only the students in their second and third years received lectures on mathematics, which lasted only half an hour each day (Fuentes 2012, 141).

Christophorus Clavius (1538–1612), one of the most famous mathematicians of late sixteenth century, studied at the *Colégio das Artes*. When he was 16 years old, he entered the Society of Jesus in Rome, and the next year he was sent to Coimbra, where he remained as a student from 1556 to 1560. It is unlikely that he was ever a student of Pedro Nuñes, and we do not know who could have been his mathematics professor at the Jesuit college, so he was probably a self-taught mathematician (Lattis 1994, 14–15; Baldini 1998, 214). In 1561, he went back to Rome, and in 1563, he began teaching mathematics at the *Collegio Romano*.

Sometimes the philosophy teachers of the *Colégio das Artes* provided their students with the necessary mathematical and astronomical knowledge by introducing the study of the *Sphaerae* in their courses on natural philosophy, between the lessons on Aristotle's (384–322 BCE) *Physica* and the study of his *De Coelo* (Baldini 1998, 205). Some manuscripts of such mixed lectures are extant, the earliest one from 1570, written by the Jesuit "Ioannis Gomesii Bracharensis," that is, João Gomes of Braga—an otherwise unknown philosophy teacher at Coimbra. A similar manuscript by another Jesuit priest, Luís de Cerqueira (ca. 1552–1614), who taught philosophy at Coimbra from 1581 to 1585, is also preserved, as are analogous manuscripts by three Jesuits teaching at Évora in the decade of 1580 (Baldini 1998, 235).

Philosophy, not mathematics, was the main focus of the professors of the *Colégio das Artes*. From 1592 to 1606, they published the eight volumes of the famous *Cursus Conimbricensis*, containing the best commentaries on Aristotle produced at that time. These underwent several editions and were used throughout Europe (Casalini 2017).

It seems that the first Portuguese Jesuit mathematician was João Delgado (1553–1612). He joined the Jesuit order in 1574 and 2 years later was sent to Rome to study at the Collegio Romano, where he learned mathematics under Clavius. He remained there for 9 years and returned to Portugal in 1585 (Leitão 2008, 46). He intended to go to Brazil as a missionary,[2] but for unknown reasons he remained in Coimbra, where he taught mathematics from 1586 to 1589. The manuscripts of his lessons corresponding to 1586/7 and 1587/9 have been preserved (Baldini 1998, 229). From that time onwards, the influence of Clavius in Portugal was remarkable (Carolino 2006).

In 1590, Delgado began to teach mathematics at the Colégio de Santo Antão in Lisbon. This Jesuit school had begun its activities in 1553 and it was initially dedicated to the teaching of Latin, Greek, rhetoric, ethics, and religion. Mathematics was taught there occasionally, but at the end of 1573, the cardinal Dom Henrique asked the Jesuits to include in their curriculum the study of subjects related to navigation. The agreement was only fulfilled many years later, under the Spanish domain. João Delgado became the first mathematics teacher of Santo Antão, starting in 1590 with the *aula da esfera* (lessons on the sphere) (Leitão 2008). He was sometimes substituted by another priest, Francisco da Costa (1567–1604). Several manuscripts of Delgado's courses have been preserved, some of them on astronomy (especially Sacrobosco and Peuerbach), but also including astrology (Silva and Ferreira 2008, 103–08). The manuscripts of Francisco da Costa that have been preserved show that he lectured on geography and hydrography—that is, the science of navigation (Silva and Ferreira 2008, 113–20). The mathematician Christoph Grienberger (1564–1636), who had also studied under Clavius, also taught at the Colégio de Santo Antão in the period 1599–1602, returning afterwards to Rome. Hence, from 1590 onwards, the Jesuit Colégio de Santo Antão in Lisbon became an outstanding center for the study of mathematics, and especially astronomy (Leitão 2008).

4 André Do Avelar's *Reportorio dos tempos*

After the disappearance or death of King Sebastian I (1554–1578) during the battle of Alcácer Quibir, his great-uncle, Cardinal Henry (1512–1580), assumed the throne. His death led to a succession crisis, and the Portuguese crown was seized by the Spanish king Phelippe II (1527–1598). He was called Phelippe I in Portugal, where he was king from 1581 to his death. During his reign, the University of Coimbra received new statutes.

They were signed by the king in 1591, with immediate effect. This decree declared as vacant the mathematics chair and established the *modus operandi* for the election of a new professor (called "lente," or reader). The new statutes of the

[2] Jesuit priests sent to distant places as missionaries were required to study mathematics (astronomy, geography, cartography etc.). For more information, see (Baldini 1998, 196–203).

university required that candidates for the chair of mathematics should read two lessons, one about Euclid and the other on the theory of the planets (Universidade de Coimbra 1593, fol. 77r).

There was only one candidate for the job: André do Avelar. In January 1592, he was examined and presented his lectures, and the selection commission decided to bestow him the chair of mathematics, because there was no other candidate ("opositor," that is, opponent) and because he was "one of the greatest men in this science" (Almeida 1967, 47–48). The university Dean ("Reitor"), who played a part in the decision, was Dom Fernão Martins Mascarenhas (1548–1628); he later became the General Inquisitor of Portugal in 1616 (Martins 2011).

Who was André do Avelar at this time? The documentation of his selection only states his name (penned as "Andre dauellar"), that he was poor (so that the university would have to pay his taxes) and a "foreigner" (probably meaning that he was not from Coimbra). Also, the selecting commission knew that he was knowledgeable in mathematics (de Almeida 1967, 47–48).

Very little is known about his life previous to his entrance into the University of Coimbra.[3] Most of the available biographical information about him comes from the documents of the Inquisition.[4] He was born in Lisbon in 1546 to a family of converted Jews ("cristãos-novos," or New Christians). His father was called Galás do Avelar, and his mother was Violante Fernandes. He had four brothers and three sisters. He studied in Salamanca and Valladolid, obtaining the degree of Master of Arts; he also studied theology. He married Luiza de Faria, and they had six children, three sons, and three daughters (Baião 1919, 134–35).

We have no information about his way of living before he became professor of the University of Coimbra, and at that time he was 45 years old. There is, however, a relevant event in this period: in 1585, he published his first book (Fig. 10.2), called *Reportorio dos tempos*—an untranslatable title; it means something along the lines of a repertory or collection of information concerning time.

This is a peculiar work, belonging to a category of books on astronomy, calendars, and astrology, which became common in Spain and Portugal from the late fifteenth to the early seventeenth century (da Costa 2007, 75–79). They usually received the names *Chronographia* or *Reportorio dos tempos* (or *Reportorio de los tiempos*, in Spanish). They were different from the almanacs found all over Europe in several respects. They were large books (sometimes with a few hundred pages), not cheap and disposable items; they were intended to function as general nontechnical treatises of permanent significance. They were always written in vernacular in order to reach a wide public. The oldest example was Andrés de Li's *Reportorio de los tiempos*, published in Zaragoza in 1495. This small work was printed several times in Spain and in Portugal, where it was translated and improved by Valentim

[3] There is only one publication describing Avelar's life (de Almeida 1967) and, although it transcribes relevant documents, it is very sketchy and does not deal with Avelar's religious involvement and the Inquisition process.

[4] Avelar presented his biographical information to the Inquisition on March 23, 1620 (Braga 1892–1902, vol. 2, 603–04; Baião 1919, 134–35).

REPORTORIO
DOS TEMPOS, O MAIS CO
pioſo que ate agora ſahio a luz,
conforme à noua reformaçào do
ſancto Padre Greg. XIII.
Anno 1582.
FEITO POR ANDRE DO AVELAR.
natural de Lisboa.
DIRIGIDO AO ILLVSTRISSIMO
ſenhor dom Manoel de Caſtelbranco.
¶Chegão as taboas dos Lunarios, & Eclypſes, ate o
anno de 1610.

NON FRANGIMVR SED FLECTIMVR

COM LICENÇA.
¶Impreſſo em Lisboa por Manoel de Lyra. Anno de 1585.
Com Priuilegio Real, por dez annos.
Taxado a rees em papel.

Fig. 10.2 Title page of André do Avelar's first book, *Reportorio dos Tempos*. Notice that it contains an emblem very similar to the one that was published in his *Sphaera*, in 1593. From (da Costa 2007)

Fernandes (died ca. 1518). It inspired successively larger works with similar titles and content, the most famous being Jerónimo de Chaves's (1523–1574) *Chronographia o reportorio de los tiempos*. The last known works of this type were composed in Portugal by Manuel de Figueiredo (1568–1630) and in Spain by Rodrigo Zamorano (1542–1620).

This highly popular class of works represents a nice example of a blend combining various approaches and interests (Cardoso and Martins 2018). Its chronological part included information required by priests to calculate the religious calendar and the correlation between the hours of each weekday and the planets, according to astrologers; it described the division of day and night in natural and artificial hours; the heavenly spheres, and the astrological influence of each planet; the Zodiac was described both from the point of view of its influence on the sublunary world and regarding the details of the sun's motion along the ecliptic, and the duration of the days and nights. This kind of work included a fairly detailed description of the universe and its parts, of the heavenly circles and their terrestrial counterparts, introducing the zones and climes. The religious calendar, with the names of Catholic saints connected to each day, was combined with recommendations concerning what should be done or avoided in each lunar phase, for each month. A considerable part of the book was devoted to astrological medicine, explaining the four humors and temperaments, and the astronomical choice of treatments according to the temperament and the configuration of the heavens. Eclipses were described both as purely astronomical phenomena, providing tables of their occurrences, and as astrological causes of storms, wars, and other calamities. The treatises dealt with the influence of the moon on the weather and on crops, and they sometimes presented information concerning navigation and tides.

The encyclopedic character of works of this kind shows that it was not aimed at a particular public with a specific interest—indeed, any person of that time could profit from reading some part of the book. This circumstance probably explains the wide circulation and popularity of those works.

André do Avelar's *Reportorio dos tempos* was not an exception; it was typical for that kind of literature. As with many other works of this type, it drew heavily from earlier *Reportorios* without citing them. However, the author also consulted many other calendrical, astronomical, and astrological works, besides books on geography and navigation, to update and improve his own composition. In the successive editions of his book, Avelar always added new information. The first edition (do Avelar 1585) had 297 pages, from the title page to the end of the work; the fourth edition (do Avelar 1602) had 575 pages, with a slighter larger mean number of characters per page. Hence, the fourth edition was about twice the size of the first one.

Avelar's *Reportorio dos tempos* was described by some authors as a mere Portuguese translation of a similar Spanish book called *Chronographia, o, Reportorio delos tiempos, el mas copioso y preciso que hasta agora ha salido a luz*, by Jerónimo or Hieronymo de Chaves. This was the most successful book of its kind. It was first published in 1548 and was reprinted twelve times in the sixteenth century, seven times as a posthumous publication. There are, of course, many similarities in content between Avelar's book and Chaves's work, as was the case with analogous publications of the time. The charge of plagiarism was dismissed, however, by a careful comparison between the two works done by Adalgisa Botelho da Costa (da Costa 2007, 81–143).

The full title of Avelar's book contained a misleading advertisement: the most copious that has hitherto been brought to light ("o mais copioso que ate agora sahio a luz"). As a matter of fact, it was smaller than Chaves's *Reportorio*.

The first edition of the book was dedicated to Dom Manoel de Castelbranco, or Castello Branco (1560–1614), Second Earl of Villa Nova de Portimão (Avelar 1585, fol. IIr). Several members of this family, including Manoel, participated in the Battle of Alcácer Quibir (1578) when the King of Portugal Sebastião I (1554–ca. 1578) was killed or disappeared. Manoel was then 17 years old and was captured by the moors. Afterwards he was released and returned to Portugal. His father, Dom João, and his grandfather, Dom Martinho, died in this battle (Bayão 1737, 706).

Why did Avelar dedicate his book to this young earl? The inscription found on the book does not elucidate this. It only mentions that the Castelbranco family had "many titles of honor and greatness." We may presume that Avelar knew Dom Manoel personally, because it was not acceptable to dedicate a work to a person without his or her permission. It is also likely that the subject of the book had some significance to the earl. A single source describes Dom Manoel de Castelbranco as a "distinguished mathematician" (*Encyclopedia e Diccionario Internacional* 1933, vol. 4, 2264) and Diogo Barbosa Machado stated that "he applied himself exceedingly to the study of mathematical disciplines" (Machado 1741–1759, vol. 3, 217–18).[5] Was he a private student of André do Avelar? Was Avelar his astrologer? Any further interpretation of their relationship would be mere guesswork.

The context of the composition and publication of Avelar's *Reportorio dos tempos* is not evident. In his preface, he tells a tale about Diogenes the Cynic (ca. 412–323 BCE) to justify why he wrote the book:

> There was a time when it was said that King Philip of Macedon would come with his army against the city of Corinth. All the dwellers of that city, in a hurry, became busy with all the things that were necessary for its defense. Some refurbished arms, others carried stones, some mended old walls, others strengthened towers and places to fight. But Diogenes, seeing the hustle and care of that people, and having nothing to do because nobody gave him any occupation, put his cloak over his arms and, in a hurry, began to roll his tub up and down the Craneum mount.[6] One of his friends asked him why he was doing this, and he answered: I also move and roll my tub, so that among all those who are busy and hurried, I do not remain the only idle one, having nothing to do. I [André do Avelar] say this because facing this multitude of writers and books that every day come to light, I should not be the only one who does not do his part, as Diogenes intended with his tub. Farewell. (Avelar 1585, fol. IIv)

This story about Diogenes was not invented by André do Avelar; it was reported by Lucian of Samosata (ca. 125–180), in his work "The way to write history," where the anecdote is introduced to provide Lucian's own support for writing that work (Lucian 1905, 110). Of course, this is not an adequate justification for writing and publishing a book; it could only be an excuse for some useless occupation. Why did Avelar present Diogenes as his prototype? Did he see himself as a philosopher? Was he really as idle as Diogenes, having no job and nothing to do? No, we know that he was married and had already one daughter (Tomásia) to take care of (Carvalho 1990, 333). Perhaps the true motive for writing the book was an attempt to get some money. And the book was successful indeed, as it underwent four editions (do Avelar 1585, 1590, 1594, 1602).[7]

[5]The Jesuit priest António de Castello Branco, who imparted lessons on the *Sphaera* at Évora in the decade of 1580 (Baldini 1998, 235) was not his relative.

[6]Avelar wrongly understood the *Craneum* as a mount. It was a gymnasium, that is, a place for physical exercise.

[7]The website of the University of Coimbra described and made available for download a book published in 1590, with the title: "Reportorio dos tempos muito curioso acrescentado e emendado de nouo...conforme a noua constituição do sanctíssimo Papa Sixto quinto que tirou os abusos," attributing it to André do Avelar. The book was published in Coimbra by António de Barreyra, the

money. And the book was successful indeed, as it underwent four editions (do Avelar 1585, 1590, 1594, 1602).[7]

A large part of the book was concerned with astrology. Judicial astrology, that is, the prediction of future events in the life of a single person, had been condemned by the Roman Church because it was understood as conflicting with human free will (Chap. 3). Avelar was well aware of that, and at the beginning of the book he added a proem, which stated: "understanding that everything that be said of the properties of the signs and planets will not remove the freedom of man, nor win over free will, nor bring necessity to human activities; and the whole [content of the book] is subject to correction in obedience to the sacred motherly Roman Church" (do Avelar 1585, fol. IIv).

As stated before, Avelar's *Reportorio dos tempos* is not a mere translation of the similar work published by Jerónymo de Chaves. It seems to have drawn information from a variety of sources, and this leads us to a relevant question: How did Avelar have access to the books he needed, since he and his family were not rich? There were no public libraries in Portugal at that time. The university library had moved to Coimbra. In Lisbon, where Avelar lived, there were only libraries belonging to wealthy families, rich professionals (such as physicians), and those of religious institutions. Therefore, we may assume that he had access to some of these. From the dedications of his books, we may guess that he had access to the libraries of noblemen.

The second edition of the *Reportorio dos tempos*, published in 1590, was dedicated to Dom Álvaro de Lencastre (1540–1626), third Duke of Aveiro. At the begining of Avelar's inscription to the Duke of Aveiro, there is a relevant piece of biographical information:

> Last year (when I bade farewell to your excellency to read on the chair of mathematics at the University of Coimbra by order of his Majesty) I promised to your excellency that any of my works that was first brought to light would be put under your protection and support. However, I had a different intent, that of [dedicating to you] the books on the *Sphaera* and its use, that I have [ready] to print at the said University. Nevertheless, in the mean time, as I should not be idle to the curiosity of the mathematical studies [não me soffrendo estar occioso a curiosidade do studo Mathematico], and, on the other hand, desiring to show to your excellency a first gesture [towards fulfilling] this offer and promise, I made in this recast of the treatise on the times [tractado dos tempos] a particular [addition] on the prognostic of the changes of the air, with some principles related both to natural philosophy and to rural astrology, and some brief but compendious rules for sowing, the culture of trees, and the breeding of animals. Besides those reasons, I was impelled [to dedicate this book to you] by seeing that your highness is so fond of the countryside and of its care. (do Avelar 1590, fol. IIv)

The same dedication is reproduced in the third edition of the *Reportorio dos tempos* (do Avelar 1594, fol. IIr). The fourth edition is also offered to the Duke of Aveiro, but it contains a different inscription (do Avelar 1602, fol. A2v).

Hence, André do Avelar was already teaching mathematics at the University of Coimbra in 1589 (and maybe before that year), although he only became a regular professor ("lente") in 1592. The above citation also shows that in 1590 Avelar had already written his *Sphaera*, probably intending to use it at the university, since it was ready to be printed there.

The citation also exhibits a friendly relationship between Avelar and the Duke, once more suggesting that he was on good terms with some noblemen and could probably profit from their favors and access to their libraries. That a dedication to Dom Álvaro de Lencastre appeared in the three editions from 1590 to 1602 (and the last one was not a reproduction of the first) proves that the relationship between them was enduring. It is possible that he also received patronage from the Duke of Aveiro during that period.

Around 1600, Avelar was better known for his *Reportorio dos tempos* than for his *Sphaera.*

5 "Flectimur Sed Non Frangimur"

The first three editions of the *Reportorio dos tempos* and the only edition of Avelar's *Sphaera* have in the title page a peculiar emblem (Fig. 10.3). The same woodcut was used in the three editions of the "Reportorio." The woodcut used for the *Sphaera* (Fig. 10.1) was a bad copy of the Lisbon original.

Venâncio Deslandes (Deslandes 1888, 127) described the emblem and suggested that it was a representation of the four elements. He wrongly transcribed the phrase appearing on the emblem as "non frangimur sed flectimur." (Fig. 10.4).

The emblem of reeds in the water, being bent but not broken, with the motto "Flectimur non frangimur" (we are bent but not broken) or "Flectimur non frangimur unde" (we are bent but not broken by the waves) had been used since the fourteenth century by two Italian families, Colonna and Acquaviva (Gelli 1916, 286–87). In the fifteenth century, the Pope Alexander VI, born Rodrigo de Borja (1431–1503), banished from Rome the Cardinal Giovanni Colonna (1456–1508) and other members of the family, who took refuge in Naples and Sicily. They then assumed the emblem again (Fig. 10.4), and this was meant as a warning: although the Colonna family was bent by pontifical persecution, the Pope should not consider them dead and buried; they were still alive and powerful, and hoped sooner or later to grow even stronger than before and get pay back (Gelli 1916, 287; Palliser 1870, 73–74). Both the emblem and its history were well known in the sixteenth century and were included in the work "Symbolorum & emblematum" of Joachim Camerarius (1500–1574) under number "95" (Camerarius 1590, fol. 97r).

CHRONOGRAPHIA 150
OV REPORTORIO
DOS TEMPOS O MAIS CO-
PIOSO QVE TE AGORA SAYO
A LVZ, CONFORME A NO-
ua reformação do ſancto Papa
Gregorio XIII.
FEITO POR ANDRE DE AVELLAR
natural de Lisboa, lente das Mathematicas na
Vniuerſidade de Coimbra.
Neſta terceira impreſſaõ reformado & acrecentado pello meſmo Author, com hum tratado do Pronoſtico da mudança do ar, & algũs principios que tocão, aſsi à Philoſophia natural, como a Aſtrologia ruſtica, & com hũas breues, mas muy proueitoſas regras pera as ſementeiras, & cultura das aruores, & criação dos animaes.

Em Lisboa com licença da ſancta Inquiſição, & Ordinario.
Em caſa de Simão Lopez. Anno
M. D. XCIIII.
Com Priuilegio Real por dez Annos.

Fig. 10.3 Title page of the third edition of André do Avelar's *Reportorio dos tempos*. From (do Avelar 1546). Courtesy of the National Library of Portugal

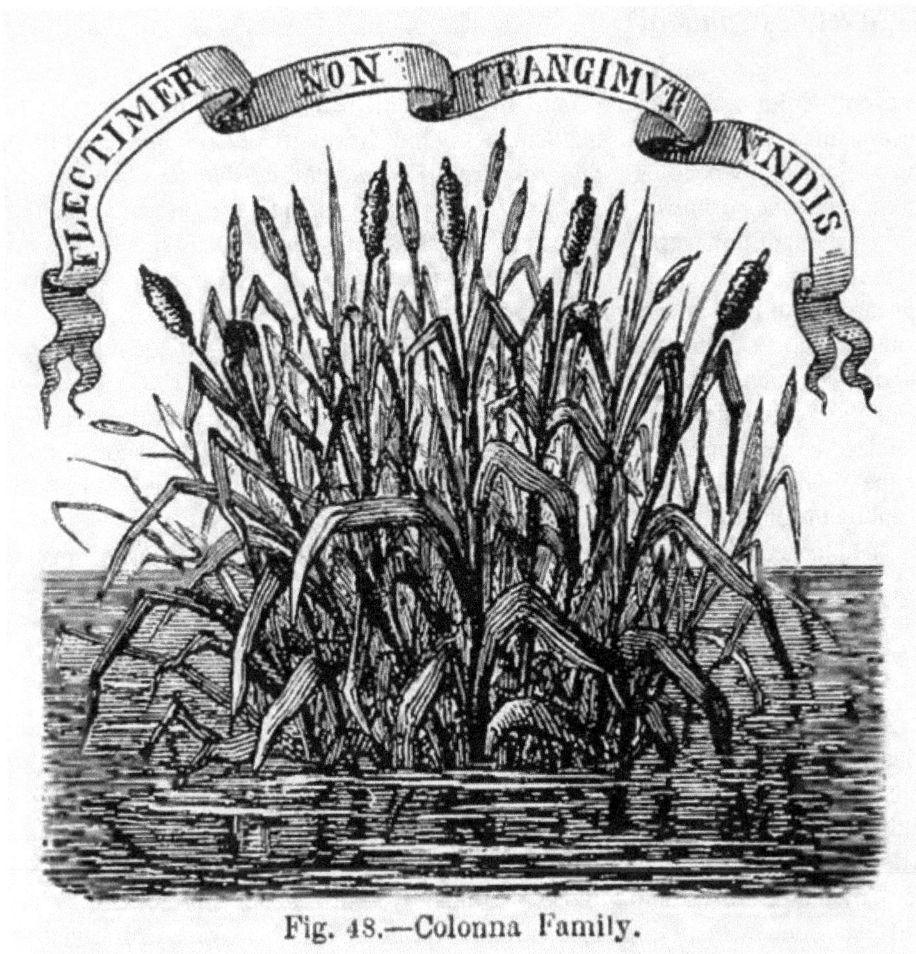

Fig. 10.4 The emblem of the Colonna family, with the motto "Flectimur non frangimur undis." According to Mrs. Bury Palliser, the emblem was created by Jacopo Sannazaro (1458–1530). From (Palliser 1870, 74). Public Domain Mark

Why did Avelar incorporate this emblem in his books? Had he suffered some personal persecution? Did he hope to rise from a defeated, humble condition to a higher one? Lack of knowledge about his personal history prevents us from giving any well-founded answer. However, one may conjecture that the emblem was related to his family situation: they were of Jewish origin, and there was a social stigma oppressing this class. As is well known, all Jews had been expelled from Spain in 1492, and from Portugal in 1497. Any Jew willing to remain in Portugal after that time was compelled to become a Christian. The pejorative name "New Christian" was applied to those Jewish converts and to their posterity. Perhaps Avelar yearned for a time when the New Christians would rise once more to their deserved social status or would be allowed to become Jews again.

6 Avelar's *Sphaera*[8]

The only edition of Avelar's *Sphaera* was published in 1593, the year after he became the official reader of mathematics at the University of Coimbra. The title of the book is highly peculiar: *Sphaerae vtrivsq[ue] tabella, ad sphaerae huius mundi faciliorem enucleationem*, meaning "Tablet of both spheres, for easier explanation of the sphere of this world." Here, "tabella" does not mean "table," but a little board or tablet. In the aforesaid title, it meant a summary or abridgment. The word appeared with the same meaning, for example, in the title of one of the books of Louis de Blois (1506–1566): *Tabella spiritualis alias sacellum animae fidelis nuncupata: continens brevem institutionem vitae Christianae* (de Blois 1583). However, Avelar's book cannot be interpreted as a summary of Sacrobosco's work, because it is indeed larger than the medieval treatise. It could be understood as a short version of the *Sphaera* only when compared to the large commentaries that had become usual by the end of the sixteenth century.

The phrase "both spheres" (*sphaerae utriusque*) used in this book was unusual. The only astronomical work I have found with a title that might have inspired Avelar was *De mundi et sphaerae: ac utriusque partium declaratione cum planetis et variis signis historiatis* by Gaius Iulius Hyginus (ca. 64–17 BCE), which was printed several times in the sixteenth century.

Avelar's *Sphaera* was dedicated to Dom Ferdinando (or Fernão) Martins Mascarenhas, the Dean (*Reitor*) of the University of Coimbra and adviser of King Phelippe who became General Inquisitor two decades later. After the dedication, the author addressed the students of mathematics that comprised his intended public. After that, we find several poems dedicated to the readers and to the author of the book by the priest António Velloso of Braga; by Pedro Mendes, professor of Latin and Greek; and by Dom Jorge de Castro (do Avelar 1593, fol. IIr–VIr). Unfortunately, it was impossible to find information about them.

Avelar's *Sphaera* does not cite the name of Sacrobosco at any point, although he closely follows the medieval text. Of course, every reader knew that the book was a new version of Sacrobosco's, therefore Avelar could not have held the expectation of being regarded as the original author of the ideas contained in the book.

Many of the commentaries on the *Sphaera* contained Sacrobosco's full original text, set in a clearly distinct typography and intermixed with elucidations. That was the style followed, for instance, by Pedro de Espinosa,[9] a professor of the University of Salamanca, who published a Latin commentary on Sacrobosco that underwent two editions (1535, 1550), a commentary that might have been read by Avelar when he was a student (de Espinosa 1550). The famous and massive commentary produced by Christophorus Clavius (Lattis 1994) also followed the same structure. Avelar, however, did not follow the example of those predecessors. It is simply

[8] This following section is an expanded and corrected version of (Martins 2003b).

[9] The name "Pedro de Espinosa" was rather common in Spain in the early sixteenth century. There is no reliable information about the dates of birth and death of the author of this commentary on Sacrobosco (Díaz Díaz 1987, v. 3, 68–69; Delgado 1983).

impossible to identify, by any typographical hint, where he was simply transcribing Sacrobosco's text and where he was presenting his own contribution. Indeed, the style of Avelar's book is that of a paraphrase of Sacrobosco's *Sphaera*.

Sacrobosco started his first chapter with definitions of the geometric sphere by Euclid and Theodosius (347–395). Avelar referred initially to the "material sphere," that is, the armillary sphere, the instrument used to explain the structure of the universe. Immediately afterwards, however, Avelar applied Theodosius's definition to that sphere and introduced the concepts of axis and pole exactly as Sacrobosco had done. In the citations shown below, the corresponding parts are underlined.

Avelar	**Sacrobosco**
Sphaera materialis est instrumentum quoddam rotundum, compositum ex variis circulis, quibus coelorum motus, totiusque mundis situs, commodissime explicantur, <u>in cuius medio punctum est, à quo omnes lineae rectae ad circunferentias circulorum ductae sunt aequales: & illud punctum dicitur centrum sphaerae, linea vero recta transiens per centrum, applicans extremitates suas ex utraque parte ad circunferentiam circa quam sphaera voluitur dicitur axis sphaerae: duo vero puncta axem terminantia, dicuntur poli mundi.</u> (do Avelar 1593, fol. 1r–v)	Sphaera etiam a Theodosio sic describitur: sphaera est solidum quoddam una superficie contentum, <u>in cuius medio punctus est, a quo omnes lineae ductae ad circumferentiam sunt aequales: & ille punctus dicitur centrum sphaerae. Linea vero recta, transiens per centrum sphaerae, applicans extremitates suas ad circumferentiam ex utraque parte, circa quam sphaera voluitur, dicitur axis sphaerae. Duo vero puncta axem terminantia dicuntur poli sphaerae.</u> (Vinet 1561, fol. 9v)

The underlined parts show that Avelar, for the most part, just copied Sacrobosco's text without substantial changes. Differences concerning the order of the words, punctuation, and even changes to some terms are not noteworthy, since similar differences appeared in different printed editions of Sacrobosco's own text.

After the definition of the sphere, Avelar presented the main circles of the heavenly sphere, a subject that appeared only at the second part of Sacrobosco's text. The explanation of the greater and smaller circles is slightly different.

Avelar	**Sacrobosco**
De circulis sphaera materialis. Cap. II Sphaera autem materialis componitur ex decem circulis, quarum haec sunt nomina scilicet aequinoctialis, zodiacus, colurus solstitiorum, colurus aequinoctiorum, meridianus, horizon, tropicus Cancri, tropicus Capricorni, circulus Arcticus, circulus Antarcticus. <u>Horum vero circulorum, quidam sunt maiores, quidam minores</u>: priores sex maiores dicuntur, seu maximi: posteriores quatuor minores appellantur, sive non maximi. <u>Maior circulus in sphaera is dicitur, qui idem centrum cum sphaera obtinet: ipsamquae, in duo hemisphaeria dividit</u>: minor vero circulus est ille, qui diversum centrum a sphaera centro possidet, ipsa sed in duo segmenta inaequalia partitur. (do Avelar 1593, fol. 2r–v)	Capitulum secundum. De circulis, ex quibus sphaera materialis componitur: & illa supercoelestis, quae per istam imaginatur, componi intelligitur <u>Horum autem circulorum quidam sunt maiores, quidam minores</u>, ut sensui patet. <u>Maior autem circulus in sphaera dicitur, qui descriptus in superficie sphaerae super eius centrum, dividit sphaeram in duo aequalia.</u> Minor vero, qui descriptus in superficie sphaerae, eam non dividit in duo aequalia sed in portiones inaequales. (Vinet 1561, fol. 20v)

André do Avelar did not follow the text of Sacrobosco and did not treat it as a sacred or immutable astronomical revelation. The original *Sphaera* was, of course, his point of departure, but he felt free to produce his own work. In this respect, his approach was uncommon, although there had been several precedents, including Galileo Galilei's (1564–1642) unpublished textbook on the *Sphaera* (Galilei 1656; Martins and Cardoso 2008; Martins and Cardoso 2017).

Avelar's book was divided into four parts like Sacrobosco's work, and it included essentially the same content, but with an inversion of the order of the first and second parts. The first part of Avelar's *Sphaera*, "On the material sphere and the circles of which it is composed," containing twelve chapters and one table and occupying nine numbered leaves, presented the main circles of the celestial sphere and their terrestrial counterparts. The sequence and content are almost identical to the second part of Sacrobosco's *Sphaera*, with some slight differences. For instance, Avelar used 23° 30′ for the obliquity of the ecliptic, a better estimate, at that time, than Sacrobosco's value of 23° 33′.

The second part, called "On the sphere of the world, or natural [sphere], and its parts," was divided into eighteenth chapters and one table, filling forty-five numbered leaves. It is similar to the first part of Sacrobosco's book, but some parts were put into a different order. For instance, Avelar discussed the size of the earth before presenting the celestial structure (ether, motions, and form of the heavens). He mentions the theory of trepidation and the need for ten heavenly orbs, instead of the nine mentioned by Sacrobosco, following, at this point, the proposal that had been introduced by King Alphonso X, el Sabio (1221–1284). Besides this change, Avelar added some quantitative data and some tables. One of them provides the declination of the sun throughout the year, essential information for the determination of latitudes. There are other relevant differences. Avelar introduced several mentions of the *Astronomica* of Marcus Manilius (fl. 1st century), a book that was seldom cited in commentaries on Sacrobosco's *Sphaera*. The outlook of Manilius's work was mainly astrological, and references to his contribution betray Avelar's deep involvement with astrology. Accordingly, in the second part of his *Sphaera* he presented some astrological information about the signs and established relationships between the four elements, the four humors, the seasons, and the four ages of mankind. This part of the book also mentions new information that became available because of the Iberian navigations, such as knowledge of the Southern Cross (Fig. 10.5), dark nebulae, the Magellanic Clouds, and a new evaluation of the size of the earth.

The third part of Avelar's *Sphaera*, containing fourteen chapters and several tables encompassing thirty-nine leaves, closely follows the structure and the content of Sacrobosco's third chapter. It deals with the rising and setting of the stars and constellations, the yearly motion of the sun, the diversity of duration of days and nights, and the climates. Avelar added to this part some quantitative tables, one of them showing the degrees of the equinoctial corresponding to each of the zodiacal signs, both in the case of the right sphere and the oblique sphere and, in particular, for the latitude of Lisbon. In his discussion of the climates, Avelar stated that Claudius Ptolemy (died 160) had described twenty-one parallel circles and seven climates, but the recent authors had introduced twenty-three climates and fourty-nine parallel circles. He provides a table with the data for those climates and the

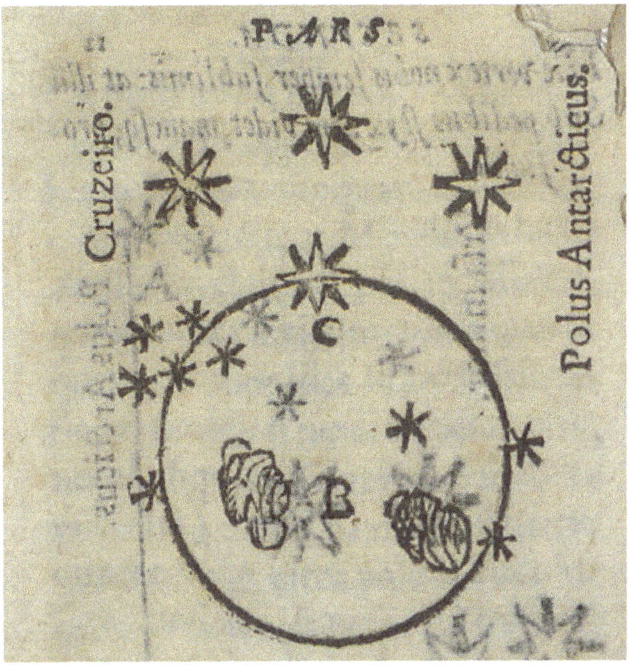

Fig. 10.5 Representation of the Southern Cross ("Cruzeiro") and the Magellanic Clouds, according to Avelar. From (do Avelar 1593, fol. 12v). Courtesy of the Biblioteca Universidad de Sevilla

variation of daylight length during the year. Following Sacrobosco, Avelar presents copious literary citations from Publius Vergilius Maro (70–19 BCE), Publius Ovidius Naso (43 BCE–18), and Marcus Annaeus Lucanus (39–65).

The fourth and last part of Avelar's *Sphaera* is quite short: six chapters and a few tables, encompassing eleven leaves. Like Sacrobosco's fourth chapter, it deals with the motions of the sun and the moon, of the planets, and of the eclipses. Avelar provided a more detailed theory than Sacrobosco, mentioning the contributions of Alphonso X and presenting data relative to the motion of the moon.

7 The Sources of Avelar's *Sphaera*

André do Avelar certainly consulted other works besides Sacrobosco's to write his *Sphaera*. It is difficult, however, to find out which sources he used. Part of this difficulty is due to Avelar himself, who did not have an inclination to name his authorities. It is possible to notice this tendency in his *Reportorio dos tempos*: Chaves had provided a number of references in his *Chronographia o reportorio de los tiempos*, but Avelar did not reproduce this information, even in the cases where he was clearly translating Chaves's book (da Costa 2007, 91–93). Note that in the late sixteenth century, several authors of academic books provided specific references to authors, the title of their works, and their relevant parts or chapters. Sometimes those references appeared inside the text, but there was an increasing use of marginal notes, the antecedent of our footnotes. One may find a profuse employment of this kind of resource in the *Comentarii Conimbricensis*, the commentaries about Aristotle's works produced by the Jesuit teachers of Coimbra (Casalini 2017).

In his *Sphaera*, Avelar usually only reproduces the meager information contained in Sacrobosco's work: vague allusions to Aristotle, Ptolemy, Alfarganus (805–870), and other authors, without any hint as to which books the stated ideas are drawn from. He did, however, mention a few other authorities. One of them was king Alphonso X (1221–1284) (do Avelar 1593, fol. 26v, 96v), who was never cited by Sacrobosco, since the *Sphaera* was written earlier. Because of the importance of Alphonso X as a reformer of medieval astronomy, he was cited in many of the commentaries on the *Sphaera*. It was natural, therefore, that Avelar should include this authority in his book.

Another author cited by Avelar but not by Sacrobosco was Marcus Manilius, who is explicitly referred to in four distinct passages (do Avelar 1593, fol. 10v, 21r, 30r, 34r). The sixteenth century commentaries on the *Sphaera* seldom cited Manilius.

Since Avelar's book was composed for use at the University of Coimbra, it could completely avoid subjects related to navigation such as the tables of declination of the sun, used for ascertaining the latitude. This was an expected subject in the teaching of the *cosmógrafo-mor* or at the *aula da esfera*, but not at the universities. However, he did include such information, as well as data provided by navigators. He did not point out, however, which sources he used to describe the Southern Cross and the Magellanic Clouds (do Avelar 1593, fol. 11v–12v). At another point, he remarks on the visible signs of Earth's roundness and the new measurement of the geographical degree, amounting to 17.5 leagues, as determined by the "navigators of Spain" (do Avelar 1593, fol. 16r–17v).[10] He also referred to "more recent" authors who had expanded Ptolemy's division of the earth from seven to twenty-three climes (do Avelar 1593, fol. 88v), but nowhere did he present the source of this information.

Although Avelar was a successor of Pedro Nuñes at the mathematics chair of the University of Coimbra, he never cited his writings (Leitão 2002, 27). In contrast, the many editions of Élie Vinet's *Sphaera Joannis de Sacro Bosco Emendata* contained Nuñes's analysis of the climes (Valleriani 2017); and the Portuguese mathematician was also cited in other commentaries, such as Clavius, where his name appeared in nine different places (Knobloch 1999, 347).

8 Literary Citations

Sacrobosco's *Sphaera* contained a significant number of poetic quotes from Virgil, Ovid, and Lucan. One of the reasons for the introduction of those classical citations was the need to elucidate the literary works themselves: many passages of those

[10] At this time, Portugal was part of Spain, under Phelippe I.

Latin authors could not be understood without an adequate knowledge of astronomy (Martins 2003a).

Avelar's *Sphaera* kept seventeen of the twenty-one original poetic references. He dismissed only three from the *Georgica* of Virgil and one from Lucan's *Pharsalia.* It is quite clear, however, that he did cherish those literary quotations. Indeed, in several cases he presented a larger number of verses than Sacrobosco did. For instance, when he quoted Ovid's *Metamorphoses*, Avelar reproduced seven verses (do Avelar 1593, fol. 54r); in most of the editions of Sacrobosco's work we find only four verses. He also added to his work three quotations from Ovid and one from Lucan (do Avelar 1593, fol. 30r, 36v, 56v, 78v) that were not used by Sacrobosco. The several quotations of Manilius presented by Avelar were also probably introduced for aesthetic reasons.

Avelar also incorporated in his *Sphaera* five additional literary quotes that have no counterpart in Sacrobosco's work. Only one of them was identified by Avelar himself (do Avelar 1593, fol. 35v): a citation of *In sphaeram Archimedis*, from Claudius Claudianus (ca. 370–404). Sacrobosco did not include any quotation from Claudianus in his *Sphaera*, but he did so in his *De anni ratione*, or *Computus ecclesiasticus* (de Sacrobosco 1550, fol. L 4r).[11]

What do those additions tell us about Avelar and his sources? At first sight, they suggest that the author was familiar with Latin poetry and that this allowed him to add several quotations to those that Sacrobosco had used in his *Sphaera.* There is, however, another explanation. Maybe Avelar just made use of some commented editions of the *Sphaera* where those passages can be found. Indeed, several of the extracts used by Avelar can be found in the very popular *Sphaera Joannis de Sacro Bosco emendata*, edited by Élie Vinet (1509–1587). Observe that Élie Vinet was professor of the Colégio das Artes in Coimbra from 1547 to 1549. He was a friend of Pedro Nuñes and helped to make some contributions of the Portuguese mathematician known in Europe, introducing them in his own edition of the *Sphaera.*

Some specific instances will be examined in the following section. The seven lines of Ovid's *Metamorphoses* transcribed by Avelar were contained in Vinet's edition (Vinet 1561, fol. 87r). There we can also find one of Avelar's citations from Ovid that does not appear in Sacrobosco's work (do Avelar 1593, fol. 56v; Vinet 1561, fol. 72v) and the additional quotation from Lucan (Avelar 1593, fol. 78v; Vinet 1561, fol. 33r). Two of the four citations of authors who were not identified by Avelar were also contained in Vinet's *Sphaera* (do Avelar 1593, fol. 26r, 57r; Vinet 1561, fol. 20r, 61v).

Hence, we see that Avelar did not require considerable erudition to insert the additional poetic quotations of his *Sphaera.* Some of those literary addenda were not taken from Vinet's work, but they might have been copied from some other commentary on Sacrobosco's *Sphaera.*

---

[11] Unlike Sacrobosco's *Sphaera*, his treatise on the calendar called *Computus ecclesiasticus* or *De anni ratione* has been scarcely studied by historians of science. Some information about that work and its genre can be found in papers published by Lynn Thorndike (1954) and Jennifer Moreton (1994).

The quotation of Claudianus had been published by Francesco Giuntini (1523–1590) in his commentary on the *Sphaera* (Giuntini 1578, 347). Clavius's commentary also contains the Claudianus's verses that were used by Avelar (Knobloch 1999, 9). The same quotation also appeared in the first edition of Clavius's commentary (Clavius 1570, 24).

One particular literary addition by Avelar should be emphasized: it was part of a poem by Philip Melanchthon (1497–1560) on the rising and setting of the stars: "Mane vehit supra terram tibi Cosmicus ortus Sidera, sed Phoebi lumine tecta latent. Mane dat Heliacus quaedam subvecta videre Astra, sed Achronicus nocte videnda trahit" (Avelar 1593, vol. 26r; Melanchthon 1579, fol. Q5v). Melanchthon was an outstanding intellectual leader of the Lutheran Reformation and an author prohibited by the *Index librorum prohibitorum* since 1559 (Lenard 2006, 55–56) (Chap. 9).

Notice that before the publication of the *Index*, many commentaries on Sacrobosco's *Sphaera* contained as a preface a piece written by Melanchthon in 1531 in praise of astronomy (Gingerich 1988). That was also the case of the early editions of Élie Vinet's *Sphaera* (the 1552 edition of Elie Vinet's *Sphaera*, for instance, contained Melanchthon's preface.)

After the proscription of Melanchthon's works by the Inquisition in 1559, the editions of the *Sphaera* containing his preface were partially censured in Catholic countries. In several cases, only Melanchthon's name was effaced, cut or crossed over while the preface was kept. In some later Italian editions, the preface was kept but as an anonymous text (Gingerich 1988, 269).

Philipp Melanchthon's epigram quoted by Avelar had been included, without any allusion to its author, at the end of the fourth chapter of Sacrobosco's *Sphaera* in the first edition printed in Wittenberg, which also included Melanchthon's preface (Sacrobosco 1531, fol. F IIr). It was reproduced in successive publications of the book in several cities, usually at the same point (although its subject had nothing to do with the fourth chapter of the *Sphaera*) and without any declared authorship (see, for instance, de Sacrobosco 1547, fol. E [V]r; de Sacrobosco 1550, fol. G2r).[12] Sometimes it was moved to the third chapter, where its theme would fit more naturally (de Sacrobosco 1545, fol. D [V]r).

Philipp Melanchthon's verses were also published in some editions of the *Sphaera Ioannis de Sacro Bosco emendata* that contained Francesco Giuntini's comments (de Sacrobosco 1564, 116; de Sacrobosco 1573, 88) at the end of the fourth chapter of the *Sphaera*, as in the first Wittenberg edition. It was later included in Francesco Giuntini's long commentary on the *Sphaera* that was part of his *Speculum astrologiae* (Giuntini 1581–1583, vol. 2, 767).[13] In the later work, Giuntini moved the quotation to the third chapter of the *Sphaera*, where its content is more appropriate.

Most of the editors and authors who reproduced this epigram were probably unaware of its authorship; and it was not explicitly disapproved by the Inquisition,

[12] I am grateful to Matteo Valleriani for pointing out those occurrences of Melanchton's citation.

[13] I am grateful to Elio Nenci for pointing out this occurrence of Melanchthon's quotation.

although all of Melanchthon's works had been condemned. Indeed, although Giuntini's *Speculum astrologiae* had been heavily censured, as witnessed by a copy that belonged to a Jesuit institution—that now can be found at the Library of the University of Sevilla (A Res. 03/3/13–14)—the epigram was not crossed out by the person who "corrected" that book, although several lines on the same page and on the following one were scored out (Fig. 10.6). It is possible that Avelar copied Melanchthon's verses without knowing their authorship.

9 Manilius's *Astronomica* and the *Sphaera*

Marcus Manilius's *Astronomica* is the earliest surviving classical astrological work. It is a lengthy Latin poem, written between the year 10 and 20 under the reign of Augustus and/or Tiberius (Goold 1977, xii; Folk 2009, 4). It is likely that part of the work was lost. In its present form, it contains five books. The first one presents a general description of the universe, the celestial circles, stars, constellations, the Milky Way, planets, and comets. The subject of the second book is the Zodiac, its movement relative to the earth, its signs and their influence upon the human body. The third one teaches how to draw a horoscope and describes the lots, the ascendant, and the influence of the stars on man. The fourth book presents the decans of the Zodiac and contains an astrological geography, providing a description of the heavenly influences upon places and people. The last book explains the variable influence of the planets according to their position and motion, describing the influence of constellations that do not belong to the Zodiac.

Sacrobosco did not quote and was probably unaware of Manilius's work, since the *Astronomica* and his author were never explicitly cited by extant Roman authors and it was barely known during the Middle Ages. Manilius's poem was retrieved from obscurity in the early fifteenth century by Poggio Bracciolini (1380–1459), who obtained a copy of a manuscript version; it was published for the first time around 1473 by Johannes Regiomontanus (1436–1476) in Nuremberg (Folk 2009, 2). In 1474, it was also published in Bologna, and 10 years later it was commented on by Lorenzo Buonincontro (1410–ca.1500). It was reprinted several times in the fifteenth and sixteenth centuries, becoming a quite popular book. The first critical edition was published in 1579 by Joseph Justus Scaliger (1540–1609).

The *Astronomica* was therefore available to Avelar when he wrote his *Sphaera*, both directly and through indirect sources, as will be described. Let us first present four passages of that work quoted by Avelar.

Avelar's first quotation of Manilius appeared at the place where he described the axis of the universe (Avelar 1593, fol. 10v). These are the corresponding lines of the *Astronomica* (I.279–282) in a recent edition (Manilius 1977, 26):

> Aera per gelidum tenuis deducitur axis
> libratumque regit diverso cardine mundum;
> sidereus circa medium quem volvitur orbis
> aetheriosque rotat cursus, immotus....

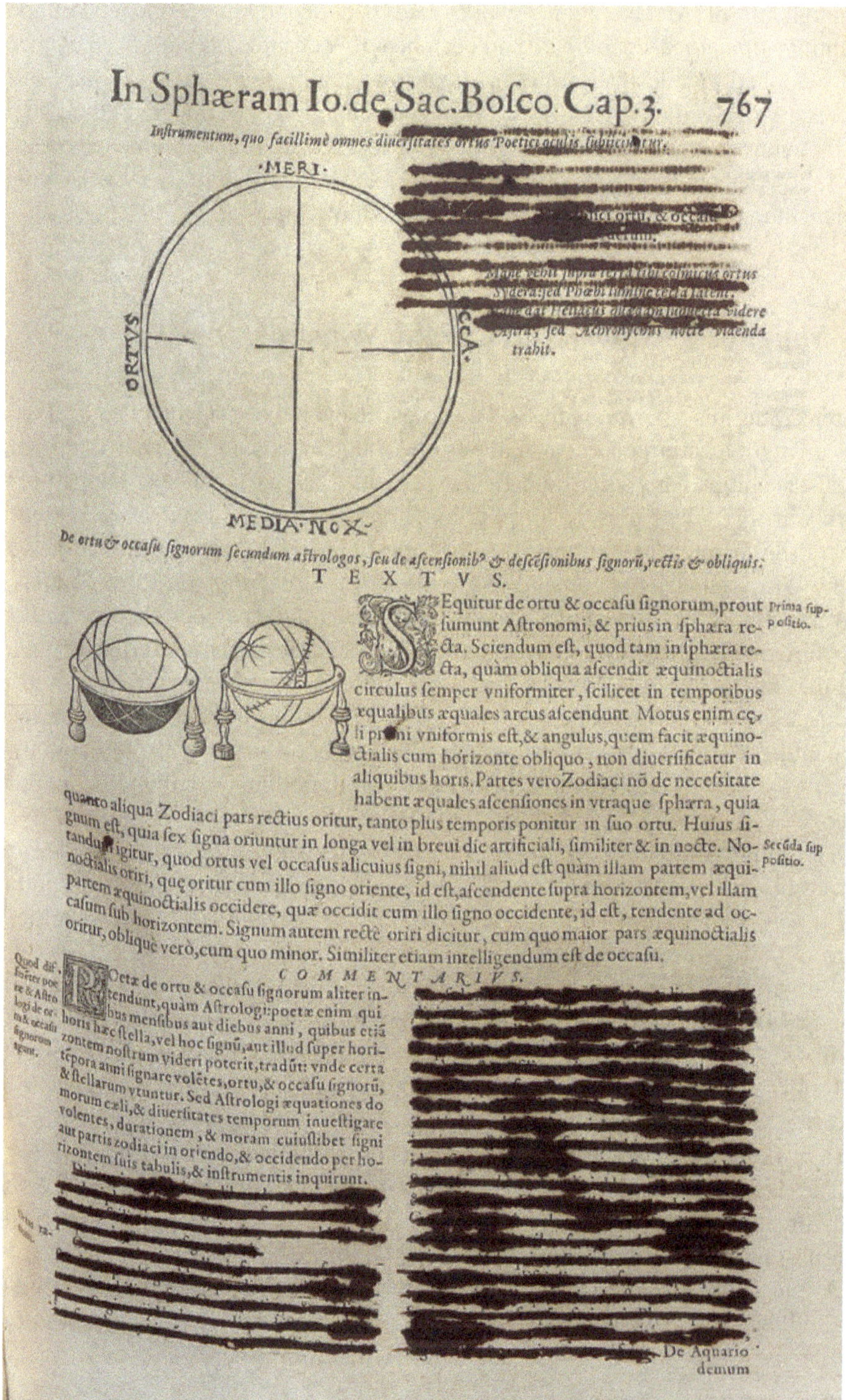

In Sphæram Io. de Sac. Boſco Cap. 3. 767

Inſtrumentum, quo facillimè omnes diuerſitates [illegible]

De ortu & occaſu ſignorum ſecundum aſtrologos, ſeu de aſcenſionib⁹ & deſcẽſionibus ſignorũ, rectis & obliquis.

T E X T V S.

SEquitur de ortu & occaſu ſignorum, prout ſumunt Aſtronomi, & prius in ſphæra recta. Sciendum eſt, quod tam in ſphæra recta, quàm obliqua aſcendit æquinoctialis circulus ſemper vniformiter, ſcilicet in temporibus æqualibus æquales arcus aſcendunt Motus enim cœli primi vniformis eſt, & angulus, quem facit æquinoctialis cum horizonte obliquo, non diuerſificatur in aliquibus horis. Partes vero Zodiaci nõ de neceſsitate habent æquales aſcenſiones in vtraque ſphæra, quia quanto aliqua Zodiaci pars rectius oritur, tanto plus temporis ponitur in ſuo ortu. Huius ſignum eſt, quia ſex ſigna oriuntur in longa vel in breui die artificiali, ſimiliter & in nocte. Notandum igitur, quod ortus vel occaſus alicuius ſigni, nihil aliud eſt quàm illam partem æquinoctialis oriri, quę oritur cum illo ſigno oriente, id eſt, aſcendente ſupra horizontem, vel illam partem æquinoctialis occidere, quæ occidit cum illo ſigno occidente, id eſt, tendente ad occaſum ſub horizontem. Signum autem rectè oriri dicitur, cum quo maior pars æquinoctialis oritur, obliquè verò, cum quo minor. Similiter etiam intelligendum eſt de occaſu.

Prima ſuppoſitio.

Secũda ſuppoſitio.

C O M M E N T A R I V S.

POetæ de ortu & occaſu ſignorum aliter intendunt, quàm Aſtrologi: poetæ enim qui bus menſibus aut diebus anni, quibus etiã horis hæc ſtella, vel hoc ſignũ, aut illud ſuper horizontem noſtrum videri poterit, tradũt: vnde certa tẽpora anni ſignare volẽtes, ortu, & occaſu ſignorũ, & ſtellarum vtuntur. Sed Aſtrologi æquationes domorum cæli, & diuerſitates temporum inueſtigare volentes, durationem, & moram cuiuſlibet ſigni aut partis zodiaci in oriendo, & occidendo per horizontem ſuis tabulis, & inſtrumentis inquirunt.

Quod differunt poetæ & Aſtrologi de ortu & occaſu ſignorum.

De Aquario demum

Fig. 10.6 One censured page of the copy of Giuntini's *Speculum astrologiae* kept at the Library of the University of Sevilla. From (Giuntini 1581–1583). Courtesy of the Biblioteca Universidad de Sevilla

The verses were translated by George P. Goold as follows: "...an insubstantial axis runs down through the wintry air and controls the universe, keeping it pivoted at opposite poles: it forms the middle about which the starry sphere revolves and wheels its heavenly flight, but is itself without motion..." (Manilius 1977, 27).

Avelar's transcription of this passage is slightly different, not only because of the chosen orthography and punctuation, but also for more significant dissimilarities (see words in italic):

Aera per gelidum tenuis deducitur axis,
Libratumque *gerit* diverso cardine mundum.
Sydereus, *medium circa* quem, volvitur orbis,
Aeternosque rotat cursus, *immotum*.... (do Avelar 1593, fol. 10v)

Avelar's second (and longest) quotation of the *Astronomica* appeared in his discussion of the position of the earth in the middle of the universe (do Avelar 1593, fol. 21r–v). These are the verses he cited (*Astronomica* I.194–206):

Nec vero admiranda tibi natura videri
pendentis terrae debet, cum pendeat ipse
mundus et in nullo ponat vestigia fundo,
quod patet ex ipso motu cursuque volantis,
cum suspensus eat Phoebus currusque reflectat
huc illuc agilis, et servet in aethere metas,
cum luna et stellae volitent per inania mundi,
terra quoque aerias leges imitata pependit.
est igitur tellus mediam sortita cavernam
aeris, e toto pariter sublata profundo,
nec patulas distenta plagas, sed condita in orbem
undique surgentem pariter pariterque cadentem.
haec est naturae facie:.... (Manilius 1977, 18–20)

Goold's translation of that citation is the following:

But the principle of the Earth's suspension should cause you no surprise. The firmament itself hangs thus and does not rest on any base, as is clear from the actual movement of its swift career; the Sun moves unsupported, as it wheels its chariot nimbly now this way and now that, keeping to its turning points in heaven; and the Moon and the stars wing their way through empty regions of the sky: therefore the Earth, too, in obedience to celestial laws, has hung suspended. Thus it is that Earth has been allotted a hollow space in mid-air, equidistant from every quarter of heaven's depths, not spread into flat plains but fashioned into a sphere which rises and falls equally at every point. This is the shape of nature;.... (Manilius 1977, 19–21)

Avelar's transcription of this extract is also slightly different from the modern edition (see words in italic):

Nec vero *admiranda tibi natura* videri
Pendentis terrae debet, cum pendeat ipse
Mundus, et in nullo ponat vestigia fundo:
Quod patet ex ipso motu, cursuque volantis,
Cum suspensus eat Phoebus, *cursuque* reflectat
Huc, illuc, *agiles* et servet in aethere metas
Cum luna et stellae volitent per inania mundi,
Terra quoque *aereas* leges imitata pependit.
Est igitur tellus mediam sortita cavernam

Aeris, e toto pariter sublata profundo:
Nec patulas distenta plagas, sed condita in orbem,
Undique surgentem pariter, pariterque cadentem:
Haec est naturae facies.... (do Avelar 1593, fol. 21r–v)

The third quotation of the *Astronomica* in Avelar's *Sphaera* appeared at a point where he describes the Milky Way (do Avelar 1593, fol. 30r), a subject that was not included in Sacrobosco's original work. This is a modern version of the verses he cited (*Astronomica* I.699–703):

trisque secat medios gyros et signa ferentem
partibus e binis, quotiens praeciditur ipse.
nec quaerendus erit: visus incurrit in ipsos
sponte sua seque ipse docet cogitque notari.
namque in caeruleo candens nitet orbita mundo.... (Manilius 1977, 60)

It was translated by Goold as follows:

At two points it cuts the three middle circles and the circle which carries the signs and is as often cleft itself. One need not search to find it: of its own accord it strikes the eyes; it tells of itself unasked, and compels attention. It shines like a glowing path in the dark-blue of the heavens,.... (Manilius 1977, 61).

In the case of the third quotation, Avelar's transcription is also somewhat different from the modern edition (see words in italic):

Tresque secat medios gyros, et signa ferentem
Partibus e binis, *quoties* praeciditur ipse,
Nec quaerendus erit, visus incurrit in ipsos
Sponte sua, seque ipse docet, cogitque notari:
Namque in caeruleo candens nitet orbita mundo.... (do Avelar 1593, fol. 30r)

Avelar's fourth and last quotation of the same work, containing a single line, appeared at his description of the equinoxes (do Avelar 1593, fol. 34r): "Libra Ariesque parem reddunt noctemque diemque" (Manilius 1977, 180), which may be translated as: "Libra and Aries make equal night and day" (Manilius 1977, 181) (Chap. 2).

Manilius had sometimes been cited by astronomical authors of the late fifteenth and sixteenth centuries, before Avelar's work. Jacques Lefèvre d'Étaples (ca. 1450–1536), in his commentary on Sacrobosco's *Sphaera*, had a short quotation of Manilius's *Astronomica* that partially corresponds to Avelar's first citation: "Aera per gelidum tenuis deducitur axis / Sidereus: medium circa quem volvitur orbis" (de Sacrobosco 1499, fol. m IIr). Later editions of Jacques Lefèvre's commentary contain the same two lines (de Sacrobosco 1508, fol. 57r; de Sacrobosco 1527, fol. 4v; de Sacrobosco 1531, fol. 128r; de Sacrobosco 1538, fol. 4v). Notice, however, that Avelar's quotation contained four lines: "Aera per gelidum tenuis deducitur axis, / Libratumque gerit diverso cardine mundum. / Sydereus, medium circa quem, volvitur orbis, / Aeternosque rotat cursus immotum" (do Avelar 1593, fol. 10v). It is quite odd that Jacques Lefèvre's quotation did not include the second verse (only the first and third ones); and hence that work could not be the source of Avelar's corresponding citation.

One of the authors who cited Manilius a number of times was Luca Gaurico (1475–1558), at several places in his 1531 edition of the *Sphaera*.[14] At the beginning of the book, in his "Oratio de laudibus astrologiae," Gaurico mentioned Manilius's name and quoted his *Astronomica* several times (de Sacrobosco 1531, fol. a IIr; a IIIv; a IIIIv). However, none of those quotations match Avelar's citations. At another part of the book, containing the commentaries of Prosdocimo de Beldomandis (ca. 1370/1380–1428) and of Bartolomeo Vespucci (fl. 1500) on Sacrobosco's *Sphaera*, we find a long quotation of the *Astronomica* that contains Avelar's second citation (de Sacrobosco 1531, fol. 15r). Gaurico's unconventional quotation skipped some of the verses and changed the order of other ones, without warning his readers of those changes (he reproduced *Astronomica* I.194–204, 211–13, 215–17, 228–29, 230–31, and 235). Avelar's corresponding extract (*Astronomica* I.194–206) included two verses that were not transcribed by Gaurico. Hence, this was not the source of Avelar's second quotation of Manilius.

Hartmann Beyer (1516–1577), in his *Quaestiones novae in libellum de sphaera Ioannis de Sacro Busto*, quoted a single verse by Manilius that was also reproduced by Avelar: "Libra Ariesque parem reddunt noctemque diemque" (Beyer 1549, fol. 26v).

Cornelius Valerius (1512–1578), in his *Sphaera*, quoted some verses from Manilius's *Astronomica* to illuminate the relation between the signs of the Zodiac and the parts of the human body (Valerius 1564, fol. 10r) and when describing the main winds (*ibid.*, fol. 18r). Those verses of the *Astronomica* were not quoted by Avelar.

Theodor Graminaeus (ca. 1530–ca. 1593), in his commentary on Sacrobosco's *Sphaera*, cited Manilius's *Astronomica* three times (Graminaeus 1567, 72, 91, 116). None of his quotations correspond to those used by Avelar.

Francisco Sánchez de las Brozas (1523–1600) published a Latin *Sphaera* where he presented two quotations of the *Astronomica* (Brozas 1579, fol. 3v; fol. 23v). They were different from those published by Avelar.

Those occurrences show that, before Avelar, several astronomical authors did quote Manilius; but none of those works could have been the source of the quotations found in Avelar's *Sphaera*. Of course, Avelar could have drawn his citations from one of the several editions of the *Astronomica*; but there is a better explanation: he probably copied the quotations from Clavius.

It was quite unexpected to find out that the author who supplied Avelar's citations of Manilius was Christophorus Clavius, since the Jesuit mathematician was not interested in astrology. The 1999 reprint of the 1611 edition of Clavius's commentary on Sacrobosco's *Sphaera*, edited by Eberhard Knobloch, contains a very useful addition: an index of cited authors and works (Knobloch 1999, 341–50). That index points to five quotations of Marcus Manilius's *Astronomica* (Knobloch 1999, 9, 107, 126, 134, 185–86). Avelar's four citations of Manilius are contained therein. He did not include in his *Sphaera* Clavius's fourth citation, which describes the

[14] I am grateful to Angela Axworthy for calling my attention to this occurrence.

constellations of the Zodiac. The same five citations had already appeared in the first edition of Clavius's commentary (Clavius 1570, 25, 249–50, 284, 295, 377), with a single difference: the first one is incomplete in this edition (only two verses are quoted, exactly as in Jacques Lefèvre d'Étaples's commentary on Sacrobosco's *Sphaera*). The second edition, however, contained the four verses that appeared in Avelar's *Sphaera* (Clavius 1581, 18). Although I did not check the other editions, it is probable that they also contained the same quotations.

It is likely that Avelar copied his citations of Manilius from Clavius's commentary, since all the quotations he used were contained there. I could find no other work published before Avelar's book that contained all those citations except, of course, the editions of Manilius's poem. Had Avelar consulted the *Astronomica* itself, he would probably have choosen some quotations different from those published by Clavius.

Of course, Clavius's commentary on Sacrobosco's *Sphaera* contained references to many additional authors. Besides Manilius, there were eleven other authors quoted by Clavius five or more times: Alfarganus (al-Farghānī), Aristotle, Lucan, Pedro Nuñes, Ovid, Proclus (412–485), Ptolemy, Regiomontanus, Erasmus Reinhold (1511–1553), Averroes (ibn Rushd), (1126–1198) and Virgil. Only three of them were also mentioned by Avelar: Lucan, Ovid, and Virgil, who had been quoted by Sacrobosco himself. If Avelar did take his quotations from Clavius's book, then it seems that he made a careful selection of the citations he wanted to include in his own book. Perhaps the choice was made just for aesthetic reasons, or maybe he added the quotations from Manilius's *Astronomica* to his *Sphaera* because of his own deep involvement with astrology.

10 Avelar's Manuscripts

The only book published by André do Avelar for his teaching activities at the University of Coimbra was his *Sphaera*. It is not altogether clear why he decided to write that book. For his classes, he could have used any of the several editions and commentaries on Sacrobosco's *Sphaera* that were available at that time. The most popular ones were the *Sphaera Ioannis de Sacrobosco Emendata* produced by Élie Vinet and the massive *In Sphaeram Ioannis de Sacro Bosco Commentarius* by Christophorus Clavius.[15] Perhaps he wanted to make a name for himself.

[15] There were thirty-three editions of Vinet's compilation, published from 1551 to 1620 in several countries, and 18 editions of Clavius's commentary between 1570 and 1618. See my survey of editions of Sacrobosco's *Sphaera*: http://www.ghtc.usp.br/server/Sacrobosco/Sacrobosco-ed.htm. Accessed June 2019. For the list of treatises that contain the tract of Sacrobosco respectively authored by Christophorus Clavius, see http://hdl.handle.net/21.11103/sphaera.100732 and for the list of treatises respectively authored by Elie Vinet, see http://hdl.handle.net/21.11103/sphaera.100903.

It seems that Avelar enjoyed writing scholarly treatises, some of them directly linked to his teaching activities at the University of Coimbra. Several of his manuscripts have been conserved (Fiolhais and Martins 2016, 714–15), and it is likely that he wrote a larger number of them. There are eight known manuscripts and digital copies of two of them are available on the Internet:

1. "Lectio[n]es Mathematicae," dated 1595, Biblioteca Nacional de Lisboa, FG. Cód. 1894.
2. "In Theoricas Septem Planetarum Purbachii et Octavae Spherae eiusdem Apostillae recte expplicatae ad Preceptore meo Andrea d'Auellar," Biblioteca do Escorial–Ms. RBME &–IV–9.
3. "Tratado do Uso da Esfera," Biblioteca Municipal do Porto–Ms. 249
4. "Galatas de Curiosidades Matemáticas," Arquivo Nacional Torre do Tombo, Lisboa–Manuscritos da Livraria, nr. 681.[16]
5. "Geographia Theorica & Practica," Biblioteca Nacional de Lisboa–Ms. Cod. 2138.[17]
6. "Juizo que tirou em Coimbra Andre do Avelar Cathedratico da Cadeira de Mathemattica sobre os effeitos do Cometa que apareceo no anno de 1607," Biblioteca da Ajuda, Lisboa–Ms. 51–V–11.
7. "Discurso Astronomico & Astrologico do Cometa que apareceo por novembro de 1619 em Coimbra," Biblioteca da Ajuda, Lisboa, Ms. 46–VIII–16.

José Pereira Osório cited another manuscript (Biblioteca Municipal do Porto, Ms. 669) that he did not describe, in which Avelar referred to Copernicus's astronomical theory, rejecting it (Osório 1986, 120).

Those manuscripts display the intense scientific production of André do Avelar. Item (1), in Latin, contains notes of Avelar's Coimbra lectures in the years 1594 and 1595. Item (2), also in Latin, contains notes made by a student concerning Avelar's teaching of Peuerbach's *Theorica Planetarum*. Item (5), in Latin, is a large manuscript containing notes on geography and astronomy, probably also included in Avelar's teaching activities at the university. The other tracts, in Portuguese, could not be intended for use at the University of Coimbra.[18]

A detailed study of all available manuscripts produced by André do Avelar is a desideratum. It will greatly broaden our perception of this Portuguese author.

[16] http://digitarq.arquivos.pt/details?id=4461821 Accessed June 2019.

[17] http://purl.pt/24925 Accessed June 2019.

[18] Although item (4) is usually attributed to André de Avelar, it is unlikely that he wrote it. The description of the manuscript on the website of the Torre do Tombo Archive states that the word "Galatas" in the manuscript title was a pseudonym of Avelar. However, I was unable to find any other source referring to this pseudonym; and at the end of the prologue of that manuscript, one finds the name "Frei Joseph de Jesus Maria Olisiponense [from Lisbon] Galatas." Hence, instead of Avelar, the author of the manuscript could have been that priest. On the other hand, although Diogo Barbosa Machado, in his *Bibliotheca Lusitana*, mentioned several authors called "José de Jesus Maria" (Machado 1741–1759, vol. 2, 863–65), none of them could be the author of that work.

11 Avelar's Career at the University of Coimbra

Besides his teaching and research activities, Avelar became involved in several other occupations at the University of Coimbra. It seems that he had financial difficulties. The annual wage of a mathematics professor was 80,000 reis, according to the statutes of the university (de Almeida 1967, 48). One year after being admitted to the university, however, he asked for a supplementary yearly payment of 20,000 reis to help support his children. The king accepted his request and ordered the dean to pay this additional amount (de Almeida 1967, 49, 51). In 1593, Avelar sold to the university an astrolabe made of wood and a large world map in color for 2700 reis; two years later he sold another map for 3400 reis (de Almeida 1967, 50, 52). The astrolabe and maps were used in his classes and are directly related to his manuscript on practical and theoretical geography.

The statutes of the University of Coimbra had determined that there should be a proof reader for the books produced by its staff. In addition to this post, the proof reader would also become the keeper of the university's library, receiving for this work 30,000 reis each year (Universidade de Coimbra 1593, fol. 69r). In 1595, Avelar attempted to become the director of the library and corrector of the books published by the university, but his request was denied on dubious grounds: the university, it was claimed, had no library at that time (de Almeida 1967, 39, 53). It is likely, however, that the official document describing the decision meant to say that there was no place to keep the books. There is reliable information that the university did have a library. In 1536, before moving to Coimbra, it had 151 volumes, all of them attached to chains to avoid being stolen (do Amaral 2014, 20). After the transfer of the university to Coimbra, access to the books was made public to any person who wished to consult them (do Amaral 2014, 33). Around 1600, it already had 739 volumes, and many others were ordered by the university. The writer Pedro de Mariz (ca. 1550–1615) was afterwards chosen to organize the books, and in 1598 Avelar replaced him (de Carvalho 1990, 169). In the same year, Avelar was also granted the position of keeper of the notary's office, adding 12,000 reis to his annual income (de Almeida 1967, 40, 55; Braga 1892–1902, vol. 2, 245). In 1598, because of this additional job, Avelar produced the first catalog of the archive of the university (Brandão 1927).

Avelar's wife Luiza de Faria died around 1600. The six children of the couple were born in the following years: Tomásia 1581; Mariana 1591; Ana 1592; Luiz 1593; Pedro 1596; and Violante 1596 (de Carvalho 1990, 333). In 1600, five of them were under ten. Avelar, however, turned his widowhood to his own advantage. During his youth he had studied theology in Salamanca or Valladolid. Now, after the demise of his wife, he took religious vows and became a Catholic priest (Machado 1965–1967, vol. 1, 137–38; vol. 4, 15–16). After that, he applied for a position at the See of Coimbra.

The annual income of the See of Coimbra was 45,000 cruzados (Botelho 1873, 30). Around 1600, the cruzado (a gold coin with a weight corresponding to 3.55 g) was equivalent to about 400 reis (Fernandes 1856, 124; Lima 2005, 171). Hence, the income of the See of Coimbra was 18,000,000 reis. Part of this huge revenue was divided between the priests connected to the See.

There were thirty-five canons attached to the See, and some were appointed by the university. One of them, who should be a Master of Arts, was "tercenário," that is, he received one third of the full annual payment of an ordinary canon (Botelho 1873, 30). Miguel Vaz Pinto, the university professor who formerly held this position, died in 1601, and three candidates applied to replace him. One of them was André de Avelar, who was elected on account of his age and capacities (de Almeida 1967, 56). However, there were some doubts whether he should ultimately assume this position, because he was the descendant of Jews. The King consulted the Pope, who in 1603 authorized the appointment (de Carvalho 1990, 184; de Almeida 1967, 55; Braga 1892–1902, vol. 2, 492–93). It seems that this additional income solved his financial problems.

In 1612, after teaching for twenty years at the university, Avelar (now 66 years old) had the right to retire and receive two thirds of his wages (de Almeida 1967, 59). However, two months later he was reappointed to the chair of mathematics for an additional interval of four years, receiving during that time the full corresponding wages (de Almeida 1967, 60–61). He was employed again not only because of his good services and the satisfaction of his students, but because there was nobody who could replace him at that time (de Carvalho 1990, 211).

Notice that the benefits successively obtained by Avelar at the University of Coimbra suggest that he was on good terms both with the successive deans and with his colleagues.

In his old age Avelar had reached a privileged status. His poverty belonged to the remote past. He had published a popular book (*Reportorio dos tempos*) and another one not so successful, his *Sphaera*. He was a respected university professor and was still producing an abundance of manuscripts on mathematics, astronomy, and astrology. He had beneficial relationships with some noblemen who probably helped to boost his career. There was, however, a dark cloud lingering over him. His last years were to become a nightmare.

12 Inquisition in Portugal

The first aim of the Catholic Council of Trent, held between 1545 and 1563, was the development of strategies for fighting the growing influence of Martin Luther (1483–1546). In Portugal, Protestantism did exist (Bodian 2002) but never became popular. The rising inquisitorial wave found another target there: crypto-Judaism among the "New Christians," the descendants of the Jews who King Dom Manoel (1469–1521) had compelled to become Christians in 1497 (Roth 1959, 27–29). The recently converted Jews and their offspring became a separate social class that was not fully accepted by the "Old Christians," even after more than one century. There was a general suspicion that the "New Christians" or "people of the Hebrew nation" were Christians only outwardly, but kept the old practices of Judaism privately. A Jew who practiced Judaism could not become a target of the Inquisition, because it was an internal institution of the Catholic Church; but a converted Jew who practiced the religion of his ancestors was classified as a heretic and could be prosecuted. So, in the second half of the sixteenth century, the hidden practice of Judaism turned out to be the main concern of the Portuguese Inquisition.

This was a peculiarity of Portugal. In Spain, after an early preoccupation with Judaizers, the Inquisition turned its attention to sexual offences (bigamy and "sexual deviation"), blasphemy, and specific heretical propositions. Between 1540 and 1700, only 10% of the Spanish inquisitorial processes involved Judaizers. In Portugal, before 1770, over 80% of the Inquisition trials dealt with people accused as Judaizers (Schwartz 2008, 95; Salomon 1998, 132). Among the forty thousand known inquisitorial processes held in continental Portugal, there were about seven thousand denunciations regarding witchcraft, but only about one thousand cases were prosecuted, and punishment was usually soft, as compared to northern Europe (Schwartz 2008, 97).

Around 1600, the Portuguese Inquisition was engaged in the suppression of crypto-Judaism, spreading a culture of religious and racial discrimination towards the New Christians. Other reasons might also have been relevant at that time for such persecution (Schwartz 2008, 105).

The situation became particularly strained in the early seventeenth century. After a few years of negotiation, a group of New Christians obtained from King Philippe III of Spain (1578–1621) (known as Philippe II in Portugal) the publication of a General Pardon in January 1605 signed by the Pope, which abolished all past religious faults and allowed them to leave the country with their property intact. In exchange for this bill, the New Christians promised to pay the enormous sum of 1,800,000 crusados, corresponding to over 6000 kilograms of pure gold. Because of this decree, on January 16, 1605, the Portuguese tribunals liberated 410 prisoners after imposing only a formal penance (Roth 1959, 40). There was an immediate public reaction against the royal act of mercy. In Coimbra, students attacked and killed some New Christians, setting their houses on fire (Curto 1997, 188; Braga 1892–1902, vol. 2, 493). Unfortunately, the New Christians could not pay the promised amount, and the pardon was revoked in 1610. There were new attempts to negotiate with the king, but they were not successful, and the situation of the New Christians deteriorated during that decade (Schwartz 2008, 116).

13 Avelar's Involvement with Jewish Religion

Shortly after the General Pardon episode, an outstanding professor of law at the University of Coimbra, António Homem (1564–1624), began to organize a secret Judaic brotherhood.[19] António Homem was the heir of a long Jewish family tradition.

[19] Some historians of the past denied the existence of this brotherhood and stated that the Inquisition had falsely accused him. There is no doubt, however, that it was real. Those interested in this subject should consult the most careful research ever produced about António Homem and the Fellowship of St. Diogo, written by (de Carvalho 1990).

His great-great-grand father lived and died as a Jew. Some of his relatives had been sentenced as Judaizers by the Inquisition: his great-grandfather, his grandmother, and two of his uncles. His father had kept the Judaic religious tradition and taught it to all his children, without the knowledge of his wife, who was an Old Christian of noble lineage (de Carvalho 1990, i).

Up to 1607, António Homem kept his Judaic practices private. He knew several other New Christians who were also Judaizers or sympathetic to those practices. He probably thought that strong cooperation and organization were needed both to support and develop their religious faith and to defend themselves (de Carvalho 1990, 32). In September of that year, he organized at his own home a celebration of Yom Kippur, to which he invited several friends, including André do Avelar. Avelar started collaborating directly with Homem, using his astronomical knowledge to determine the right day for Judaic ceremonies (de Carvalho 1990, 37).

Avelar had been initiated in the Jewish faith at the age of sixteen by his maternal aunt, Brites Lopes (Braga 1892–1902, vol. 2, 603). However, his wife Luiza de Faria was an Old Christian, and he kept his belief to himself until her death. Only a few years later, after the General Pardon, did he disclose his faith to his children, who also became crypto-Jews (de Carvalho 1990, 169).

Under the leadership of António Homem, the growing clandestine Jewish community became the "Confraria de São Diogo" (Fellowship of St. Diogo), a name that was supposedly Christian, but was actually a tribute to a Judaic martyr, the Franciscan Friar Diogo da Assumpção (1571–1603), who spontaneously converted to Judaism in 1599. He tried to escape from Portugal to a country where he could become a Jew, but he was denounced and arrested by the Inquisition. Friar Diogo tried to convince the Inquisitors that he was right and that the Christian religion was false, and maintained this attitude for years during his imprisonment. He was finally convicted for the crimes of heresy and apostasy and was burned alive on the 3rd of August, 1603, without showing a sign of regret (Teixeira 1895, 217–64; Curto 1997, 181–88).

The underground Jewish activities of the Fellowship of St. Diogo were kept secret for a few years, and during this period Avelar suffered no harassment. The situation changed around 1620 for two reasons: official inspections of the University of Coimbra by the order of the King and the pursuits of the Inquisition.

The first royal inspection of the university in the seventeenth century was prompted by rumors of frauds and bad behavior amongst the students, professors, and other personnel of the institution (Baião 1919, 104). In 1615, the King appointed Dom Martim Affonso de Mexia (ca. 1570–1623), Bishop of Leiria (afterwards of Lamego), as "visitador e reformador" (visitor and reformer) of the university (Braga 1892–1902, vol. 2, 500). The Bishop spent several months hearing all types of allegations. António Homem was accused of bribery in the election of professors, and of protecting and favoring New Christians (Braga 1892–1902, vol. 2, 497). There was also a more serious charge. In May 1616, one of the canons of the Coimbra Cathedral named Alvaro Soares Pereira wrote a letter to an Inquisitor in Coimbra, denouncing António Homem for the sin of sodomy, pointing out the names of people who could be interrogated to confirm the accusation. He also identified one of the young lovers of António Homem: Jorge Mexia, nephew of the Bishop (de Carvalho 1990, 54–65). At that time, any homosexual activity was regarded as a crime and could be severely punished both by the civil authorities and by the Inquisition (de Azevedo 1921, 141).

The Inquisitor who received the letter was Francisco de Brito de Menezes (ca. 1570–1631), a fierce enemy of António Homem with whom he had fought over a vacant canon at the Coimbra Cathedral in 1610. António Homem was chosen, although he was a New Christian and notwithstanding the support the defeated professor received from members of the Inquisition (Braga 1892–1902, vol. 2, 499). Now, Francisco de Menezes had a nice opportunity for revenge, but nothing happened, possibly for two reasons: first, António Homem was a friend and legal adviser of the Bishop Dom Martin Affonso Mexia (de Carvalho 1990, 64); second, it was not advisable to expose the nephew of the "university visitor" and Bishop.

After a brief investigation, António Homem was formally charged in November 1616 for only corruption and bribery. Although he defended himself, he was found guilty and had to pay a fine of 100,000 reis, more than the annual remuneration received by Avelar (de Carvalho 1990, 69; Braga 1892–1902, vol. 2, 506–11). António Homem's punishment was light compared to that of other people. Another professor of law, Manuel Rodrigues Navarro, was found guilty of usury, corruption, and sodomy. In June 1616, he was expelled from the university *in perpetuum* and his name was struck off the university records. Afterwards, he was sentenced to deportation and the loss of his property. He went to Italy, where he lectured in the universities of Bologna and Naples (de Carvalho 1990, 70, 392). Many other professors received serious or light punishments.

Nothing happened to André do Avelar during the visitation of Dom Martim Affonso de Mexia, but all New Christians were regarded as highly suspicious after that time (Braga 1892–1902, vol. 2, 500). Besides that, the mood of intrigue and denunciations stimulated by the visitor would soon produce new difficulties.

14 Astrology and the Jewish Messiah

In the sixteenth century, the people who were more deeply interested in astrology and who understood its technical details were physicians (and other people involved with health issues, such as surgeons) and, of course, astrologers (sometimes called "mathematicians"). In old documents Avelar is not described as an astrologer, and he did not study medicine. However, there was a strong tradition in the teaching of astrology at the University of Salamanca, where he was a student. That might have been the origin of his deep association with this subject, as shown in his *Reportorio*.

During his involvement with the fraternity of Saint Diogo, Avelar made use of his extensive knowledge about the calendar and astronomy to choose an adequate time for the Jewish ceremonies performed by the group. We may conjecture that he was also interested in the astrological prediction of the coming of the Jewish Messiah, since this was a central interest of crypto-Jews and particularly of António Homem's group. There had already been several outstanding attempts to use astrology as an instrument for predicting the coming of the Messiah in the preceding centuries.

At least since the tenth century, there had been Jews in the Iberian Peninsula who described their hope for the coming of the Messiah (Montenegro 2017, 23). Messianic belief involved both the idea of a religious-spiritual redemption and national and political independence (Sáenz-Badillos 1999, 178). The Jews used several methods for predicting the coming of the Messiah. The most common and earliest one was the interpretation of the Book of Daniel, alone or together with other Biblical texts. In the tenth century, Sa'adiah ben Yosef Gaon (882/892–942) interpreted the prophecies of Daniel as indicating that the coming of the Messiah would occur abruptly. His advent would be accompanied by many disgraces to the enemies of Israel: fire, stones, violent deaths, etc. The Jews would gather in Jerusalem and there would follow idyllic times of peace and happiness, without diseases or sorrows (Sáenz-Badillos 1999, 189).

Another method made use of comparisons between the different periods when the Jews were kept in exile, in Egypt and in Babylon. The third one was an analysis of the Torah using special numerical techniques for the interpretation of letters and words, such as gematria and Kabbala. Astrology was another independent method (Silver 1927, 243–59).

Although many early Jew authorities denied the validity of astrology in late Antiquity, taking inspiration from the Greeks and Romans, it became a highly significant belief. Its impact increased in the Middle Ages, due to the influence of Islam (Silver 1927, 255–56).

Two medieval Jewish authors who attempted to predict the coming of the Messiah by means of astrological calculations were Solomon Ibn Gabirol (ca. 1021–1058) and Abraham Bar Ḥiyya (ca. 1065–ca. 1140) (Ruderman 2001, 26; Sela 2010, 9). Gabirol was probably the first Jewish astronomer to attempt to predict the "end of days" using astrological techniques (Goldstein 2001, 42). Using astrological methods developed by Abū Ma'shar (787–886) and especially by Māshā'allāh ibn Atharī (762–ca. 815), those Jewish astrologers attempted to explain past historical events and to predict future ones by astral causes (Rodríguez-Arribas 2011, 14). Māshā'allāh's *Book of Eclipses* contained an astrological theory of world history and associated the great conjunctions of Jupiter and Saturn to the coming of great prophets and seers (Goldstein 1996, 14). Both Gabirol and Bar Ḥiyya linked the Saturn-Jupiter conjunctions in the sign of Pisces to the advent of the Messiah, because a historical conjunction of the same type was associated with the exodus of the Jews from Egypt. Actually, the detailed analysis of the astrological situation concerning the exodus was much more detailed and intricate (Rodríguez-Arribas 2005, 146–48). According to Bar Ḥiyya's calculations, the new deliverance of the Jews could happen in 1358 (the last of twelve great conjunctions) or in the years corresponding to the largest conjunction, 1448 or 1468 (Sáenz-Badillos 1999, 190; Sela 2010, 9–11).

Many Jewish authors denied the possibility of predicting the arrival of the Messiah by astrological calculation, saying only prophets could anticipate it. Two examples of scholars with this attitude are Saadia Gaon (882–942) and Abraham ben Meir ibn Ezra (1092–ca.1167) (Rodríguez-Arribas 2011, 14, 17). Ibn Ezra accepted historical astrology and the relevance of the conjunctions of Jupiter and

Saturn (Goldstein 1996, 14), but criticized its use for predicting the coming of the Messiah (Rodríguez-Arribas 2005, 149–53).

One important Spanish kabbalist was rabbi Moshe ben Shem-Tov (ca. 1240–1305), known as Moses de León, who is regarded as the composer or compiler of the *Zohar*. Messianism is an important subject of that work, and the coming of the Messiah is described as the way to redeem mankind from evil, bringing harmony to the world. The coming of the Messiah would be accompanied by the death of the Pope and great wars, followed by the reconstruction of Jerusalem (Montenegro 2017, 33).

In the following centuries, the religious relevance of the Saturn-Jupiter conjunctions was discussed by other Jewish astrologers such as Levi ben Gershon, or Gerson (1288–1344), who studied the event of 1345 (Goldstein 1996, 18–19). He agreed with the prevailing Jewish view that the Messiah would come in 1355 or 1358. Christian authorities were also anticipating some dangerous religious upheaval. The cardinal Pierre d'Ailly (1351–1420) associated the coming of the Antichrist with a Saturn-Jupiter conjunction (Goldstein 1996, 19).

In the fifteenth century, the famous Portuguese-born Jewish philosopher Isaac ben Judah Abravanel or Abarbanel (1437–1508) used the same method employed by Bar Ḥiyya. The first conjunction of Saturn and Jupiter in Pisces had occurred in the Jewish year 2365, three years before the birth of Moses and eighty-three years before the exodus; the new conjunction, in the year 5224 (1464), might announce the birth of the Jewish Messiah (Cohen-Skalli 2006, 196–97). The redemption of the Jews would occur a few decades after that time, between 1503 and 1571, perhaps in 1531 (Lawee 2001, 18). There would be terrible wars involving the Jews, the Christians, and the Muslims; the Messiah would defeat their enemies and the Jews would return to Jerusalem (Rodrigues 2010, 98).

In the transition from the fifteenth to the sixteenth century, the Jewish astrologers Bonet de Lattes (1450–1514) and Abraham ben Samuel Zacut (1452–ca. 1515)[20] (Chap. 3) also made messianic predictions (Ruderman 1992, 308). Zacut made use of the calculation of eclipses and planetary conjunctions and predicted that the salvation of Israel would begin in 1503 or 1504 (Goldstein 1998, 182). In the early sixteenth century, the expectation was so strong that in 1521 an inquiry was sent from Rome to Jerusalem asking whether the signs of the approaching Messiah had not already appeared there (Silver 1927, 149).

Possibly because of the failure of those astrological forecasts, during the sixteenth and seventeenth centuries most of the Messianic predictions were made using biblical interpretation and gematria, not astrology (Silver 1927, 116–43, 183–92). Moses de Léon, the author of the *Zohar*, suggested a handful of dates: 1300, 1306, 1324, 1334, 1340, 1608, and 1648. In the seventeenth century, several Kabbalists and non-Kabbalists accepted the year 1648 as the date of Redemption (Silver 1927, 151); others, however, predicted different years, such as 1713 or 1725 (Patai 1988, 56). In 1648, the last date predicted by the *Zohar*, Shabbatai Zevi (1626–1676), a

[20] Samuel Zacut or Zacuto, born in Spain, was a highly influential astronomer and astrologer. His astronomical tables were used by Portuguese navigators (Goldstein 1998).

Jew born in Smyrna, became convinced that he was the Messiah and produced a strong commotion (Merrill 1973, 161; Patai 1988, xlv). That famous episode, however, happened after Avelar's passing.

Did Avelar get involved in astrological predictions concerning the Messiah? His published works display no clear hint about this subject. Luís Miguel Carolino and Carlos Ziller Camenietzki studied the two known manuscripts left by Avelar containing astrological predictions (Carolino and Camenietzki 2006). Both manuscripts described the effects of comets that appeared respectively in 1607 and 1618. Comets, in astrology, are usually a warning of bad occurrences, and Avelar's analysis of those comets was no exception. The 1607 event, according to Avelar, would produce the following effects:

> ...a king will almost die of a sudden death, and princes and nobles of the highest rank will die. A war will be waged in a nearby land, or rather many wars will occur in many places, and there will be the destruction of cities, shipwrecks, and great dangers and raids by pirates at sea, and their fleets will spread fear and do great ill at seaports. There will be great changes in religion and political affairs. There will be different and unexpected journeys and embassy missions. (Carolino and Camenietzki 2006, 38)

Predictions of the "end of days" usually described wars and the violent deaths of key persons, together with political changes, before the dawn of the new age brought by the Messiah. Therefore, Avelar's description of the effects of the 1607 comet was compatible with this scenario. However, although Avelar did mention great political and religious changes, there was no direct allusion in those manuscripts to the coming of the Jewish Messiah.[21] If Avelar made any astrological prediction about this subject, it was probably kept secret.

Luiz de Avelar, son of André do Avelar, published a short dialogue on the 1618 comet (do Avelar 1619). He mentioned the usual injurious effects of comets, but he did not suggest any influence of that celestial occurrence on religious affairs.

15 Avelar and the Inquisition. First Proceedings

Meanwhile, the Coimbra Inquisition was busy receiving denunciations, arresting, interrogating, and judging hundreds of suspects. The first known allusion to André do Avelar in a process of the Inquisition arose in 1612. A prisoner called Antonio Gomes accused Avelar and three other canons of the Coimbra See who were New Christians of practicing Judaism (Antonio Dias da Cunha, Chrispim da Costa and Fernão Dias). The second mention, much more serious, appeared early in 1619 in the process of Miguel Gomes, called "o Manco" (the lame one). In his deposition we find the first description of the Brotherhood consecrated to Friar Diogo, in which Gomes identified some of its members, including himself. He mentioned António

---

[21] Luís Miguel Carolino, private communication, 22nd of June 2018. Carolino conjectures that Avelar would be very careful in his public prognostics because of his position as professor at the University of Coimbra and his status as a New Christian.

Homem, André do Avelar, Diogo Lopes da Rosa, António Dias de Almeida, Diogo Lopes de Sequeira, António Corrêa, Luiz de Sá, André Vaz Cabaço, Manuel Gomes Tasquinha, Francisco da Costa, Pêro Cabral, Francisco de Aguiar, Miguel da Fonseca, Fernão Dias da Silva, José Coutinho, and António de Oliveira (Braga 1892–1902, vol. 2, 511–12). In the following months, other members of the Brotherhood were successively arrested and interrogated.

By coincidence, in 1619, Francisco de Brito de Menezes, the opponent of António Homem, had a new opportunity to take vengeance: he was appointed the new visitor and reformer of the University of Coimbra as well as substitute dean, or intervenor (Braga 1892–1902, vol. 2, 516). Francisco de Menezes began his work in May 1619. He summoned students and professors at random and questioned them about the cases of immorality and corruption they knew. In a few weeks he had already obtained several depositions against António Homem, who was accused of bribery, corruption, and homosexuality (de Carvalho 1990, 82–85).

In October 1619, the promotor of the Coimbra Inquisition requested the arrest of António Homem and the confiscation of his estate, since there were three separate reliable testimonies against him by relatives; he was a New Christian and it was known that he had other relatives that had been convicted as Judaizers; and he had also been accused of moral offenses with male boys. He was arrested on Sunday, the 24th of November, and three weeks later was transferred to the Lisbon headquarters of the Inquisition (de Carvalho 1990, 90–93; Braga 1892–1902, vol. 2, 555; Baião 1919, 105). He would never be set free.

André do Avelar knew that he was also going to be arrested, but it was impossible to escape. Although his name had already been mentioned by several members of the Fellowship that had been seized previously, the formal charge against André do Avelar was made on the 1st of March, 1620, by the canon Crispim da Costa, who stated that once, when they were leaving the Coimbra See, they had confessed to each other that they followed the Jewish faith (Baião 1919, 134; de Carvalho 1990, 98). Avelar was arrested by the Inquisition on the 20th of March. He was seventy-four years old at that time. The next day, he asked to be heard and made a detailed confession. He stated that he had also disclosed his Jewish belief to other Judaizers and named António Homem, Miguel Gomes, Henrique de Aredes, Francisco de Almeida, Francisco Gomes, and others (Baião 1919, 134; de Carvalho 1990, 99; Braga 1892–1902, vol. 2, 602–03). The process of André do Avelar was unbelievably fast. A few days later, on the 29th of March, his sentence was read and he was ordered to abjure his mistakes. After doing so, he was "imprisioned" in Coimbra on the next day, though this meant in effect his release from detainment (Baião 1919, 135–36; Braga 1892–1902, vol. 2, 604).[22]

It is difficult to understand the unusually gentle and speedy treatment received by Avelar on this occasion, but Avelar was not the only one to receive such mild treatment. Crispim da Costa, who had denounced him and António Homem, and Miguel

---

[22] In the case of mild offenses, the sentence of incarceration *ad arbitrium* was interpreted as a temporary obligation to live in one's own town, during a given number of months (de Azevedo 1921, 145).

da Fonseca, who had also denounced António Homem, were also allowed to abjure on the same day (29th of March 1620) and were afterwards released (de Azevedo 1921, 174). Maybe, at that time, the main interest of the Inquisition was obtaining additional evidence to convict António Homem.

16 Avelar and the Inquisition. Second Proceedings

Avelar only remained free until the next year. There were successive arrests of members of the Fellowship and the mathematics professor was now accused of being one of the heads of the group (Baião 1919, 136). On the 11th of February 1621, his son Pedro Homem de Faria (twenty-five years old) was arrested by the Inquisition at Lisbon. He had enlisted to serve in India and it was suspected that he was running away to avoid prosecution (de Carvalho 1990, 112). He was questioned but denied any involvement with Judaic practices and made no denunciation. The other son, Luiz de Avelar (twenty-eight years old), who was a Master in Arts, was arrested on the 10th of September (de Carvalho 1990, 115). He soon admitted his own guilt and accused his father, brother, and four sisters.

Three of André do Avelar's daughters were nuns at the Monastery of Santa Ana in Coimbra: Ana de Faria (twenty-eight years old), Mariana do Deserto (thirty), and Violante de Faria (twenty-five). They were all arrested by the Inquisition on the 17th of October 1621, in the morning. Soon afterwards, their sister Tomásia de Faria (fourty years old), who was married and had six children, was arrested at her home; the same happened to André do Avelar (de Carvalho 1990, 118). The whole family was now in prison. The youngest of the daughters, Violante de Faria, immediately confessed (de Carvalho 1990, 281), stating that she became a Judaizer under the influence of her father. She also disclosed the existence of a larger group of crypto-Jews at the Monastery of Santa Ana; they were arrested a few days later.

After being seized, André do Avelar first requested some time "to ponder his faults" because he was very disturbed by his arrest. One week later, on the 25th of October, he was heard and said that everything he had confessed in the previous year was false, and that he had never accepted the Judaic faith (Carvalho 1990, 119). Of course, given the weight of the accusations against him, this attitude could not be accepted by the Inquisitors (Baião 1919, 136). Two weeks later, on the 8th of November, he asked for an audience with the Inquisitor and declared that he was guilty and that his former statement was untrue. He knew that his sons had been arrested before him, but he may not have known that his daughters were also in prison; he possibly thought that he had been denounced by them and that he should disclose that the whole family followed the Judaic faith. He told the Inquisitor that he had withheld some information because of his love for his children, and declared that sometime after the General Pardon he had told them that he accepted the Law of Moses and instructed them accordingly (Baião 1919, 136–37). A few days later, two other daughters, Mariana do Deserto and Ana de Faria, made their confessions, blaming the father for their apostasy. They also confirmed the Jewish cult meetings at the Monastery (de Carvalho 1990, 121).

A few weeks later, on the 28th of November, there was an *Auto de Fé* in Coimbra where 174 people were sentenced, including three of André do Avelar's children: Ana de Faria, Tomásia de Faria, and Luiz de Faria. They had confessed, their faults were not severe, and they were condemned to prison and penitential dress *ad arbitrium* (de Carvalho 1990, 123–24). As a matter of fact, they were "detained" in Coimbra. Ana de Faria, like the other nuns who were pronounced guilty, could not return to the monastery, and the family estate had been confiscated by the Inquisition. She was entrusted to a widow and afterwards to her mother's relatives (de Carvalho 1990, 277).

Up to the beginning of 1623, André do Avelar did not confess to his active participation in the Fellowship of St. Diogo. Evoking his old age, Avelar alleged that his memory was worn out and that he did not remember the things they asked to describe (Baião 1919, 137–38; Braga 1892–1902, vol. 2, 605). He only admitted his prominent role in that Brotherhood on the 21st of February, naming António Homem as its High Priest (de Carvalho 1990, 131–32). He accused over twenty other people and made a detailed description of the ceremonies of the Fellowship, and of his own role in those meetings (Baião 1919, 139; Braga 1892–1902, vol. 2, 605–07).

However, his new confession was deemed incomplete. On the 20th of May 1623, he was admonished and then taken to the torture chamber, where he was stripped of his clothes and tied up (Baião 1919, 139). However, before the beginning of the torment, he asked to be heard and incriminated Father Diogo da Mata and his mother, but he had nothing else to add. On the same day, his sentence was decided: he should be kept for life in the penitential prison of the Inquisition in Lisbon to avoid his contact with other "people of his nation," as he could corrupt them (de Carvalho 1990, 134–35).

Avelar, his two daughters Mariana do Deserto and Violante de Faria, and 136 other people were sentenced in the *Auto de Fé* that occurred in Coimbra on the 18th of June 1623 (de Carvalho 1990, 137). The two nuns were condemned to prison and penitential dress *ad arbitrium*, as were her other sisters (de Carvalho 1990, 285).[23] Avelar was condemned to perpetual imprisonment without remission, with the use of penitential dress with the insignia of fire (de Carvalho 1990, 137), meaning that his faults were so severe that he deserved to be burned, but he repented and was acquitted.

On the 20th of June 1623 André do Avelar was transferred from Coimbra to the Inquisition prison in Lisbon. At that time, he was 76 or 77 years old. There is no additional information about him after this day. It is likely that he died there, after a few years (de Carvalho 1990, 285).

António Homem, as the leader of the crypto-Jews of Coimbra, could have informed the Inquisition about everyone involved in his Fellowship. He made, however, a firm decision: he never confessed anything, he never accused anybody, he always denied every charge against him, and declared that he was innocent and a good Christian. He maintained this consistent attitude over more than three years,

[23] António Baião wrongly stated that he was sent to the Inquisition prison in Lisbon (Baião 1919, 147).

but the evidence against him was overwhelming. He was finally sentenced by the Inquisition, excommunicated, and "released to the secular arm." At the last moment, before being burned at the stake, António Homem made a public declaration of Christian faith and chose to die as a Catholic. He was garroted, and his body was burned on the 5th of May, 1624 (de Carvalho 1990, 142–47).

The younger son of André do Avelar, Pedro Homem de Faria, did not cooperate with the Inquisition. He refused to admit any fault. In March 1624, three years after being arrested, he continued to deny his guilt, although several witnesses had presented depositions against him, including his relatives (de Carvalho 1990, 127). On the 3rd of May, Pedro Homem was informed that he would be handed over to the secular arm, that is, he would be burned alive. The next day, on the eve of his execution, he asked to be heard and confessed. His life was spared. In the same *Auto da Fé* in which António Homem died (Sunday, 5th of May 1624), Pedro Homem abjured, received absolution, and was sentenced to perpetual imprisonment and penitential dress with the insignia of fire, without remission, and penal servitude in the King's galleys for five years, without payment. On the 17th of May, he embarked, with other convicted prisoners, on the ship "Madalena" (de Carvalho 1990, 142–44, 280, 288–89; Baião 1919, 153–54).

17 Concluding Remarks

André do Avelar's *Sphaerae vtrivsque tabella, ad sphaerae huius mundi faciliorem enucleationem* (1593) was the only Latin version of Sacrobosco's *Sphaera* published in Portugal. It was written to be used by the students of the University of Coimbra, unlike Pedro Nuñes's Portuguese translation of the *Sphaera*. The composition of Avelar's *Sphaera* was similar to that of his *Reportorio dos tempos* in several respects. In both cases he took a successful, previously existing book and rewrote it, abandoning some parts, following a different arrangement, and adding new information without identifying his sources. Although the contents of the *Reportorio* and the *Sphaera* are quite different, Avelar introduced in his version of Sacrobosco's work some information useful for navigators and also some astrological knowledge. The creation of both works required access to a variety of books that were possibly consulted by Avelar at the private libraries of the noblemen to whom he dedicated the *Reportorio dos tempos*. Avelar's *Sphaera*, on the other hand, was dedicated to Dom Fernão Martins Mascarenhas, the dean of the university, who ironically became the General Inquisitor at the time when the mathematician was arrested and condemned.

It seems that Avelar's Jewish ancestry was not a hindrance to his professional work at the University of Coimbra up to his retirement, because he obtained several new appointments that helped to complement his income, including the duty of a canon at the Coimbra cathedral. He led an exterior life of a Catholic, but he was really a Jew, to most effects, suffering the prejudices of the Portuguese against the New Christians and following the Law of Moses in secrecy. This situation was conceivably the motivation for his emblem with the motto "Flectimur sed non frangimur." However, he was finally bent and broken by the Inquisition.

The known association between Avelar's astronomical knowledge and the activities of the Fellowship of St. Diogo consisted in the need to settle the adequate time for Jewish ceremonies. There is, however, another conceivable link between Avelar's religious faith and his interest in astronomy and astrology. There was a general trend of Messianic thought among the Portuguese crypto-Jews in the sixteenth and seventeenth centuries, and it was even thought that the Messiah could be born among the New Christians at that time. António Homem was one of those believers who were preparing for the coming of the Messiah and who, in turn, would save the Jews from the oppression of Christians. We do not know whether Avelar personally tried to predict the time of the Messiah's appearance by astrology, but it would be meaningful to search for that connection.

References

Baião, António. 1919. *Episódios Dramáticos da Inquisição Portuguesa. Volume I. Homens de Letras e de Sciência por Ela Condenados*. Porto: Renascença Portuguesa.

Baldini, Ugo. 1998. As Assistências ibéricas da Companhia de Jesus e a actividade científica nas missões asiáticas (1578–1640). Alguns aspectos culturais e institucionais. Trans. Bernardino Fernandes. *Revista Portuguesa de Filosofia* 54 (2): 195–246.

Bayão, José Pereira. 1737. *Portugal Cuidadoso e Lastimado com a Vida e Perda do Senhor Rey Dom Sebastiaõ, o Desejado de Saudosa Memoria*. Lisboa Occidental: Officina de Antonio de Sousa da Sylva.

Bensaude, Joaquim. 1912. *L'Astronomie Nautique au Portugal a l'Epoque de Grandes Découvertes*. Bern: Akademische Buchhandlung von Max Drechsel.

Beyer, Hartmann. 1549. *Quaestiones novae in libellum de sphaera Ioannis de Sacro Busto*. Frankfurt: ex officina Petri Brubachii. http://hdl.handle.net/21.11103/sphaera.100645; http://hdl.handle.net/21.11103/sphaera.100644.

Bodian, Miriam. 2002. In the cross-currents of the reformation: Crypto-Jewish martyrs of the inquisition 1570–1670. *Past & Present* 176: 66–104.

Botelho, Bernardo de Brito. 1873. *Historia Breve de Coimbra: sua Fundaçam, Armas, Igrejas, Collegios, Conventos, e Universidade*. 2nd ed. Lisboa: Imprensa Nacional.

Braga, Teophilo. 1892–1902. *Historia da universidade de Coimbra nas suas relações com a instrucção publica portugueza*. Lisboa: Academia Real das Sciencias.

Brandão, Mário. 1927. Contribuições para a história da Universidade de Coimbra: A livraria do P. Francisco Suarez. *Biblos* 3 (4–5): 325–349.

Brozas, Francisco Sánchez de las. 1579. *Sphaera Mundi, ex variis autoribus concinnata*. Salmanticae: ex officina Ildefonsi a Terranoua.

Camerarius, Joachim. 1590. *Symbolorum & Emblematum ex Re Herbaria Desumtorum Centuria Una*. Noribergae: Hofmann.

Cardoso, Walmir Thomazi, and Roberto de Andrade Martins. 2018. Iberian approaches to astronomy during the sixteenth century. *Mediterranean Archaeology and Archaeometry* 18 (4): 265–271.

Carolino, Luís Miguel. 2006. João Delgado SJ e a 'Quaestio de Certitudine Mathematicarum' em inícios do século XVII. *Revista Brasileira de História da Matemática* 6 (11): 17–49.

Carolino, Luís Miguel, and Carlos Ziller Camenietzki. 2006. Tokens of the future: Comets, astrology and politics in early modern Portugal. Cronos. Cuadernos Valencianos de. *Historia de la Medicina y de la Ciencia* 9: 33–58.

Casalini, Cristiano. 2017. *Aristotle in Coimbra: The cursus conimbricensis and the education at the college of arts*. Trans. Luana Salvarani. London: Routledge

Clavius, Christoph. 1570. *In Sphaeram Ioannis de Sacro Bosco Commentarivs*. Romae: apud Victorium Helianum. http://hdl.handle.net/21.11103/sphaera.100365.

———. 1581. *In Sphaeram Joannis de Sacro Bosco Commentarius*. Roma: ex officina Dominici Basae. http://doi.org/21.11103/sphaera.101117.

Cohen-Skalli, Cedric. 2006. Authorship in the age of early jewish print: Isaac Abravanel's Ma'aynei ha-Yeshu'a and the first printed edition in Ferrara 1551. In *Tradition, heterodoxy and religious culture: Judaism and Christianity in the early modern period*, ed. Chanita Goodblatt and Howard Theodore Kreisel, 185–201. Beer-Sheva: Ben-Gurion University of the Negev.

Curto, Diogo Ramada. 1997. The stranger within in the time of Quijote. *Portuguese Studies* 13: 180–197.

da Costa, Adalgisa Botelho. 2007. *O "Reportorio dos Tempos" de André do Avelar e a Astrologia em Portugal no século XVI*. Rio de Janeiro: Booklink.

da Fonseca, Fernando Taveira. 2001. Pedro Nuñes na Universidade. II – Coimbra. *Revista Portuguesa de História* 35: 297–333.

———. 2004. Pedro Nuñes na Universidade. I – Lisboa. In *Estudos em Homenagem a Luis António de Oliveira Ramos*, ed. Francisco Ribeiro da Silva et al., vol. 1, 537–544. Porto: Faculdade de Letras da Universidade do Porto.

da Mota, Avelino Teixeira. 1976. Some notes on the organization of hydrographical services in Portugal before the beginning of the nineteenth century. *Imago Mundi* 28: 51–60.

de Albuquerque, Luís Mendonça. 1965. *Os Guias Náuticos de Munique e de Évora*. Lisboa: Junta de Investigações do Ultramar.

de Almeida, Manoel Lopes. 1967. Apontamentos para a Biografia de André de Avellar: Professor de Matemática na Universidade. *Revista da Faculdade de Ciências da Universidade de Coimbra* 29: 31–72.

de Azevedo, João Lucio. 1921. *Historia dos Christãos Novos Portugueses*. Lisboa: Livraria Clássica Editora.

de Blois, Louis. 1583. *Tabella Spiritualis aliâs Sacellum Animae fidelis Nuncupata. Continens brevem Institutionem vitae Christianae*. Pataviae: Matthaeus Nenninger.

Carvalho, João Manuel Almeida Saraiva de. 1990. *The fellowship of St. Diogo: New christian judaisers in Coimbra in the early 17th century*. PhD diss. University of Leeds.

de Espinosa, Pedro. 1550. *Sphera Ioannis de Sacro Busto. Cum Commentariis Petri a Spinosa Artium Magistri, Celeberrimiq[ue] Praeceptoris Salmanticensis Gymnasij, Aeditis*. Salmanticae: Ioannes Iunta. http://hdl.handle.net/21.11103/sphaera.100199.

de Sacrobosco, Johannes. 1499. *Sphaera mundi cum tribus commentis*. Venezia: per Simonem Papiensem dictum Biuilaquam. http://hdl.handle.net/21.11103/sphaera.100021.

———. 1508. *Nota eorum quae in hoc libro continentur*. Venezia: Iuncta de Iunctis. http://hdl.handle.net/21.11103/sphaera.100915.

———. 1527. *Textus de sphaera Ioannis de Sacrobosco*. Paris: Simonem Colinaeum. http://hdl.handle.net/21.11103/sphaera.100998.

———. 1531. *Liber Iohannis de Sacro Busto, de sphaera. Addita est praefatio in eundem librum Philippi Mel. ad Simonem Gryneum*. Wittenberg: Josephum Clug. http://hdl.handle.net/21.11103/sphaera.100138.

———. 1538. *Textus de sphaera Ioannis de Sacrobosco*. Paris: Simonem Colinaeum. http://hdl.handle.net/21.11103/sphaera.101011.

———. 1545. *Sphaera Ioannis de Sacrobosco typis avctior*. Paris: apud Guilielmum Richardum. http://hdl.handle.net/21.11103/sphaera.101054.

———. 1547. *Sphaera Iohannis de Sacrobosco*. Antuerpiae: Iohannem Richardum. http://hdl.handle.net/21.11103/sphaera.100917; http://hdl.handle.net/21.11103/sphaera.100143.

———. 1550. *Ioannis de Sacro Busto Libellus de sphaera: Accessit Eiusdem Autoris Computus Ecclesiasticus*. Wittenberg: Iohannem Cratonem. http://hdl.handle.net/21.11103/sphaera.100157.

———. 1564. *Sphaera Ioannis de Sacro Bosco emendata*. Lugduni [Lyon]: apud Haeredes Iacobi Iunctae. http://hdl.handle.net/21.11103/sphaera.101071.

———. 1573. *Sphaera Ioannis de Sacro-Bosco, emendata*. Antuerpiae: apud Ioannem Bellerum. http://hdl.handle.net/21.11103/sphaera.100460.

Delgado, Vicente Muñoz. 1983. Pedro de Espinosa (+1536) y la Lógica en Salamanca hasta 1550. *Anuario Filosofico* 16: 119–208.

Deslandes, Venancio Augusto. 1888. *Documentos para a Historia da Typographia Portugueza nos Seculos XVI e XVII*. Lisboa: Imprensa Nacional.

Dias, Nuno, Elza Amaral, and José M. Cobos. 2012. Influência dos Jesuítas no Ensino em Portugal. *Llull* 35 (75): 111–128.

Díaz Díaz, Gonzalo. 1987. *Hombres y Documentos de la Filosofía Española*. Madrid: Consejo Superior de Investigaciones Cientificas.

do Amaral, António Eugénio Coelho Maia. 2014. *Os Livros em sua Ordem: para a História da Biblioteca Geral da Universidade (antes de 1513–2013)*. Coimbra: Imprensa da Universidade de Coimbra.

do Avelar, André. 1546. *Chronographia ou reportório dos tempos o mais copioso que te agora sayo a luz conforme a noua reformação do santo Papa Gregorio XIII*. Lisboa: Simão Lopez.

———. 1585. *Reportorio dos tempos. O Mais Copioso que Ate Agora Sahio a Luz, Conforme à Noua Reformação do Sancto Padre Greg. XIII. Anno 1582*. Lisboa: Manoel de Lyra.

———. 1590. *Reportorio dos Tempos. O Mais Copioso qve Ate Agora Saio a Luz Conforme a Noua Reformação do Sancto Papa Gregorio XIII. Nesta segunda impressam reformado & acrescentado pelo mesmo author*. Lisboa: Manoel de Lyra.

———. 1593. *Sphaerae Utriusq[ue] Tabella ad Sphaerae huius Mundi Faciliorem Enucleationem. Autore Andrea d'Avellar Olysiponensi, Artium, ac Philosophiae magistro, & publico in Conimbricensi Academia Mathematum professore*. Conimbricae: Antonius Barrerius. http://hdl.handle.net/21.11103/sphaera.100534.

———. 1594. *Chronographia ou Reportorio dos Tempos. O Mais Copioso que te Agora Sayo a Luz Conforme a Noua Reformação do sancto Papa Gregorio XIII. Nesta terceira impressaõ reformado & acrecentado pello mesmo author*. Lisboa: em casa de Simão Lopez.

———. 1602. *Chronographia ou Reportorio dos Tempos. O Mais Copioso que te Agora Sayo a Luz. Conforme a Noua Reformação do Santo Papa Gregorio XIII. Nesta quarta impressam reformado & acrescentado pelo mesmo author*. Lisboa: Iorge Rodriguez.

do Avelar, Luís. 1619. *Nox attica. Hoc est Dialogvs de Impressione Metheorologica, et Cometa Anni Domini 1618*. Coimbra: Nicolau Carvalho.

Encyclopedia e Diccionario Internacional. 1933. *Organizado e redigido com a collaboração de distinctos homens de sciencia e de lettras brasileiros e portuguezes*. Rio de Janeiro: W. M. Jackson.

Fernandes, Manuel Bernardo Lopes. 1856. *Memoria das Moedas Correntes em Portugal, desde o Tempo dos Romanos, até o Anno de 1856*. Lisboa: Academia Real das Sciencias.

Fiolhais, Carlos, and Décio Martins. 2016. Reportório da Ciência Matemática em Portugal entre 1582 e 1623: André de Avelar em Coimbra e os Jesuítas da Aula da Esfera em Lisboa. In *Concilio de Trento: innovar en la tradición: historia, teología y proyección*, ed. José Eduardo Franco et al., 711–718. Alcalá de Henares: Universidad de Alcalá.

Folk, Katharina. 2009. *Manilius and his intellectual background*. Oxford: Oxford University Press.

Fuentes, Jesus Luis Paradinas. 2012. Las Matemáticas en la Ratio Studiorum de los Jesuitas. *Llull* 35 (75): 129–162.

Galilei, Galileo. 1656. In *Trattato Della Sfera Ovvero Cosmografia*, ed. Urbano d'Aviso. Roma: N. A. Tinassi.

Gelli, Jacopo. 1916. *Divise, Motti e Imprese di Famiglie e Personaggi Italiani*. Milano: Ulrico Hoepli.

Gingerich, Owen. 1988. Sacrobosco as a textbook. *Journal for the History of Astronomy* 19 (4): 269–273.

Giuntini, Francesco. 1578. *Commentaria in Sphaeram Ioannis de Sacro Bosco Accuratissima*. Lugduni: Philippum Tinghius. http://hdl.handle.net/21.11103/sphaera.100393.

———. 1581–1583. *Speculum astrologiae*. 2 vols. Lugduni [Lyon]: Simphorianum Beraud. http://hdl.handle.net/21.11103/sphaera.101207.

Goldstein, Bernard R. 1996. Astronomy and astrology in the works of Abraham ibn Ezra. *Arabic Sciences and Philosophy* 6 (1): 9–21.

———. 1998. Abraham Zacut and the medieval hebrew astronomical tradition. *Journal of the History of Astronomy* 29 (2): 177–186.

———. 2001. Astronomy and the jewish community in early islam. *Aleph* 1: 17–57.

Goold, George P. 1977. Introduction. In *Astronomica*, ed. Marcus Manilius, xi–cxxiii. Cambridge, MA: Harvard University Press.

Graminaeus, Theodor. 1567. *Uberior enarratio eorum, quae a Joanne de Sacro Bosco proponuntur: ita ut adjecta difficilioribus locis commentarii vicem supplere possit*. Coloniae: apud Maternum Cholinum.

Hopkins, Marjorie Irene. 2016. *Humanism as civic project: The Collège de Guyenne 1533–1583*. Ph.D. Dissertation. University of Guelph.

Knobloch, Eberhard, comp. 1999. *Christoph Clavius. In Sphaeram Ioannis de Sacro Bosco Commentarius*. Olms-Weidmann: Hildesheim.

Lattis, James M. 1994. *Between Copernicus and Galileo: Christoph Clavius and the collapse of ptolemaic cosmology*. Chicago: The University of Chicago Press.

Lawee, Eric. 2001. The messianism of Isaac Abarbanel, *father of the [jewish] messianic movements of the sixteenth and seventeenth centuries*. In *Millenarianism and Messianism in early modern European culture*, ed. M. Goldish and R.H. Popkin, 1–39. Dordrecht: Springer.

Leitão, Henrique. 2002. Introdução. In *Pedro Nuñes (1502–1578): novas terras, novos mares e o que mays he: novo ceo e novas estrelas*, ed. Henrique Leitão and Lígia de Azevedo Martins, 15–28. Biblioteca Nacional: Lisboa.

———. 2008. *A Ciência na Aula da Esfera no Colégio de Santo Antão: 1590–1759*. Lisboa: Comissariado Geral das Comemorações do V Centenário de S. Francisco Xavier.

Leite, Serafim. 1963. *Estatutos da Universidade de Coimbra (1559)*. Coimbra: Universidade de Coimbra.

Lenard, Max. 2006. On the origin, development and demise of the *Index Librorum Prohibitorum*. *Journal of Access Services* 3 (4): 51–63.

Lucian. 1905. *The works of Lucian of Samosata*. Trans. Henry Watson Fowler, and Francis George Fowler. Oxford: The Clarendon Press.

Lima, Fernando Carlos G. de Cerqueira. 2005. Uma análise crítica da literatura sobre a oferta e a circulação de moeda metálica no Brasil nos séculos XVI e XVII. *Estudos Econômicos* 35 (1): 169–201.

Machado, Diogo Barbosa. 1965–1967. *Bibliotheca Lusitana Histórica, Crítica e Cronológica*. Coimbra: Atlântida.

Manilius, Marcus. 1977. *Astronomica*, ed. and trans. George P. Goold. Cambridge, MA: Harvard University Press.

Martins, Roberto de Andrade. 2003a. Las Fuentes Literarias del *Tratado de la esfera* de Sacrobosco. In *Epistemología e Historia de la Ciencia. Selección de Trabajos de las XIII Jornadas*, ed. Víctor Rodríguez and Luis Salvatico, 307–314. Córdoba: Universidad Nacional de Córdoba.

———. 2003b. O Tratado da *Esfera* de André do Avelar (1593). Paper presented at the 9º Seminário Nacional de História da Ciência e da Tecnologia and 2º Congresso Luso-Brasileiro de História da Ciência e da Tecnologia, Rio de Janeiro, October 08–10, 2003. http://www.ghtc.usp.br/server/pdf/ram-Sacro-Avellar.pdf. Accessed June 2019

Martins, Maria Teresa Payan. 2011. O Índice Inquisitorial de 1624 à Luz de Novos Documentos. *Cultura. Revista de História e Teoria das Ideias* 28: 67–87.

Martins, Roberto de Andrade, and Walmir Thomazi Cardoso. 2008. O *Trattato della sfera ovvero cosmografia* de Galileo Galilei e algumas cosmografias e tratados da esfera do século XVI. *Episteme* 13 (27): 15–38.

———. 2017. *Galileo's Trattato della Sfera ovvero Cosmografia* and its sources. *Philosophia Scientae* 21: 131–147.

Melanchthon, Philipp Petrus Vincentius. 1579. In *Epigrammatum Libri Sex*, ed. Petrus Vincentius, 1579. Wittenberg: Haeredes Iohannis Cratonis.

Merrill, Eugene H. 1973. Sabbatai Zvi and Jewish Messianism. *Journal of the Evangelical Theological Society* 16 (3): 159–165.

Montenegro, Enrique Cantera. 2017. La esperanza mesiánica en el judaísmo hispano medieval. *Estudios de Historia de España* 8: 11–57.

Moreton, Jennifer. 1994. John of Sacrobosco and the calendar. *Viator* 25: 229–244.

Nuñes, Pedro. 1537. *Tratado da Sphera com a Theorica do Sol & da Lua. E ho Primeiro Liuro da Geographia de Claudio Ptolomeo Alexãdrino*. Lixboa: per Germão Galharde. http://hdl.handle.net/21.11103/sphaera.101010.

———. 1940. *Obras*. Vol. 1. Lisboa: Academia das Ciências de Lisboa.

Osório, José Pereira. 1986. Sobre a história e desenvolvimento da astronomia em Portugal. In *História e desenvolvimento da ciência em Portugal. I colóquio–até ao século XX*, ed. José Pinto Peixoto, vol. 1, 111–142. Lisboa: Academia das Ciências de Lisboa.

Palliser, Bury. 1870. *Historic devices, badges, and war-cries*. London: Sampson Low, Son & Marston.

Patai, Raphael. 1988. *The messiah texts. Jewish legends of three thousand years*. Detroit: Wayne State University Press.

Reportorio dos Tempos Muito Curioso. Acrescentado e Emendado de Nouo Conforme a Noua Constituição do Sanctíssimo Papa Sixto Quinto que Tirou os Abusos. 1590. Coimbra: António de Barreira impressor da Universidade

Rodrigues, Manuel Augusto. 2006. Notas sobre a Universidade de Coimbra desde as origens (1290) até à fixação definitiva em Coimbra (1537). In *Estudos em homenagem ao Professor Doutor José Marques*, vol. 3, 111–128. Porto: Faculdade de Letras da Universidade do Porto. 4 vols.

———. 2010. O judeu português Isaac Abravanel no espaço ibérico e europeu. *Revista de Historia das Ideias* 31: 77–115.

Rodríguez-Arribas, Josefina. 2005. Historical horoscopes of Israel. In *Horoscopes and public Sphaeras: Essays on the history of astrology*, ed. Günther Oestmann, H. Darrel Rutkin, and Kocku von Stuckrad, 145–164. Berlin: Walter de Gruyter.

———. 2011. The terminology of historical astrology according to Abraham Bar Hiyya and Abraham Ibn Ezra. Aleph. *Historical Studies in Science & Judaism* 11: 10–54.

Roth, Cecil. 1959. *A history of the Marranos*. New York: Meridian Books.

Ruderman, David B. 1992. Hope against hope: Jewish and christian messianic expectations in the late middle ages. In *Essential papers on Jewish culture in renaissance and Baroque Italy*, ed. David Ruderman, 299–323. New York: New York University Press.

———. 2001. *Jewish thought and scientific discovery in early modern Europe*. Detroit: Wayne State University Press.

Sáenz-Badillos, Ángel. 1999. Milenarismo y Cábala entre los judíos hispanos. In *Milenarismos y milenaristas en la Europa Medieval*, 177–202. Nájera: Instituto de Estudios Riojanos.

Salomon, Herman Prins. 1998. Crypto-judaism or inquisitorial deception? *The Jewish Quarterly Review* 89 (1/2): 131–154.

Schwartz, Stuart B. 2008. *All can be saved: Religious tolerance and salvation in the Iberian Atlantic world*. New Haven: Yale University Press.

Sela, Shlomo. 2010. *Abraham Ibn Ezra, the book of the world: A parallel hebrew-english critical edition of the two versions of the text*. Leiden: Brill.

Silva, Ana Cristina, Ferreira, Teresa Duarte. 2008. Catálogo. In Sphaera Mundi*: A ciência na* Aula da Esfera. *Manuscritos científicos do Colégio de Santo Antão nas colecções da BNP*, ed. Henrique Leitão, 99–247. Lisboa: Biblioteca Nacional de Portugal.

Silver, Abba Hillel. 1927. *A history of messianic speculation in Israel. From the first to the seventeenth century*. New York: Macmillan.

Teixeira, Antonio José. 1895. *Antonio Homem e a Inquisição*. Coimbra: Imprensa da Universidade.

———. 1899. *Documentos para a Historia dos Jesuitas em Portugal*. Coimbra: Imprensa da Universidade.

Thorndike, Lynn. 1954. Computus. *Speculum* 29 (2): 223–238.

Universidade de Coimbra. 1593. *Estatutos da Vniuersidade de Coimbra. Confirmados por el rey Dom Phelippe primeiro deste nome, nosso Senhor em o anno de 1591*. Comibra [sic]: por Antonio de Barreira.

Valerius, Cornelius. 1564. *De Sphaera, Et Primis Astronomiae Rvdimentis, Libellvs vtilissimvs*. Antverpiae: Silvius.

Valleriani, Matteo. 2017. The tracts on the *Sphere*. Knowledge restructured over a network. In *The structures of practical knowledge*, ed. Matteo Valleriani, 421–473. Dordrecht: Springer.

Vinet, Élie. 1561. *Sphaera Joannis de Sacro Bosco Emendata*. Lutetiae: apud Gulielmum Cavellat. http://hdl.handle.net/21.11103/sphaera.100211.

4

Prolegomena to the Study of Early Modern Commentators on Johannes de Sacrobosco's *Tractatus de sphaera*

Matteo Valleriani

Abstract By way of introduction to the present volume, a corpus of 359 treatises is described that was used in early modern educational institutions for introductory classes on cosmology and that is referenced by the following contributions. Following a taxonomy of early modern commentaries, central characteristics are analyzed in detail such as the rate of production of the treatises, the places where they were produced, and their various languages and formats. The focus then turns to the balance between the temporal dynamics of production of the treatises and the lifespans of their commentators. This reveals how the early modern textbooks first amplified medieval scientific knowledge and only slowly began to support and spread the echoes of scientific debate among contemporary scholars. The institutional and intellectual profiles of the commentary authors are then described on the basis of the results presented in the contributions to this volume. The commentators are described by referring on one hand to their relations to the universities, religious orders, and commercial institutions, and on the other to their engagement with disciplines both inside and outside the conceptual framework of the quadrivium. Finally, a quantitative summary of the results achieved by this volume is presented along with outlines for future research endeavors that will focus, consequently, on the role of the printers and publishers of the same commentaries.

Commentaries on the *Sphaera* of Johannes de Sacrobosco (died ca. 1256) constitute a peculiar genre in the *mare magnum* of medieval and early modern scientific commentaries. They are peculiar for a number of reasons. First and foremost, they do not comment on an ancient text but rather on a late medieval textbook, compiled for coursework at the University of Paris (Thorndike 1949, 76–142). Sacrobosco's work, a short qualitative introduction to geocentric cosmology, was not ambitious in its treatment of mathematical astronomy. It is actually best defined as a manual for using a scientific instrument, namely, the armillary sphere. It is a piece of deictic writing and it was probably used for deictic teaching—its main purpose was to make students familiar with this specific instrument. A second peculiarity of these commentaries is the fact that their history of publication, the precise run of editions between the thirteenth and the mid-seventeenth century,[1] exhibits such exceptional continuity that the usual historical periodization that divides the Middle Ages from the early modern period does not seem significant in the least.

The original text does not introduce any relevant innovation from the point of view of cosmological knowledge. It is clearly based on the geocentric conception of the universe found in Ptolemy's *Almagest*, but it is also influenced by later works, especially from the Islamic scientific tradition, as Lynn Thorndike has clearly shown (Thorndike 1949, 1–75). Nevertheless, the textbook cannot be considered a simple paraphrase or abridgment, first of all because it contains a short but significant passage at the end to contextualize it in the general frame of Christian theology, and secondly because it has an original textual structure—it was designed to function in a particular context: the newly conceived university, built around the scheme of the quadrivium.

Retracing the history of the treatise over four centuries is a fascinating project that can only be accomplished by splitting it up into specific aspects and time windows. The exceptional continuity of this commentary tradition is due to the central role of cosmology in the general scientific and cultural framework that emerged in the late medieval time. The cosmological worldview was the nucleus around which the European knowledge system—with its constellation of scientific subjects—was organized. This central role remained unchanged until the mid-seventeenth century, when it mutated not only because of the emergence of alternative worldviews but also because of the progressive specialization of subjects previously attached to cosmology, which in turn achieved the status of new, independent disciplines. The matter of which scientific subjects were seen as kindred to cosmology had in fact been evolving throughout the centuries in question. In the thirteenth and fourteenth centuries, for instance, the study of cosmology at the universities was certainly instrumental to the study of *computus ecclesiasticus*. The association between these two subjects—cosmology and calendric—is readily apparent to even a cursory analysis of the many *Sphaera* manuscripts deposited in archives of medieval sources. They very often appear in collections of university textbooks that contain treatises on both subjects. Yet this association can scarcely be found in later periods. Meanwhile, the text of Sacrobosco in the early modern period was often written or printed together with texts whose subjects were previously either non-existent, such as cosmography, or not associated with cosmology.

[1]After 1650, some further editions of the commentaries on Sacrobosco's *Sphaera* were published. However, the role of this text, for instance in the context of university teaching, declined to such an extent that those late editions can no longer be considered representative of the scientific debate in Europe in that period.

If the tradition of commentaries on the *Sphaera* of Sacrobosco is seen in light of the role of cosmology within the entire scientific knowledge system, what appears are four centuries characterized by a profound continuity. If, however, the same tradition is analyzed with the aim of determining which scientific subjects pivoted on cosmology in the same knowledge system, this continuity also seems to accommodate a more dynamic development, one which might allow us to reconstruct how scientific knowledge, as it was imparted at European universities, fundamentally changed over time.[2]

This characteristic opens up the possibility for a temporal segmentation of this long commentary tradition, a segmentation which is further supported—and perhaps made necessary—by three important contextual factors and changes that took place in the first phase of the early modern period.

Firstly, commentaries on the *Sphaera* got on board the exciting new medium of the printed book in 1472. Two 1472 editions, one printed in Ferrara and the other in Venice (de Sacrobosco [1472a]; 1472b), opened a long, spectacular series of 359 different editions of treatises on the *Sphaera* up to 1650.[3] As book historians (recently, Angela Nuovo in particular) have well demonstrated, the market for printed books became a transnational European market very quickly. Large printer's workshops set up a selling network able to cover large geographic areas, which smaller workshops could also tap into. The European book market was already well established before the turn of the fifteenth century, and great central nodes such as book fairs supported the continuous development of this market during the whole early modern period. An elaborate texture of printers, publishers, and booksellers emerged all over Europe and continuously expanded. These central nodes—the book fairs, and centers such as Paris, Venice, and Basel—attracted a myriad of printer's workshops, either because they were appropriately situated on the lines of distribution or because of the presence of authorities and institutions relevant and necessary to their economic activity (Nuovo 2013).

The second factor relates to the history of the universities. From the same period of the expansion of book printing activity toward the end of the fifteenth century, the so-called third phase of university foundation took place. Especially the sixteenth century is marked by an impressive increase in the number of universities in Europe. Because of the high mobility of lecturers and students, the universities as such can be seen as another network of interconnected nodes sharing similar aims, structures, and, most importantly, knowledge (Rüegg 1994–2011, Vol. 2; Grendler 2002).

The increasing homogenization of scientific knowledge over the continent, also detectable in the corpus of the early modern commentaries on the *Sphaera*, might be

---

[2] The reconstruction of a shared scientific identity in Europe between the late medieval and early modern periods, based on an analysis of the evolution of its underlying knowledge system, is the overarching goal of the project in whose context the present study was conducted. For more information.

[3] The number of known books constituting the corpus of the printed commentaries on the *Sphaera* of Sacrobosco can obviously increase if new texts are found. The database displaying the current status of research is available through the website of the *Sphaera* research project (see footnote 2).

explained by the combination of these first two factors. However, homogenization does not mean stagnation or lack of innovation. In fact, homogenization was a process that could only take place through the continuous input of innovations by the actors of networks—innovations such as additional *scholia*, new images for the same textual content, or a description of a new solar clock. Innovations were the motor that justified exchange and connectivity in the network, and the higher the level of exchange, the more homogenous the knowledge became—the shared scientific knowledge of Europe that manifested itself in the teaching of young students in their first years at the faculties of liberal arts.

The third factor is represented by great epochal processes and events in general. Although their consequences are often difficult to grasp by looking directly at the *Sphaera* commentaries, the journeys of exploration and events such as the Reformation in 1517 clearly determined "waves of knowledge" that are detectable in the content of these treatises—for instance, they had bearing on the appearance of new subjects and the disappearance of others.

Because of these factors, it appears justifiable to segment the analysis of the commentary tradition into at least two great temporal phases, the first dating from the authorship of the treatise by Sacrobosco's own hand until the first appearance of the treatise as an incunable in 1472; the second from this same year until 1650, when the commentary tradition of this specific text and its scientific relevance came to an end. What follows will refer to the second phase of this editorial history.

This segmentation into two epochs, however, does not yet fully consider the medium through which the knowledge was disseminated. The diffusion of print technology did not suddenly nullify the habit of using handwritten material for study purposes. It is well known that, at least during the first phase of the history of printed books, manuscripts remained predominant as a medium. Therefore, when an editorial history of this temporal length is investigated, one has to deal with two different time scales, each associated with a specific medium. This would imply that the investigation of the corpus of early modern commentaries on the *Sphaera* of Sacrobosco should not be limited to the printed books only, but needs to take into consideration the manuscript tradition as well.[4] Unfortunately, however, no census of the manuscripts of this era containing the text of Sacrobosco with or without commentary has ever been compiled. This means, first, that research can only be carried out based on the exclusive analysis of printed sources, which in contrast to handwritten manuscripts have been systematically and completely identified and collected; and second, that research results, especially concerning the phase during which the printed book established itself as the academic standard, have to be considered temporary until further source analyses can complete the picture.

[4] Research that systematically combines the analysis of printed sources with the handwritten ones for a long period during the fifteenth and sixteenth centuries would require additional temporal scaling because of regionally different speeds of decline for the manuscript and emergence for the printed book as the principal medium for scientific and teaching purposes.

1 The Corpus of the Printed Commentaries

The 359 printed treatises, which cover a time span of 178 years, offer the possibility to investigate multiple aspects of this editorial history. On one hand, there are basic changes to the knowledge system (even as it remained grounded in cosmology), and on the other there is the way these changes came to be "represented" through the evolution of printed books. These two points of focus can be investigated on the basis of their mutual relationship. In this respect, it is useful to sub-divide the corpus into a taxonomy that reflects the book or edition typology. In particular, five categories can be identified by following a bottom-up approach, that is, by analyzing the corpus in its entirety and focusing only on textual elements (Fig. 1.1).

The first type of book that can be easily identified contains the short original tract of Sacrobosco, and nothing more. This is the smallest group of books in the corpus and it mostly appears in the first phase of the history of production. A close look at the texts nevertheless shows that they cannot be considered truly identical. In these books as well as in all the others that contain a reproduction of the original text, deviations of all sorts can be detected. Indeed, a census of all these variations is still a *desideratum*. Often, such changes concern linguistic or syntactic aspects. Sometimes they are made explicit—for instance, on title pages or colophons (*sphaera revisa, sphaera emendata*, etc.)—particularly under the influence of the humanistic requirement for a more elegant Latin, but many times they were just applied by the correctors employed in the printer's or publisher's workshops, who mostly remained anonymous. Thorndike indeed noticed that different variants of the text were circulating already during the first phase of its history in the thirteenth and

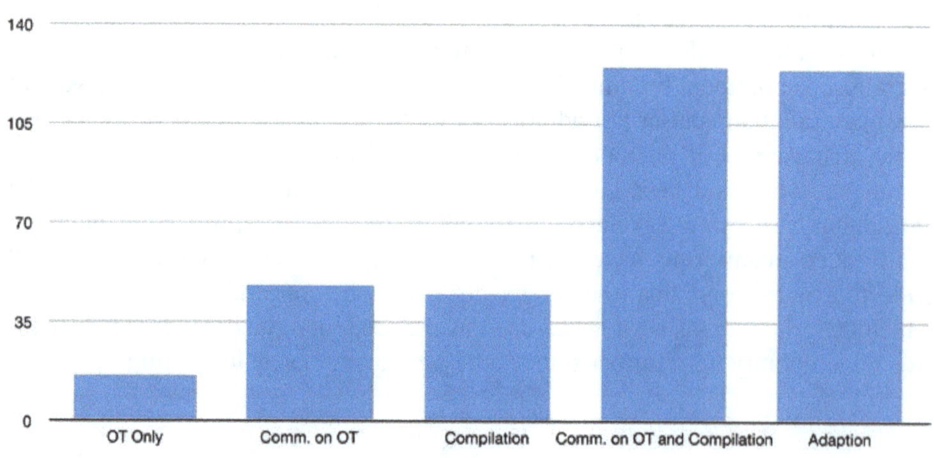

Fig. 1.1 A typology for the editions constituting the corpus of early modern printed commentaries on the *Sphaera* of Johannes de Sacrobosco: editions that contain the original medieval tract (OT) only; those that contain the original treatise with commentary; those that contain the original treatise and other treatises (compilations); those that contain the original treatise, commentary, and other texts; and adaptions

fourteenth centuries (Thorndike 1949, 1–76). Other differences are more content related. They can concern the citation of sources and even aspects of cosmology. Finally, new variants were created through the process of translation into local tongues, as will be shown later. Generally speaking, it can be said that the reproduction of the original text certainly was faithful, but not made with the historical and philological sensitivity that would be expected nowadays. In the spirit of the time, the continuous variations compiled until the very end of this series of editions did not encounter any scientific objection, as the main goal was scientific and not philological.

The second type of book displays the original treatise plus a commentary on it. This is not, however, the most relevant of the sub-groups of treatises, as one might expect. It also appears quite early in the history of this corpus, but in total only accounts for 48 editions. The identity of the corpus is characterized by a combination between the second and the third type of edition, namely, compilations.

Compilations are volumes that contain the original tract (without comments) alongside other texts. The most studied example of this is the compilation of Sacrobosco's text with Georg von Peuerbach's (1423–1461) *Theorica planetarum* (Chap. 6). In this case, the compilation of the two texts clearly aimed to enlarge the spectrum of subjects the volume dealt with: from the description of the *machina mundi* to a method for determining the positions of the planets. Often, however, commentaries on the *Sphaera* were added that might or might not refer to specific passages of the original text—though they were never directly integrated into its textual flux. A classic example of this kind of commentary addition is represented by the compilation of Sacrobosco's text with the *Questions* of Pierre d'Ailly (1340–1520). The latter text itself is not a commentary on the *Sphaera*, but placing the two texts together nevertheless creates a commentary, a way to deepen certain aspects of the original text.

Identifying compilations, which account for 45 editions in the corpus, is made particularly difficult and "risky" for source analysis because of a series of issues related to the characteristics of the early modern book market. In particular, the practice of producing bound books developed very slowly and did not become a market standard until after the period in question.[5] Quaternions were printed, stored in the bookshops as piles of sheets, and bound in the shops themselves, together with other books, selected by the customer at the moment of purchase. However, this does not imply that printers and publishers were producing only single texts and avoiding the design and manufacture of more elaborate books like compilations. For a researcher with some knowledge of the quadrivium tradition—the teaching

[5] During the second half of the sixteenth century, some publishers and printers introduced the *corezze*, a sort of light binding whose main function was to facilitate the transportation to, and storage at bookshops. They did not represent normative guidelines concerning the content of a final volume, which was bound at the shop by the seller according to the selection of books made by the customer (Nuovo 2013, 136).

framework to which these books mostly belong—recognizing the printer's intention is often quite simple indeed.[6]

As mentioned, a combination of the second and third book types gives us the most essential representation of the corpus of editions. These are volumes that contain at least one commented edition of the *Sphaera* of Sacrobosco and further content-related texts as well. The great Venetian anthologies of 1508 and 1531, published by Giuntino Giunta (1477–1521) and Lucantonio Giunta (1457–1538), respectively, are among the most impressive examples of this mixed typology. They contain commentaries and texts of authors such as Bartolomeo Vespucci, Francesco Capuano (died ca. 1490), Jacques Lefèvre d'Étaples (ca. 1455–1536), Johannes Regiomontanus (1436–1476), Pierre d'Ailly, and Robert Grosseteste (ca. 1175–1253) (de Sacrobosco et al. 1508, 1531), and they were produced with the clear intent to give an up-to-date survey of the relevant subjects. This group amounts to 125 editions and is therefore the most significant for understanding both the role of cosmology and the way knowledge was systematized over time.

The last type of treatise can be called 'adaption.' Identifying the books in this category requires the most background knowledge. These books, together a considerable group of 124 editions, do not actually contain Sacrobosco's text. They are seen as related to it, above all because they share a similar design, that is, they discuss the same subjects and in the same order. They also often make use of the same illustrative apparatus, at least partially. Finally, adaptions also include the so-called *Quaestiones*, books written in the form of questions and answers and intended for the students as an auxiliary means to prepare for an examination. While the identification of adaptions is very simple in certain cases, such as in the case of the works of Alessandro Piccolomini (1508–1578),[7] in many others the application of this category requires a close reading and a final decision on each individual source at hand. This type of edition serves as a limit category to distinguish the corpus from other works produced in the frame of early modern cosmology. It is necessary for the corpus to maintain a well-defined identity, and at the same time to keep the number of sources under examination at a manageable level.

Apart from the first type (a simple reproduction of Sacrobosco's treatise), all the *Sphaera* editions can be considered forms of commentaries. Beginning from the kind of comment that is created by breaking the original text into sections for the insertion of additional text, or "atomizing" the text, as Anthony Grafton calls it

[6] Besides a content-related analysis of the texts, there are several further ways to distinguish between compilations of texts as conceived by the printers and those created by individual customers in the bookshops. The principal ones are: (a) if it exists, an analysis of the table of contents, (b) the position of the colophon, (c) an examination of binding, paper, types, and ink, and (d) a comparison with other copies of the same volume. Nevertheless, a certain degree of uncertainty remains sometimes in reference to some editions, in particular to those which were produced in the early phase of the diffusion of printing technology, during which the design and conception of 'book' as a final product was still under development.

[7] Alessandro Piccolomini authored fourteen editions that belong to the corpus of the printed commentaries on the *Sphaera* of Sacrobosco. For more information about these editions.

(Grafton 2010), these typologies can be used to understand the historical developments that slowly brought the practice of commentary in science to an end, giving way to the emergence of a new genre, which in turn became a forerunner to the scientific monograph. Moreover, the compilation of texts, despite being a practice that always existed on an individual level (any medieval manuscript is testament to this), almost became a new genre of its own when it entered the world of the printed book—namely, when it exited the realm of individual research and interests to become an object of public scientific discourse, a process motivated in large part by the printer's will to conquer or maintain sectors of the academic book market.

If other elements aside from the textual aspects of the treatises are considered, such as the visual apparatus, the panorama becomes even more sophisticated. Ongoing research is showing that the books constituting this corpus contain roughly 16,000 content-related scientific illustrations, most of them astronomic and cosmological diagrams. Together with other sorts of images—frontispiece, illustrations from title pages, printer marks, decorations, and initials—they amount to about 20,000. Scientific visual material does not simply display knowledge that is otherwise written. It conveys it in a different way and can even display different scientific content. The development of the visual material used in this corpus (Chap. 9) attests to the increasing relevance of this medium for the transmission and transformation of knowledge. In other words, the insertion of images into the texts can be seen as an additional way of commenting on it.

Textual typology and visual apparatus are aspects of the book whose materiality calls for the analysis of further aspects, including those related to the printing process and those concerned with selling the book as a product. Although not all this data is available, some further considerations pertaining to the internal dynamics of the corpus and especially to the role of the commentators within the framework of this corpus can be easily added at this stage, namely those concerned with the places of production, the formats of the books, and the languages in which they were published.

In Europe, the production of these commentaries took place in forty-one different cities whose geographical distribution basically covers the entire continent, from Krakow to Lisbon and from London to Rome. Nevertheless, the production was not equally distributed and if only those centers of production are considered where at least ten different editions were printed, only nine cities remain (Fig. 1.2). As one might expect, the two most relevant centers are Venice and Paris, both hubs of the transnational European market for printed books. The third position of Wittenberg is all the more relevant if one considers that it entered this market as late as 1531, forty-one years after the first printed edition. As the intellectual center of the Reformation, its political and pedagogical influence over Europe is reflected by this position.[8] The fourth position of Leipzig, on the contrary, is based on the production of books that took place before 1520, at which point its role as the main producer of university textbooks was completely taken over in the course of only a

[8] For a study that shows how Wittenberg's treatises shaped the content of the textbooks on cosmology in Europe, see (Valleriani et al. 2019).

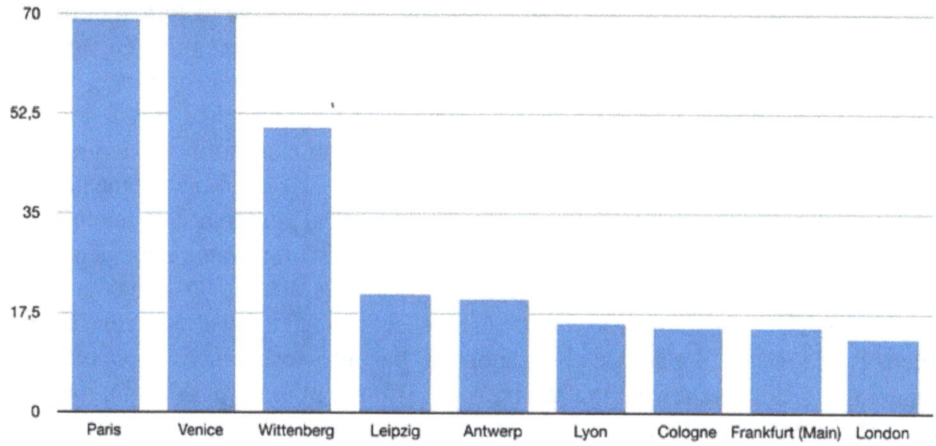

Fig. 1.2 The top nine places where early modern printed commentaries on the *Sphaera* of Johannes de Sacrobosco were produced

few years by Wittenberg after the onset of the Reformation. Antwerp deserves a special mention, too; its position at number five is due to production that, though beginning in 1543, was prolific only between 1560 and 1585, when the intensity of production of this commentary was at its peak across the continent. If the book-producing cities are considered nodes of a network, what results is a wide circulation of knowledge that is mostly determined by a few centers. In other terms, the centers of production of the treatises collected in the corpus analyzed here are the same major centers of production of printed books in general, except perhaps for the low relevance of Basel.[9]

That these books were mostly destined for the student market can be easily shown by considering their prevalent formats (Fig. 1.3). Among the three major format categories—in order of decreasing size and price: folio, quarto, and octavo—[10] the in-octavo books are the models that shape the material identity of the corpus most. Moreover, the folio format was mostly produced during the first 50 years of the history of *Sphaera* editions, that is, in the period for which the data at our disposal are less representative (because reading material still regularly came in manuscript form).

[9]For an interesting study concerned with the diffusion of the commentary on Sacrobosco's *De sphaera* in the region culturally influenced by Prague, see (Hadravovi and Hadravovi 2019).

[10]The editions of the corpus were printed twice in sextodecimo format otherwise always either folio, quarto, or octavo. We disregard the folding of the sheets. The precise dimensions of the books, moreover, are more differentiated, and certain editions would more accurately be defined as small folios or small quartos. Furthermore, the three categories do not denote any universally accepted book dimension, as libraries tend to follow slightly different organizing principles. Nevertheless, the use of the categories (disregarding those small differences) still accounts for how printers and publishers made investments, and therefore represents an indication of the final price of the books.

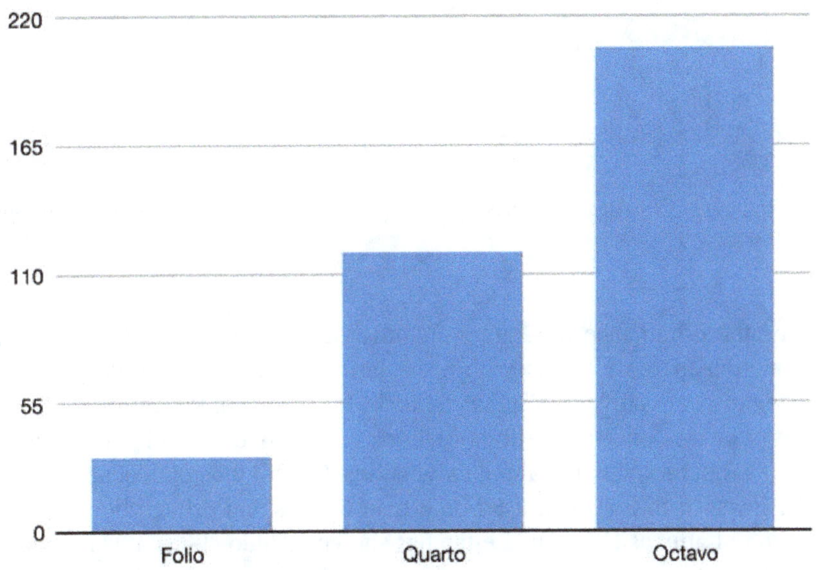

Fig. 1.3 Book formats of the early modern printed commentaries on the *Sphaera* of Johannes de Sacrobosco

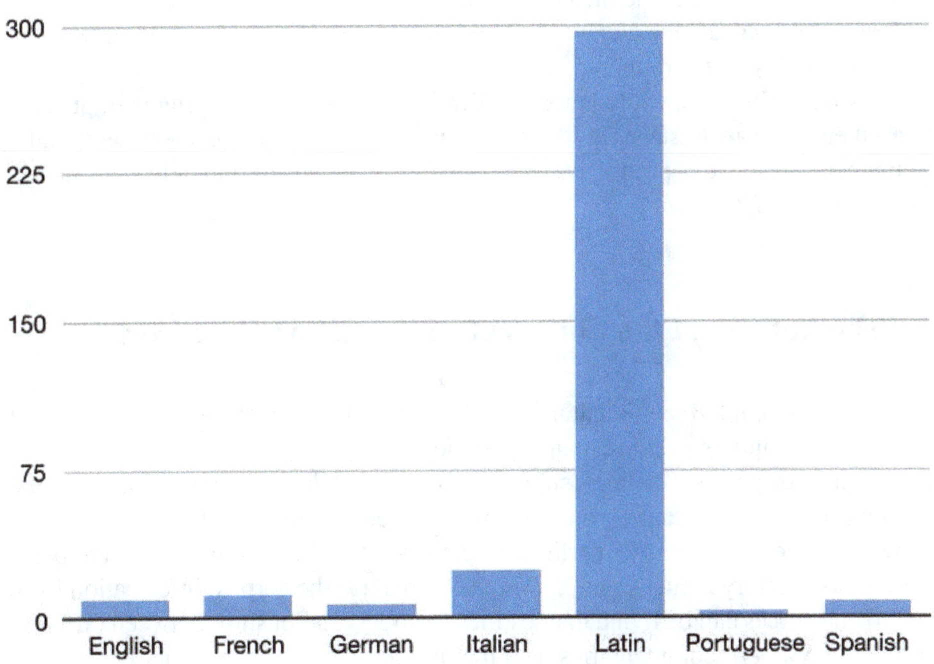

Fig. 1.4 Languages in which the early modern commentaries on the *Sphaera* of Johannes de Sacrobosco were published

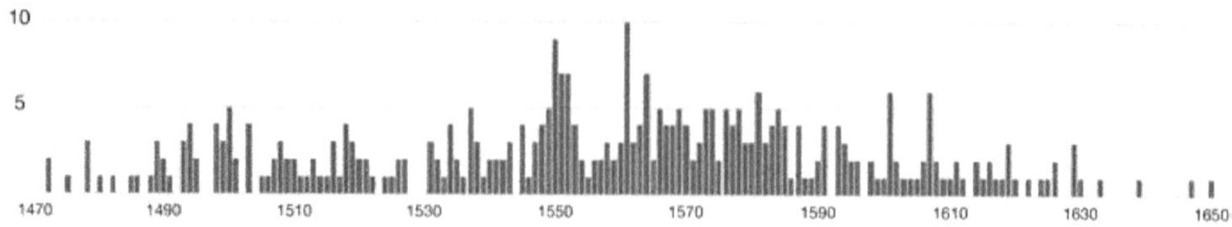

Fig. 1.5 Temporal distribution of the production of the early modern commentaries on the *Sphaera* of Johannes de Sacrobosco

Finally, the dominance of Latin over other languages is the clearest indicator of the function of these books within the university teaching context (Fig. 1.4). This does not necessarily imply that treatises compiled in different languages, in many cases one-to-one translations of books published in Latin, were not used for teaching, but probably not at the universities. Other kinds of educational institutions, for instance in Portugal, used both manuscripts and printed books that were not written in Latin, as Henrique Leitão has shown (Leitão 2008). It is interesting to notice that, in spite of the fact that the majority of books produced in local tongues are in Italian, the first Italian book did not appear on the market before 1537, at which point books had already been produced in Portuguese (first edition: 1509), German (first edition: 1516), and French (first edition: between 1525 and 1529), with a printed version of the medieval commentary of Nicole Oresme (1320–1382).

Finally, concerning the dynamics of production, this is not constant over time but it relevantly increases around 1550 and keeps this peak until about 1580 (Fig. 1.5). In this specific interval, while Paris and Venice remain dominant centers of production, their relevance strongly decreases in favor of other centers, especially Wittenberg—almost as productive as Venice but by means of only a few publishers and printers—and Antwerp.

2 The Authors of the Commentaries as a Social Network

The corpus of *Sphaera* commentaries lends itself to being analyzed as a series of at least two overlapping networks: an economic one, whose protagonists are the printers, publishers, and booksellers (supported by a network of distribution all over the continent), and a "semantic" one, wherein the circulation and transformation of knowledge are dynamically interlinked. The role of authors in this context appears to be fairly aleatory at first sight. Yet some external (to the corpus) information from historical sources can be found that testifies to direct relationships between different authors of *Sphaera* commentaries, and the outlines of a social network can be partially reconstructed from this information. For instance, it is known that Élie Vinet (1509–1587) and Pedro Nuñes (1502–1578) were in contact with each other.[11]

[11] For the connection between Élie Vinet and Pedro Nuñes, identified as a weak tie, see (Valleriani 2017).

However, it is neither realistic nor methodologically possible to reconstruct all the relevant social relationships by following this procedure. It is possible, however, to approach the subject by applying limitations to the research field as well as taking steps to identifying where the author of a given commentary is positioned within the social network and then trying to profile who they were.

Authors in the network are people who must have had a connection with a particular publisher to publish their text. The publisher or printer is conceived of here as a point of convergence in the network; his workshop represents a sort of social micro-region populated by employees—press-men, correctors, and engravers—in addition to many relevant external figures, such as creditors, book-sellers, representatives, and, in the case of the academic book market, representatives of educational institutions.[12] The advantage of taking the printer's workshop or the publisher's office as a starting point is that one can systematically approach the question of whether a social network indeed exists amongst the authors by looking at a set of bibliographic metadata for the editions constituting the corpus.

A first limitation to the study is represented by the fact that only the so-called 'credited authors' can initially be considered. Credited authors are those whose names are detectable on the title pages, frontispieces, colophons, or in the long incipit that characterizes the fifteenth-century incunables. This means that some authors will be left out. A cursory analysis of the treatises that have been defined as compilations reveals namely that those editions contain more texts (by more authors) than what the title pages or even the tables of contents declare. The anthological work mentioned above published in Venice in 1531 by Lucantonio Giunta, for instance, mentions 13 authors in the table of contents but contains texts authored by fifteen scholars—among them Campanus of Novara (1220–1296), who is not mentioned.

The case of Campanus, moreover, points to a second vital limitation, namely, in linking the dates of birth and death of the authors to the date of publication of the specific edition bearing their names. Obviously, if the credited authors are dead, no social relationship can be assumed between them and the publisher. This research is therefore executed in reference to all editions of the corpus and not in reference to the publishers themselves—that is, their lifetimes or periods of activity.

A search for pairs of authors who were both mentioned in one edition and both alive at the time of its publication results in 120 relationships, which appear in the impressively small number of 36 editions. In other words, the publication process of only 36 out of 359 editions could have involved some sort of exchange between more than one scholar and a publisher, and therefore between the scholars by way of the publishers. One way to explain this astonishing result is through a phenomenon well known to (especially economically-oriented) historians of the early modern book, although systematic studies have not yet been completed. It seems that the first generations of printers and publishers were particularly reluctant to publish

[12] Isabelle Pantin has been able to demonstrate how early modern publishers of academic textbooks were producing their books in accordance with or by means of an exchange with the scholars and representatives of educational institutions that were within their market. For more information, see (Pantin 1986, 1998) and (Chap. 9).

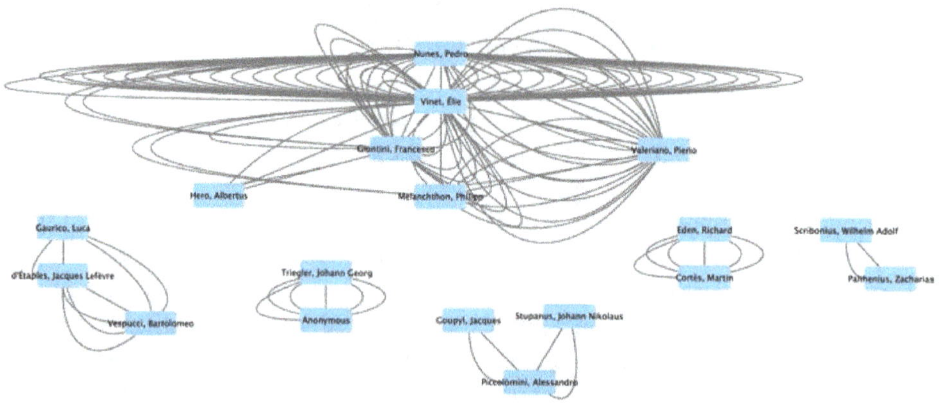

Fig. 1.6 Potential social relationships among the authors of the commentaries on the *Sphaera* of Johannes de Sacrobosco reconstructed by choosing authors who were alive when the editions containing their texts were published and when more than one author simultaneously contributed to the same edition. The humanist Élie Vinet comes out as the most central figure

texts by contemporary authors and scholars.[13] This tendency, which makes sense if one assumes that there was no demand for contemporary authors, can be recognized in the editorial history of the commentaries on Sacrobosco's *Sphaera*, too. Indeed, until 1550, only two works appeared in which more than one living scholar appeared as a credited author, and these were in fact the two peculiar anthologies printed in Venice in 1508 and 1531 mentioned above. The living authors involved were Bartolomeo Vespucci, Jacques Lefèvre d'Étaples, and Luca Guarico (1475–1558). This tendency suddenly changed thanks to the almost herculean efforts of basically one man, whom Isabelle Pantin calls the first publisher specialized in scientific books in Western history (Pantin 1998): Guillaume Cavellat (ca. 1500–1576), active in Paris. The first followers of Cavellat's style arrived 11 years later in London (Richard Jugge, ca. 1514–1577) and twelve years later in Venice (Girolamo Scoto, 1505–1572).

The resulting social network inferred in this way is made up of six components (Fig. 1.6). Three are pairs of authors, two are between three authors, and the final one involves six scholars: Élie Vinet, Francesco Giuntini (1523–1590), Albertus Hero (1549–1589), Pierio Valeriano (1477–1560), Pedro Nuñes, and finally Philipp Melanchthon (1497–1560), whose social contacts with the rest of those mentioned through the Catholic publisher Cavellat should be considered improbable. If the

[13] The number of publications of texts of contemporary authors quickly increased, beginning in the 30s of the sixteenth century. The change of tendency was due to the new legal protections means for printed books that favored new books and did not consider the so-called 'common books,' a category of books to which academic works such as Sacrobosco's belonged. In order to obtain the legal protection, the books had to display substantial novelties which finally resulted in the publication of a quickly increasing number of commentaries written by contemporary scholars. For more information on such normative aspects, see (Nuovo 2013, Chap. 6, esp. 212–14).

parameter of centrality of this network of eighteen actors is then calculated, it turns out that Élie Vinet is positioned at the center of all potential exchanges.[14]

If the analysis is expanded to consider all those authors who were alive at the time of publication of their texts but who appear to be mentioned only together with Johannes de Sacrobosco (with whom there is no possible relationship) or just alone, the numbers suddenly acquire a different quality. The first case (author mentioned with Sacrobosco) involves another twenty-four scholars whose texts were published and re-published a total of 76 times. Within these new parameters, one sees that the intensity of publication enters its peak in the 1540s, although the number of publications of the previous period is much more significant. The major publisher until 1520 is Martin Landsberg (died ca. 1523) in Leipzig. In the second case, which mostly concerns the authors of adaptions for which the name of Sacrobosco is no longer mentioned on the title page but rather in the *proemium* (if at all), there are fourteen scholars appearing in a total of twenty-eight editions.[15]

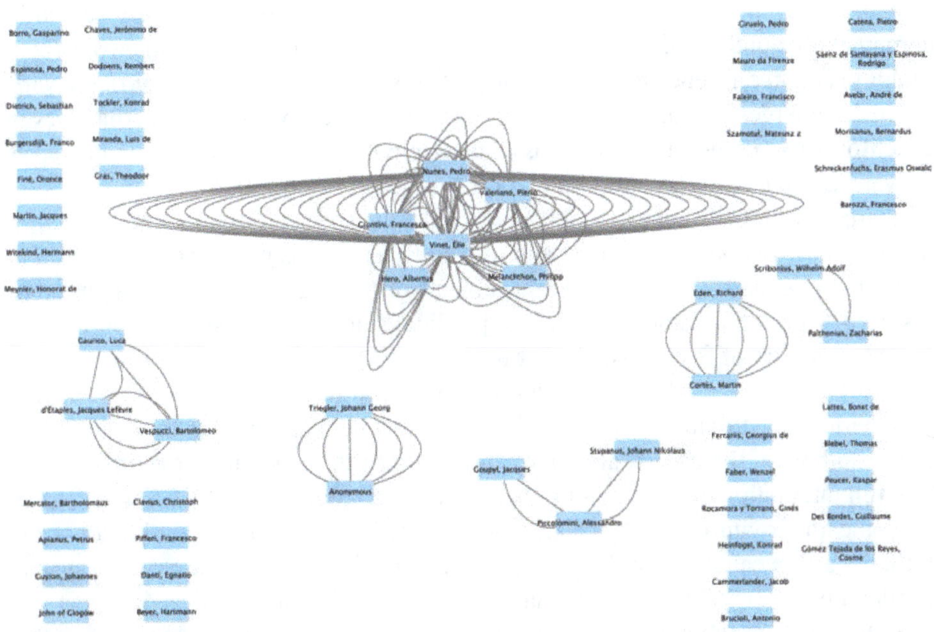

Fig. 1.7 Representation of all authors of commentaries on the *Sphaera* of Johannes de Sacrobosco who were alive when the editions containing their texts were published (fifty-eight authors) and potential relationships among them (involving only eighteen of them)

[14] A more precise reconstruction of the possible social network among the authors of the commentaries could be achieved by segmenting the editorial history into time windows of 40 or 50 years in order to be more sensitive to generational changes. This step, which would be necessary if the amount of resulting data had been significantly bigger, is not strictly necessary at this stage, as the few authors discussed here can be temporally situated very easily.

[15] If one looks at the data from another angle, it appears that the three groups of publishers—those who acted as social intermediaries between scholars, those who worked one-to-one with a single

To sum up, only fifty-eight of 164 credited authors were (at least potentially) in active contact with their respective publishers over the course of the 178-year editorial history of commentaries on the *Sphaera*, and a very small portion of these authors might have been in contact with each other through the common engagement with the same publishers (Fig. 1.7).

Such social segmentation, in spite of the basic similarity of their work in terms of content and intention, can now be further investigated by trying to reconstruct these authors' intellectual profiles. The following relies on the papers in this book.

3 Profiling the Authors

A quick overview of the authors of scientific commentaries in the early modern period makes clear that their profiles cannot be placed in simple and well-defined categories. Even against the background of a smaller field, as the only scientific commentaries under consideration here were written in the same framework (cosmology) and in reference to the same original text (Sacrobosco's *Sphaera*), the variety that emerges certainly cannot be circumscribed by simple—perhaps even anachronistic—dichotomies such as natural philosophers versus astronomers, as Edward Grant has suggested (Grant 2009, 46–59).

Starting with their educational backgrounds, the spectrum immediately enlarges to encompass mathematics, logic, and astrology, as well as medicine, theology, *and* natural philosophy. Thus, all possible subjects of the academic curriculum are represented, with the notable exception of juridical studies. This aspect displays once more the vitality of Sacrobosco's text from an intellectual point of view, as this text came to be considered the test bench for entering the scientific discourse during this long period. As a matter of fact, the centrality of the *Sphaera* was imparted to the future authors of the commentaries when they were still students at the universities. The full integration of this text into the leading pedagogical model in Europe at that time, the quadrivium,[16] made it a kind of crossroads in the pursuit of all fields—and fittingly so, because cosmology was the pivot of the entire knowledge system (Valleriani 2017). While it is true that contemporary disputes such as disagreement over the reality of the planetary orbs (Chap. 6) can only be explained by the fact that

---

scholar and remained close to Sacrobosco's text, and those who worked one-to-one with a single scholar but published adaptions—do not significantly overlap, though they were involved in the same book market.

[16]The quadrivium, as conceived by Boethius, was followed in the European universities of the early modern period with a degree of flexibility (Moyer 2012). Especially the subject of music (in its original form, which included the *musica sphaerarum*) had quickly declined and perhaps disappeared altogether by the beginning of the sixteenth century. Nevertheless, the classic Boethian division was still considered an efficient taxonomy even as the nature of the content of the individual disciplines went through profound transformations.

both astronomers (i.e., mathematicians) and natural philosophers were involved, this still does not exhaust the potential reach of the commentaries, which contained knowledge ranging from technological innovations on clocks and observational instruments to theological disquisitions.

Against such an educational background, it is easy to imagine that the institutional activities in which the authors of the commentaries were involved over the course of their lives were at least as heterogeneous as the educational background itself, if not even more so, albeit with some common denominators. Among the common denominators, the most relevant is the institutional engagement of the authors: As it turns out, they were all university lecturers. However, they were not only lecturers in astronomy and/or cosmology. Clearly, most of the authors of the commentaries were active in the frame of the quadrivium, and as such they were often employed to teach subjects related to mathematics. Many of them, however, were teaching different subjects, too. Pedro Ciruelo (1470–1554), for instance, also taught dialectics and theology after having been professor of astrology; Francesco Capuano was also lecturer in philosophy; John of Glógow (1445–1507) taught logic and philosophy, as did Franco Burgersdijk (1590–1635), who additionally gave classes in ethics. Turning our attention to the more practical undertakings related to geocentric cosmology, for instance, cosmography, geography, cartography, and nautical astronomy, plenty of *Sphaera* commentators, especially from the Iberian Peninsula (Chap. 7), can be identified who used both their publications and their classrooms as forums to discuss the latest geographic discoveries, which unfolded continuously. Working with the *Sphaera* was therefore also a means to reflect upon the most groundbreaking of early modern activities, namely, the exploration of the earth's surface.

Moreover, as active academicians, it is not rare to find the same figures involved in the process of reforming the curriculum of studies. Famous in this respect are the French scholars Jacques Lefevre d'Étaples and Oronce Fine (1494–1555) (Chaps. 2 and 8). Both of them insisted on a mathematical turn in the pedagogical framework of their time. However, it was not only about more or better mathematics; increasingly, it concerned more practical applications of mathematics, forging in this way a common spirit, or a shared scientific identity, between them and many of the Iberian authors of *Sphaera* commentaries. Also involved in a process of re-shaping curricula was Burgersdijk, who did not, however, act in favor of any particular subject but rather worked to create a textbook that accorded with the political and religious conditions of a Calvinist society (Chap. 11).

While some of them, such as Francisco Faleiro and Jerónimo de Chaves (1523–1574), were working with agencies like the Spanish House of Trades, and others, such as Capuano and Pedro Avelar, entered religious orders, another significant activity of the authors of the commentaries outside the universities seems to has been their involvement with printing workshops. As Isabelle Pantin has shown, printers were fundamentally connected to these university lecturers and authors of commentaries because they received from them the information and content necessary to design whatever editions they planned to put on the market next (Chap. 9).

But from another perspective, for instance, by looking at the activities of Lefèvre d'Étaples and Fine, it becomes evident that such a connection was so close that these roles sometimes overlapped. Looking at the census of treatises on the *Sphaera*, there are several people who appear to be both the printers or publishers and, at the same time, authors of the textbooks. The most relevant among them is certainly Petrus Apianus (ca. 1501–1552).

The intellectual profiles of the authors of the commentaries are best investigated if the content of their commentaries is also scrutinized. True, the scientific content was influenced by a series of external factors, such as the economic interest of the printers, as already mentioned, as well as by the kind of institution in which the authors were teaching. The example of the two editions of the commentary on the *Sphaera* by Capuano, for instance, shows that the content of his texts changed according to his audience: first the students of the faculty of arts, and later the clerics of the religious order of St. Augustine (Chap. 4).

In spite of this difficulty, however, an overview of the commentaries reveals that the variety of the topics they cover perfectly matches the heterogeneity of the educational backgrounds of the scholars and of their teaching subjects.

The commentaries here examined delve into astronomy, cosmography, geography, geometry and arithmetic, medicine, astrology, theology, logic, philosophy, as well as technological applications of these knowledge domains, especially the design of mathematical instruments. It is not easy to say which of these subjects were prevalent, if any. The content of individual commentaries depended very much on their particular authors and on the context (time and space) in which they were active.

Certainly, many commentaries facilitated better connections between the standard disciplines of the quadrivium (Chap. 5). The most relevant examples to this effect are given by commentaries that dealt with geometry. Specific and different classes were dedicated to the study of Euclidean geometry, but it was nevertheless a subject to which the commentators on the *Sphaera* of Sacrobosco dedicated increasing space in their editions. As exemplified by the case of John of Glogów, authors of the commentaries also devoted themselves to the establishment of a new planetary astronomy based on the *Theorica planetarum* of Georg von Peuerbach. The same question of the philosophical and scientific status of astronomy—that is, where it stood in the scientific context of the early modern period—was discussed by Capuano and Fine, for instance. Philosophical questions of this sort, associated with the ongoing process of mathematization, resulted in frequent discussions concerning the reality of the orbs and of the cosmological system delineated in these treatises (Chaps. 6, 8 and 9): The lathe depicted in Lefèvre d'Étaples's 1495 commentary is a symbol for the ontological interpretation of mathematics in general, as pointed out by Richard Oosterhoff (Chap. 2).

The discussion surrounding the reality of the orbs and the general development in favor of an ontological interpretation of mathematics was accompanied and perhaps supported by the increasing connection of cosmology and geography, a link that implied the *practical* reality of the inhabited zones, and therefore the necessity to grasp them mathematically. In this vein, cosmography emerged as a mathematically codified discipline, whose fundamental task was to map the great celestial spheres onto the earth's surface. The treatises on the *Sphaera* were objects that absorbed and furthered this tendency, beginning in 1495 (Chap. 2). An entire new genre emerged specifically dedicated to cosmography, whose texts also belong in part to the corpus of the treatises on the *Sphaera*, though they increasingly freed themselves from such a traditional background, quickly moving toward the more practically and technologically-oriented frameworks of nautical astronomy and the design of mathematical instruments (Chap. 7). Finally, the increasing relevance of cosmography (Chap. 8) might have worked to re-shape the identity of the commentary tradition of Sacrobosco's *Sphaera* to emphasize practical knowledge. Such a long transformative process thus began with an appeal for a realistic interpretation of mathematically-defined planetary orbs and resulted in diminishing the centrality of ontological lines of questioning: it became useful knowledge—it worked well, and that sufficed.

In fact, the practical turn had already been realized in the late medieval period, but at that stage it was confined to another field: medicine. The authors of the early modern *Sphaera* commentaries still associated cosmology with medicine, but their commentaries only rarely become explicit in this respect. The same authors dealt with proper medical knowledge in other works, and the *Sphaera* commentaries of this period simply refer the reader to them, if the subject is mentioned at all. Much more relevant as a subject of the commentaries is astrology, the discipline that bridged cosmology and medicine. It was astrology that allowed cosmology to be the conceptual and scientific nucleus of the whole knowledge system. Astronomy, astrology, and medicine together constituted a circle around cosmology, and all possible connections between them and with the center were exploited. Because of the 1494 prohibition of judicial astrology, which was enacted at the University of Paris, the debate about the epistemic fundament of the discipline, its scientific character, and its usefulness became very vivid at least during the first phase of the sixteenth century (Chap. 3). The availability of increasingly precise tables, useful for the mathematical workflow required to locate heavenly objects, only supported the refinement of the discipline in mathematical terms. During the first phase of the sixteenth century, this process combined perfectly with the simultaneous practical turn toward cosmography. Both frameworks made similar ontological assumptions and applied similar criteria of truth while using mathematical models and workflows, and both had the goal of asserting something useful about the real world—the earth and its inhabitants. It is conceivable that the process of mathematization that necessarily accompanied the cosmographic turn is the final explanation for this great expansion of the medieval knowledge system. In the early modern period, the circle of knowledge pivoted on cosmology expanded to involve geography, nautical astronomy, and technological developments, all harmoniously connected to each other and elegantly expressed by the tradition of commentaries on the *Sphaera*.

As is well known, however, this system later came undone. While cosmology, mathematical and nautical astronomy, and technology increasingly joined together and became the knowledge of specialized personnel such as pilots, astrological medicine was increasingly marginalized (Thomas 1971, chap. 3). Both these knowledge clusters eventually exited the universities. The first one entered new schools for the education of specific professions,[17] while astrological knowledge became increasingly insignificant in the curriculum of most universities over the course of the seventeenth century.

Most of the commentaries printed between 1472 and 1650 were destined for the academic book market, that is, they were conceived as textbooks for students. From the perspective of the authors of those commentaries, however, the function of these texts was much broader. In line with the practical turn mentioned above, the treatises also played an important political role: they offered the authors of the commentaries an opportunity to make a confessional statement (MacLean 2012), an aspect that becomes evident when the corpus is investigated in its entirety.

Writing a commentary on cosmology was a convenient occasion to profess adherence to Christianity. Authors such as Francesco Capuano and Pedro Ciruelo, for instance, mentioned explicitly that knowing the structure of the cosmos is a way to admire God's creation. For this reason, the treatise was also an occasion for dissimulation in matters of confessional belonging. The case of Pedro Avelar, who was an active member of a secret Jewish confessional group and as such brutally persecuted by the Catholic Inquisition in Portugal, is exemplary in this regard.

For him, publication was an opportunity not only to teach, but also to keep up a front of religious conformism (Chap. 10). The *Sphaera* was never under serious attack by the Congregation or the Inquisition, though several editions were indeed controlled and some of their content censored (Sander 2018). Nevertheless, the text was definitely an object of political control. Several perspectives show that this was the case. The first example chronologically is the case of the production of textbooks in Wittenberg. When Philipp Melanchthon designed the curriculum, he chose books and commentaries for confessional reasons and their function was made explicit (Chap. 9). The famous Letter to Grynaeus moreover, published 64 times in the corpus of the printed commentaries on Sacrobosco's *Sphaera*, shows that Melanchthon's commitment to astrology was appreciated well beyond the geographic boundaries of the Reformation. The second famous example is the commentary on the *Sphaera* by Christophorus Clavius (1538–1612). The book, first published in 1570 and later re-published seventeen times, was intended as part of a curriculum conceived in the frame of the Jesuit *Ratio studiorum*—that is, against a confessional background (Lattis 1994, 1–29). Thirdly, there is the example of

[17]The Portuguese *Aula de esfera*, a school dedicated to the education of pilots, was indeed operative much earlier, during the sixteenth century. However, this was an exception, as all the other political and geographic entities of Europe entered the planetary nautical market later; they systematized the necessary knowledge in the seventeenth century.

Franco Burgersdijk, who, in recompiling the treatise in the wake of Clavius's commentary, dressed it up to suit the Calvinist *School Order* (Chap. 11). Finally, the process of integrating knowledge that helped make sense of the great journeys of exploration—and on occasion helped facilitate the journeys themselves (Valleriani 2017; Crowther et al. 2015) (Chaps. 7 and 10)—not only lent credence to the claim that the treatise was useful but also represented an opportunity to celebrate "Christian knowledge." In the rhetoric of the time, this furthermore served as a justification for Christian navigators to brutally conquer the planet.

4 The Publishers and Future Research

The great changes and developments that can be detected in the corpus of the early modern printed commentaries on the *Sphaera* of Sacrobosco are clearly related to the regional, temporal, and intellectual context in which these commentaries were conceived and written. The profiles of the authors are therefore indispensable for understanding this evolution of knowledge. Against the background of the entire corpus of historical sources, the present volume considers about 43% of them (Fig. 1.8). By means of an interactive map (https://sphaera.mpiwg-berlin.mpg.de/doi-visualisation-authors-volume), moreover, it is possible to explore the way the contributions of this volume deepen the investigation of the sources. The entire period covered by the corpus is considered, and the number of sources investigated at given time intervals is indeed proportional to the dynamic of production of these treatises which increases around the half of the sixteenth century. Nevertheless, as book historians have recently disclosed (Nuovo 2013; MacLean 2009) and as historians of science are increasingly becoming aware (Valleriani 2017) (Chap. 9), the final decisions on the content of these treatises did not simply emerge from the milieu of scholars. It was in fact the publishers who had the last word on all aspects of the books' publication—their form *and* content. Indeed, these publishers engaged in intellectual, institutional, and economic negotiations with authors, professors, and other institutional representatives for each edition within the corpus—generally speaking, the same applies for every book they published. By laying more emphasis on the booksellers and printers—on the shifting commercial, material, and intellectual landscapes on which the corpus of printed *Sphaera* commentaries came into existence—it might be possible to enrich the investigation undertaken in this chapter. Future research will be undertaken to investigate the social network of the publishers as a way to account for the similarities among these editions!

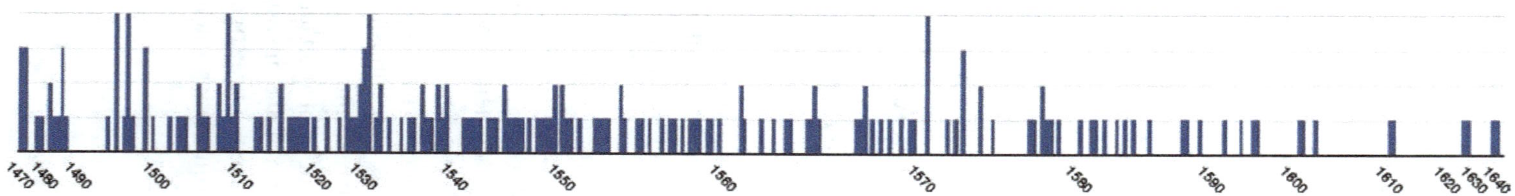

Fig. 1.8 The diagram displays the historical sources belonging to the corpus of the editions of Sacrobosco's *Tractatus de sphaera* that are mentioned in the present volume. The picks show the frequency of mentions, the size of the columns the number of historical sources in a given time windows and the size of empty spaces along the time lines the sources that are not mentioned. The timeline is not linear as the lengths of the time intervals is determined by the number of historical sources published in that specific time interval. For a more detailed and interactive map.

References

Crowther, Kathleen M., Ashley Nicole McCray, Leila McNeill, Amy Rodgers, and Blair Stein. 2015. The book everybody read: Vernacular translations of Sacrobosco's *Sphere* in the sixteenth century. *Journal for the History of Astronomy* 46 (1): 4–28.

de Sacrobosco, Johannes. [1472a]. *Tractatum de Spera*. Venice: Florentinus de Argentina. http://hdl.handle.net/21.11103/sphaera.100685.

———. 1472b. *Ioannis de Sacrobosco anglici v[iri] c[larissimi] spaera mondi feliciter incipit*. Ferrara: Andrea Bellifortis. http://hdl.handle.net/21.11103/sphaera.101122.

de Sacrobosco, Johannes, Jacques Lefèvre d'Étaples, Johannes Regiomontanus, Robert Grosseteste, Pierre d'Ailly, Bartolomeo Vespucci, and Francesco Capuano di Manfredonia. 1508. *Nota eorum quæ in hoc libro continentur. Oratio de laudibus astrologiae habita a Bartholomeo Vespucio florentino in almo Patavio Gymnasio anno.M.d.vi. TEXTUS SPHAERAE IOANNIS DE SACRO BUSTO. Expositio sphaerae Eximii artium & medicinae doctoris Domini Francisci Capuani de manfredonia. Annotationes nonnullae eiusdem Bartholomei Vespucii hic ide itersertae. Iacobi frabri stapulensis Commentarii in eandem sphaeram. Reverendissimi Domini Petri de aliaco Cardinalis & episcopi Cameracensis in eandem quaestiones subtilissimae numero xiiii. Reverendissimi episcopi Domini Roberti linconiensis sphaerae compendium. Disputationes Ioannis de regio monte contra cremonensia deliramenta. Theoricarum novarum textus cum expositione eiusdem Francisci Capuani omnia nuper diligentia summa emendata*. Venice: Giuntino Giunta. Printed by Giovanni and Bernardino Rosso. http://hdl.handle.net/21.11103/sphaera.100915.

de Sacrobosco, Johannes, Bartolomeo Vespucci, Francesco Capuano di Manfredonia, Georg von Peuerbach, Jacques Lefèvre d'Étaples, Johannes Regiomontanus, Luca Guarico, Michael Scot, Nūr al-Dīn Abū Isḥāq al-Bitrūǧi, Pierre d'Ailly, Prodoscimo Beldomandi, Robert Grosseteste, and Gerard of Cremona. 1531. *Spherae tractatus Ioannis de Sacro Busto Anglici viri clarissimi. Gerardi Cremonensis theoricae planetarum veteres. Georgii Purbachii theoricae planetarum novae. Prosdocimi de beldomando patavini super tractatu sphaerico commentaria, nuper in lucem diducta per. L. GA. nunque amplius impressa. Ioannis baptistae capuani sipontini expositio in sphaera & theoricis. Ioannis de monteregio disputationes contra theoricas gerardi. Michaelis scoti expositio brevis & quaestiones in sphaera. Iacobi fabri stapulensis paraphrases & annotationes. Campani compendium super tractatu de sphera. Eiusdem tractatulus de modo fabricandi spheram solidam. Petri cardin. de aliaco epi cameracensis. 14. Quaestiones. Roberti linconiensis epi tractatulus de sphaera. Bartholomei vesputii glossulae in plerisque locis sphaerae. Eiusdem oratio. De laudibus astrologiae. Lucae Gaurici castigationes & figurae toto opere diligentissime reformatae. Eiusdem quaestio Nunquid sub aequatore sit habitatio. Eiusdem Oratio de inventoribus & laudibus Astrologiae. Reverendissimo cardin. epo. D. Bernardo Tridentinor principi dicata. Alpetragii Arabi theorica planetarum nuperrime Latinis mandata literis a calo calonymos hebreo neapolitano, ubi nititur salvare apparentias in motibus Planetarum absque eccentricis & epicyclis*. Venice: Lucantonio Giunta. http://hdl.handle.net/21.11103/sphaera.100999.

Grafton, Anthony. 2010. Commentary. In *The classical tradition*, ed. Anthony Grafton, Glenn W. Most, and Salvatore Settis, 225–233. Cambridge: Harvard University Press.

Grant, Edward. 2009. *Planets, stars, & orbs. The Medieval cosmos, 1200–1687*. Cambridge: Cambridge University Press.

Grendler, Paul F. 2002. *The universities of the Italian Renaissance*. Baltimore/London: The Johns Hopkins University Press.

Hadravovi, Alena, and Petr Hadravovi. 2019. *Traktát o sféře. Iohannes de Sacrobosco: Tractatus de spera materiali s komentářem Václava Fabera z Budějovic*. Praha: Akropolis.

Lattis, James. 1994. *Between Copernicus and Galileo. Christoph Clavius and the collapse of Ptolemaic cosmology*. Chicago: University of Chicago Press.

Leitão, Henrique. 2008. *Sphaera Mundi: A Ciência na Aula de Esfera. Manuscriptos científicos do Colégio de Santo Antão nas colecções da BNP*. Lisboa: Biblioteca Nacional de Portugal.

MacLean, Ian. 2009. *Learning and the market place: Essays in the history of the early modern book*. Leiden: Brill.

———. 2012. *Scholarship, commerce, religion. The learned book in the age of confessions, 1560–1630*. Cambridge: Harvard University Press.

Moyer, Ann E. 2012. The Quadrivium and the decline of Boethian influence. In *A companion to Boethius in the Middle Ages*, ed. Noel Harold Kaylor and Philip Edward Phillips, 479–517. Leiden: Brill.

Nuovo, Angela. 2013. *The book trade in the Italian Renaissance*. Leiden: Brill.

Pantin, Isabelle. 1986. *Imprimeurs et libraires parisiens du XVIe siècle: Cavellat – Marnef et Cavellat*. Paris: Bibliothèque nationale.

———. 1998. Les problèmes de l'édition des livres scientifiques: l'exemple de Guillaume Cavellat. In *Le livre dans l'Europe de la Renaissance: Actes du XXVIIIe Colloque international d'Etudes humanistes de Tours*, ed. Bibliothèque Nationale, 240–252. Paris: Promodis, Editions du Circle de la Librairie.

Rüegg, Walter. 1994–2011. *A history of the university in Europe*, 4 vols. Cambridge: Cambridge University Press.

Sander, Christoph. 2018. Johannes de Sacrobosco und die *Sphaera*-Tradition in der katholischen Zensur der Frühen Neuzeit. *NTM Zeitschrift für Geschichte der Wissenschaften, Technik und Medizin* 26 (4): 437–474. https://doi.org/10.1007/s00048-018-0199-6.

Thomas, Keith. 1971. *Religion and the decline of magic: Studies in popular beliefs in sixteenth and seventeenth century England*. London: Weidenfeld and Nicolson.

Thorndike, Lynn. 1949. *The Sphere of Sacrobosco and its commentators*. Chicago: The University of Chicago Press.

Valleriani, Matteo. 2017. The tracts on the Sphere. Knowledge restructured over a network. In *The structures of practical knowledge*, ed. Matteo Valleriani, 421–473. Dordrecht: Springer Nature.

Valleriani, M., F. Kräutli, M. Zamani, A. Tejedor, C. Sander, M. Vogl, S. Bertram, G. Funke, and H. Kantz. 2019. The emergence of epistemic communities in the Sphaera corpus. *Journal of Historical Network Research* 3(1): 50–91. https://doi.org/10.25517/jhnr.v3i1.63.

5

Borrowers and Innovators in the History of Printing Sacrobosco: The Case of the In-Octavo Tradition

Isabelle Pantin

Abstract The Sacrobosco editions constitute a textual tradition wherein innovation owed to the interaction of two types of actors: borrowers, responsible for chains of related editions, and innovators, who wrote relatively original commentaries. This can be observed in the "in-octavo tradition," a series of editions printed in several European towns, which began with Apianus's 1526 *Sphaera* and endured until the seventeenth century. Different kinds of innovation were produced in it (concerning the layout, the illustration, and different kinds of textual additions). This successful attempt at standardization produced a kind of manual that always retained an essential part of its original features while remaining capable of evolution. This evolution was due to close cooperation between the publishers and mathematicians, and to the fact that the former, with a few exceptions, kept control over the process, and combined innovation with the artful practice of reusing borrowed material.

1 Introduction

The advent of print culture greatly facilitated the building of textual traditions. The great number of manuscripts of Johannes de Sacrobosco's (died ca. 1256) *Sphaera* were copied in codices containing variable sets of texts—with or without illustrations, with or without different kinds of commentary, and with numerous variants on the original text itself. Many of these manuscripts were copied from earlier ones, so it is possible to reconstruct certain lines of transmission, but a stemma would be impossible to achieve. Confronted with "legions" of manuscripts, Lynn Thorndike contented himself with collating twenty significant manuscripts and inventorying the principal commentaries (Thorndike 1949). With the printed editions, in contrast, it is possible to make a census of all those still in existence and to study their interrelations.[1] Very few of them were completely unrelated to the others, as their authors, editors, and publishers were more or less aware of the other editions available on the market. Thus, the entire body of Sacrobosco editions, with its different branches—some of which eventually became trunks of their own[2]—can be studied as a textual tradition. In this frame, we can observe more precisely how innovations occurred. I thus intend to focus on the interaction between two types of actors that contributed to the transmission of the Sacrobosco tradition: borrowers, responsible for long chains of related (if not identical) editions, and innovators, who chose to write commentaries that were, up to a certain point, original, and as a consequence signed their contributions personally.

2 The Dialectic of Tradition and Innovation: The In-Octavo Editions

In the case of the Sacrobosco editions, borrowers and innovators worked together in close collaboration. This is probably linked to the fact that booksellers were, as a rule, particularly involved in the publication of the treatises of the sphere, and that they understood the art of combining innovations with well-tried materials as crucial to commercial success. These booksellers had to respond to the specific demands of the teachers of mathematics (as well as to the vaguer demands of a larger clientele), as they depended on these teachers for achieving the mathematical part of the work; but they kept much more control over the whole process than in the case of more sophisticated scientific works. In some instances (Oronce Finé (1494–1555), Peter Apian (1495–1552)), the mathematicians were also competent in book design, book printing, and even book selling, which contributed to blurring the distinction between the roles.

Quite often, the more innovative writers of commentaries tended to bury the original text under a considerable amount of new data and arguments, or even to depart from the original pattern, so that their claims for full authorship prevailed over their linkage to the tradition. That was the case with Oronce Finé (Chap. 8), Francesco Maurolico (1494–1575),[3] and many others, notably the authors of oversized treatises intended as *summae* of cosmological knowledge rather than simple

[1] For a census of the printed editions of treatises of Johannes de Sacrobosco's *Tractatus de sphaera*, see the Database produced by the project *The Sphere. Knowledge System Evolution and the Shared Scientific Identity of Europe*.

[2] From the beginning of the printed tradition of the *Sphaera*, some editors and commentators (for instance Jacques Lefevre d'Étaples and Oronce Finé in France) took this manual as a basis for developing topics that had been perfunctorily treated by Sacrobosco, notably cosmography and mapping (Chaps. 2 and 8). Ultimately, that lead to the publication of books detached from the Sacrobosco tradition, entitled *Cosmographiae*.

[3] Maurolico's *Cosmographia*, published in Venice in 1543, was an adaption of the *Sphaera* in the form of a dialogue. There are very few studies on this part of Maurolico's work, only brief remarks in (Clagett 1974; Moscheo 1998; Rosen 1957). On the notion of adaption, see the introduction to this volume.

textbooks, which deeply transformed the field around 1570. These works were elaborated mostly for the use of teachers of astronomy, but they also originated from the desire to assert the mastery of Catholic mathematicians over the field of cosmology, which was then dominated by Lutherans.[4] Erasmus Oswald Schreckenfuchs (1511–1579), who published his commentary in 1569 (Schreckenfuchs 1569),[5] taught at the Catholic University of Freiburg im Breisgau and dedicated his work to Jacobus Curtius, Canon of the Cathedral of Constance, a patron and benefactor of the university; Christophorus Clavius (1538–1612), as is well known, published his *Sphaera* (first edition 1570) (Clavius 1570) to provide a summa of the orthodox cosmological doctrine for the use of Jesuit colleges.[6]

However, we can observe at least one long series of editions in which borrowing, copying, and imitating carried as much weight as innovating. This series, which I shall call the "in-octavo tradition," began with Apianus's 1526 edition and endured until the middle of the seventeenth century, though it dwindled down in the last decades of the sixteenth century due to the emergence of successful competing models of treatises on the *Sphaera*. Francesco Giuntini (1523–1590) was its last significant contributor, that is if we set aside the 1626 edition by Franco Burgersdijk (1590–1635), which revived the "in-octavo tradition" in extremis (Chap. 11). But Giuntini soon abandoned it to write two more original treatises on the *Sphaera* under his own name. This is a strong indicator that the tradition could not integrate highly idiosyncratic projects.

I shall describe the development of the "in-octavo tradition" both through time and space (since several printing centers in Europe were involved), and try to analyze the different kinds of innovation that were produced in it. These innovations concerned the commentary, the illustration, and the layout and editing (including the formulation of titles). It is often tough to determine whom exactly they ought to be attributed to, but the following inquest should at least improve our understanding of the type of actors involved, and of their principal aims and motives.

3 The Predecessors: The Venetian Incunabula and the Leipzig Editions

The *Sphaera* was probably first printed in Venice in 1472 or earlier (Sacrobosco ca. 1472a);[7] the first dated edition appeared in Ferrara in 1472 (Sacrobosco 1472b). These first editions were without commentaries and had no printed diagrams, but

4 On the rivalry between Catholic and Lutheran mathematicians, see (Pantin, 1999).

5 Schreckenfuchs was also the author of a huge commentary on Peuerbach's *Theoricae planetarum* with innovative diagrams. See (Mosley 2004).

6 On the original version of Clavius's commentary and on its successive revisions, see (Lattis 1994).

7 This Venetian edition is undated, but a manuscript note in the Library of Congress copy shows that it was published before 8th of May 1472. It could be anterior to the Ferrara 1472 edition, which is revised and has a more elaborate layout (with the use of subtitles).

some blank spaces (two in the Venice edition, three in the Ferrara edition) allowed the insertion of manuscript drawings. Other editions followed, and from 1478 onwards, some printed figures were added.[8] The complete set of the 'Venetian Sacrobosco diagrams' first appeared in the revised edition printed by Johannes Santritter (before 1460–after 1498) in 1488. This Santritter 1488 edition originated a tradition. It was copied by other Venetian printers and book-sellers, notably Ottaviano Scoto (ca. 1444–ca. 1499) and Melchior Sessa (active 1505–1555).[9]

The Venetian editions, from the 1488 Santritter edition onwards, contained significant corrections to the original text and at least four important additions: a description of the armillary sphere, a preliminary treatise of geometry, an excursus on the constellations at the end of chapter II, and a large note concerning domification (the construction of the *figura caelestis* for the horoscopes) in chapter III. The books were small quartos,[10] of which the *Sphaera* occupied around forty pages, followed by two other texts: Johannes Regiomontanus's (1436–1476) *Disputationes contra Cremonensia deliramenta* and Georg von Peuerbach's (1423–1461) *Theoricae novae planetarum.*[11] The long-line layout was dark and dense, and the white spaces quasi-inexistent. Some detached subtitles and, above all, numerous images helped the readers to find their way through the text. These images were small and purely descriptive (lettered geometrical diagrams were avoided). Some of them derived from a long manuscript tradition.

Indeed, a significant minority of the manuscripts of the *Sphaera* had been illustrated,[12] and a kind of iconographical tradition had even been established as early are the second half of the thirteenth century. The diagrams were always related to the same parts of the text, that is, mainly to the first and fourth chapters, and to a lesser extent the third chapter. Some of them had very ancient sources and had been transmitted through the manuscript traditions of late Antiquity texts, notably Calcidius's *In Timaeum* (fourth cent.) and Macrobius's *In somnium Scipionis*

---

[8] In 1478, Franciscus Renner from Heilbronn and Adam Burkardt from Rottweil printed in Venice two different editions with three diagrams (the elementary and celestial spheres; the climes; the solar and lunar eclipses). In 1482, still in Venice, Erhard Ratdolt published a new edition that added a fourth figure (an armillary sphere). In 1485, Ratdolt released another edition with twenty-two small new diagrams and some improvements to the existing diagrams.

[9] The Santritter edition was used as a model for the editions printed in Venice by [Bonetus Locatellus] for Octavianus Scotus, 4th of October 1490, for Melchior Sessa, 3rd of December 1501 and 3rd of December 1513, and by Jacobus Pentius for Sessa, 24th of December 1519.

[10] The dimension of the small quarto page is about 180 × 140 mm (145 × 115 mm for the printed text).

[11] Regiomontanus's and Peuerbach's texts do not begin on the recto of the first folio of a quire, which indicates that they were not meant to be sold or bound separately.

[12] Only seven of the twenty manuscripts described by Thorndike (see Footnote 1) are with diagrams and only five of these seven have what could be considered as a 'complete set.'

(first quarter of the fifth cent.), of which the earliest surviving manuscripts are from the ninth century.[13]

In the following list, we have underlined the diagrams for which prototypes can be traced back to the thirteenth and fourteenth centuries (or even earlier). Of course, for each type of diagram, there existed a quantity of variants. Further research on the fifteenth century manuscripts of the *Sphaera* would be necessary to identify the sources of the diagrams in the first illustrated editions of the *Sphaera*. This list is the brief census of the diagrams that illustrated the text of Sacrobosco in the Venetian incunabula, from the Santritter 1488 edition onwards (Table 9.1).

Even before the turn of the century, other types of treatises on the *Sphaera* appeared in which Sacrobosco's text was accompanied by commentaries, sometimes collected in massive in-folio volumes (Sacrobosco 1495b, 1498, 1499a, b). Other diagrams were conceived for these publications, often more purely geometrical than in the series above described, and new elements were added, notably samples of astronomical tables. At the same time, however, the idea of providing the students with handy textbooks was carried forward. This intention was most apparent in a series of tracts printed in Leipzig starting in 1486 (Sacrobosco 1486, ca. 1487, [1489], and so on). The volumes were small quartos, and they contained no other text than the *Sphaera*, except for some of these editions which included a commentary by Wenceslaus Faber from Budweis (died in 1518) (Sacrobosco [ca. 1495a], 1499a, b, 1500, 1501a, b, 1503a, 1505, 1508). The *Sphaera* commented by Conrad Tockler (1470–1530) (Chap. 5) was published by the same printer in Leipzig, in the same format and with the same figures (Sacrobosco 1503b, 1509).

The layout of the Leipzig quartos was quite different from that of the Venetian incunabula. Large interlinear spacing facilitated note-taking. The text was divided by pilcrows, but there were no detached subtitles (other than the titles of see Chapter II, III and IV of *Tractatus de Sphaera* by Sacrobosco). The diagrams (which were revised and augmented around 1494) illustrated almost the same topics as those of the Venetian editions, but they were different in style; they were often best suited to give the reader a simple but concrete understanding of the basic cosmological phenomena. For instance, the revised iconography provided what is probably the first attempt, in any version of the *Sphaera*, to include a map in the diagram of the terrestrial zones (*Asia*, *Africa*, and *Europa* are crudely delineated in the northern habitable zone). Besides, some diagrams had no counterparts in the Venetian editions, like a series of seven diagrams in the third chapter that showed the variation of the horizon (and its consequences) for people living at different latitudes (Fig. 9.3) (Table 9.5).

[13] Late Ancient diagrams concern the concentric spheres of the world (see Chapter I of *Tractatus de Sphaera* by Sacrobosco), the terrestrial and celestial zones, the climes (see Chapter II of *Tractatus de Sphaera* by Sacrobosco), the *figura retrogradationis* (showing how the movement of the planet on its epicycle produces, as observed from the earth, cycles of direct motion, station, and retrograde motion), and the lunar and solar eclipses (see Chapter IV of *Tractatus de Sphaera* by Sacrobosco). On this iconographic tradition, see notably (Obrist 2004; Müller 2008).

Table 9.1 The diagrams in the Venetian editions from 1488. (NB: The diagrams that had ancient models are underlined)

Preliminaries	
1.	Armillary sphere: 1 large diagram
2.	Geometry: 24 small diagrams
Chapter I	
3.	The elemental and celestial spheres: 1 large diagram that shows the four elemental spheres (earth and water are separated, the former being completely surrounded by the latter), the seven orbs of the planets, the sphere of the fixed stars, and the ninth invisible sphere, divided into the twelve zodiac signs.[a]
4.	*Sphaera recta/sphaera obliqua*: 2 diagrams showing a circle and two diagonals (in the first diagram, the equator and the perpendicular right horizon, in the second, the equator and the oblique horizon).
5.	The four elemental spheres or orbs: 1 diagram. This time, the earth is shown as a small eccentric sphere, which, at its top, emerges from the surrounding orb of water into the orb of air.
6.	*De cæli revolutione*: 1 diagram. The diurnal revolution of heaven is symbolized by a semi-circle (with the earth at its center), within which fixed stars are shown in three supposedly successive positions: at the oriental horizon (*oriens*), at the meridian (*mer*), and at the occidental horizon (*occi*).
7.	*De cæli rotunditate*: 1 diagram. A circle and three regular polygons inscribed in it (triangle, square, and pentagon) to show that the sphere is "the most capacious" of all shapes and the best suited to enclose all natural things, and that, if the world were not round, "it would follow that some place would be vacant and some body without a place" (Thorndike 1949, 120).
8.	Effects of refraction: 2 diagrams. The first one is synthetic to the point of being illegible. It shows, first, the false hypothesis of a flat sky (in which case the stars would be nearer at the zenith), and, second, the reason why the stars look bigger at the horizon than at the zenith: the interposition of vapors that disperse the visual rays. In order to illustrate this phenomenon, the second diagram shows a piece of money immersed in water, and how it is enlarged.
9.	Quod terra sit rotunda: 2 diagrams. In the first one, the earth is surrounded by the circles of the moon and of the sun. These planets are in opposite positions, so that the moon must be eclipsed. That illustrates (not quite clearly) the fact that, as the earth is round, oriental people see lunar eclipses at an earlier hour than occidental people. The second diagram (Fig. 9.1) shows the earth surrounded by the circle of the fixed stars. Four men are walking around the upper hemisphere, from north (sept) to south (mer), and lines drawn from their position at the surface of the earth to stars at different positions on the outer circles symbolize the fact that southern observers do not see the same stars as northern observers.
10.	Quod aqua sit rotunda: 1 diagram. The visual ray (Radius visualis) from an observer on a ship at the top of the mast reaches without obstacle a signal on the seacoast (*Signum littoris*), while the visual ray from an observer on the deck is intercepted by the bulge of the water.
Chapter II	
11.	The equinoctial: 1 diagram. A circle that represents the world, the axis of the world, and the equator.
12.	*De zodiaco circulo*: 1 diagram. A circle that represents the world, the axis of the world, and the zodiac divided in twelve signs. The equator is not drawn, and the inclination of the zodiac with respect to the axis of the world is not correct.

(continued)

Table 9.1 (continued)

13.	Zodiacal signs: 2 diagrams illustrate two significations of "sign." The first one is a pyramid, of which the basis is the corresponding portion of the zodiac (in the eighth sphere), and the apex is at the center of the world. The second one is a crescent-shaped portion of the eighth sphere, of which the points are at the poles and the larger part is the corresponding portion of the zodiac.
14.	*De duobus coluris*: 1 diagram. In a circle that represents the world, the axis of the world, and both colures (equinoctial and solsticial).
15.	*De meridiano et horizonte*: 1 diagram.
16.	Celestial zones and circles: 1 diagram. In a circle that represents the world, the equator, the tropics, the polar circles, the zodiac, the axes of the equator and of the zodiac.
17.	Terrestrial zones: 1 diagram. The Arctic and Antarctic zones and the zone between the tropics are marked *Inhabitabilis*. The two zones between the tropics and the polar circles are marked *habitabilis*. In the lower habitable zone (in the northern hemisphere) a landscape is drawn.
Chapter III	
18.	*Ortus/occasus cosmicus, chronicus, heliacus*: 1 diagram. The circle of the sun (with the sun in two different positions, oriental and occidental), surrounded by a circle that represents the zodiac: on its oriental part, below and above the radiating sun, several stars are represented. Without reading the text, it is not easy to guess that this is a visual definition of the cosmic, acronychal, and heliacal rising and setting of the stars.
19.	What is necessary to understand the astronomical rising and setting: 1 diagram. How the equator, the zodiac, the right horizon, and the oblique horizon mutually intersect.
20.	*Dies naturalis*: 1 diagram. A circle that represents the zodiac surrounds the earth and the circle of the sun. The symbol of Aries is marked on the right, and the position of the sun is slightly below. That corresponds probably to the definition of the natural day: "the revolution of the equinoctial with as much as the sun covers meanwhile by its own movement against the firmament" (Thorndike 1949, 132).
21.	The circles of natural days: 1 diagram. A great circle of the world, the axis of the world, the polar circles, the tropics, the zodiac, and a crude representation of the spiral movement of the sun caused by the combination of its diurnal and annual motions.
22.	The great circles of the sphere: 1 diagram.
23.	The 12 celestial houses: 3 diagrams corresponding to three different ways of drawing an astrological chart. They illustrate the addition on the methods of domification.
24.	The shadows at the equator: 1 diagram. A hemisphere, delimitated by the line of the horizon, and a semi-circle that represents the zodiac, below it, the circle of the sun, with the sun at three different positions: at the zenith, at noon, when the sun is in either equinoctial point (according to the text), when the sun is in the northern signs, and when it is in the southern signs (again according to the text). These three suns cast the shadows of a tower surrounded by ramparts: the shadow is "perpendicular" in the first case, southern in the second one (*umbra austr*), and northern (*umbra bore*) in the third.
25.	Terrestrial zones and climates: 2 diagrams. One large diagram: a circle that represents the world, with the equator, the tropics, the polar circles, the zodiac, and the axis of the equator marked. In the northern hemisphere (north is below) the seven Ptolemean climes are drawn. One smaller diagram: the great circle that passes through the poles and the equinoctial points, the equator, lines drawn from the pole to the equinoctial points, and the seven climes: the geometrical significance of this diagram is not explained in the text.

(continued)

Table 9.1 (continued)

Chapter IV	
26.	Circles of the sun: 1 small diagram shows the eccentric deferent of the sun (and the sun on it), a larger circle that represents a great circle of the sphere (probably the ecliptic), and the centers of both circles.
27.	Moon's circles, *caput* & *cauda draconis*: 1 small diagram. A larger circle that represents a great circle of the sphere, probably the ecliptic, surrounds the eccentric deferent of the moon with its epicycle, and the equant circle, which has a different center than the deferent. The equant and the deferent intersect at two points, called the "head" and the "tail" of the Dragon.
28.	The stationary, direct and retrograde motions (*figura retrogradationis*): 1 small diagram. A larger circle that represents a great circle of the sphere, probably the ecliptic, surrounds the eccentric deferent of any planet (the sun and the moon excepted) with its epicycle, the equant circle, and the three centers of these circles. Tangents to the epicycle are drawn from the center of the world: that shows why a terrestrial observer sees a succession of stationary, direct, and retrograde motions as the planet rotates on its epicycle.
29.	The lunar eclipse: 1 diagram. The traditional diagram shows simply the disposition of the three globes (Sun, Earth, and Moon) and the production of the umbra cone. Here, the eclipse is represented in a broader frame: the ecliptic circle (with stars), the horizon, the deferent of the sun, the deferent and the equant of the moon are drawn. The sun is at the nadir, the moon at the zenith (and at one node of the Dragon), and a landscape is visible on the large earth.
30.	The solar eclipse: 1 diagram shows the deferent circle of the sun, the horizon, and the earth, moon, and sun in alignment. The earth, represented with terrestrial and maritime landscapes, is so large that two men are seen, silhouetted against the sky near the eastern and western horizon. A tangent to the earth passing through the border of the umbra cone of the moon marks the limit outside of which the eclipse is completely invisible.

[a]Concerning the diagram that shows the spheres of the elements and the celestial orbs, there is a discrepancy between the diagram, which corresponds to the original text of Sacrobosco (according to which the sphere is divided into nine celestial spheres), and the revised text of the Santritter edition and its followers that lists ten celestial spheres (Chap. 5).

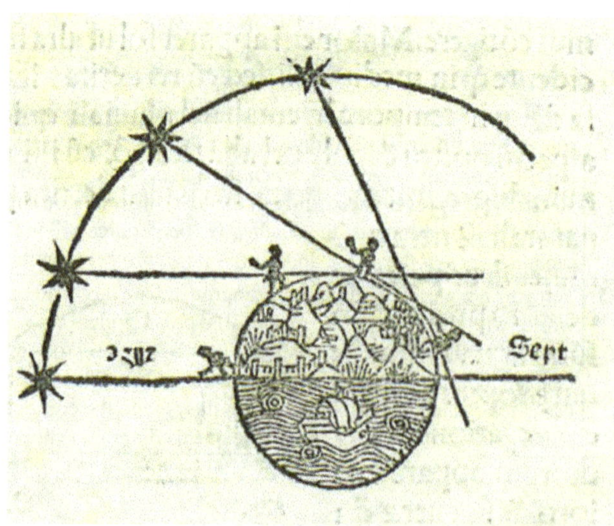

Fig. 9.1 *Quod Terra sit rotunda* (Table 9.1, n 9). From (Sacrobosco 1488). Deutsches Museum, München/Public Domain Mark

4 The Founder of the Tradition: Petrus Apianus (1526)

The new orientation towards pedagogical clarity, cosmographical realism, and interest in a more practical understanding of astronomy was accentuated in the edition printed by Petrus Apianus in Ingolstadt in 1526 (Sacrobosco 1526), which founded the in-octavo tradition. Apianus was then still at the beginning of his career. He had recently moved to Ingolstadt from Landshut where he had already published at least one world map, astrological *practica*, a book on a new kind of sundial (Apianus 1524a), and what was to become (after its revision by Gemma Frisius (1508–1555)) a best-seller, the *Cosmographicus liber* (1524), a manual based on Ptolemy's (second century) geographical knowledge that also contained abundant up-to-date information about land surveying, map projections, instrument making, and navigation (Apianus 1524b).[14] Moreover, Apianus's bias towards practicality showed itself in the way the book was printed: it was lavishly illustrated and contained volvelles (wheel charts) (Vanden Broecke 2000, 130–50). This kind of paper instrument, already used in the Middle Ages, had been introduced in printed books by Regiomontanus, followed by other mathematicians, instrument makers, and designers of astronomical books like Lazarus Behaim and Johann Stoeffler (1452–1531) (Regiomontanus (1472–1474?); Stoeffler 1514; Gingerich 1993, 63–74; Bennett 1998, 195–222; Rhodes and Sawday 2001; Kremer 2011; Schmidt 2011; Stijnman and Upper 2014).

This background must be kept in mind to better understand the new orientation around practicality given to the Sacrobosco printed tradition by Apianus's edition, which at first sight is a much less innovative book than the *Cosmographicus liber*. Indeed, Apianus refers to this earlier work in the *Ad lectorem* of his 1526 *Sphaera*. He adds that this new edition of the *Sphaera* has been conceived "because of the numerous correspondences that exist between geography and astronomy."[15] Indeed, if young students acquired a precise knowledge of the sphere and its circles via Sacrobosco, they would understand thoroughly the measures of Earth and heaven.[16]

The Apianus edition contained only the text of the *Sphaera*, slightly revised, without additions, notes, or commentary, and exhibited thorough editorial work. The layout was airier, using some paragraphs with first-line indents and blank interlinear spaces at the end of the chapters and of the sections in chapters. These chapters and sections had titles and subtitles printed in capital letters, and their set, more complete than in the Venetian incunabula (Table 9.2), better helped the reader in moving through the succession of topics in the text, and in grasping the logical organization of the treatise.

---

[14] On Apianus, see (Röttel 1995).

[15] (Sacrobosco 1526, A1v):"...cum Geographia plurimum comercii habeat cum Astronomia...."

[16] (Sacrobosco 1526, A1v):"...visum est mihi haud inutile fore, si ingenuis adolescentibus primum omnium Astronomiae rudimenta praelegerem, Sphaeram IANI de Sacrobusto accuratissime interpretarer. Futurum tandem existimans ut ex sphaerae circulorumque ejus attenta cognitione, spaciorum terrae coelique absoluta notitia proveniret."

Table 9.2 Titles and subtitles in the Venetian editions from 1488 and in the edition of Apianus (1526)

Venetian editions	Apianus edition (1526) (NB: new subtitles are underlined)
 Divisio secundum accidens sphaerae	*Prooemium auctoris* *Diffinitio sphaerae. Cap. 1*
Quae forma sit mundi *Quod aqua sit rotunda*	*Quae forma sit mundi* *De coeli revolutione* *De coeli rotunditate* *Quod Terra sit rotunda* *Quod aqua sit rotunda* *Quod Terra sit centrum mundi* *De immobilitate Terrae* *De quantitate absoluta Terrae*
De circulis ex quibus sphaera materialis componitur: et illa supercaelestis quae per istam imaginatur componi intelligitur. Capitulum secundum *De zodiaco circulo* *De duobus coluris* *De meridiano et horizonte* *De quatuor circulis minoribus* **De caracteribus sphaerae decimae vel nonae et stellationibus octavae* ** Quae signa quibus circulis dividantur* ** Sequuntur figurae septentrionales* ** Figurae Australes*[a]	*Capitulum secundum De circulis ex quibus sphaera materialis componitur: et illa supercaelestis (quae per istam imaginatur) componi intelligitur* *De zodiaco circulo* *De duobus coluris* *De meridiano et horizonte* *De quatuor circulis minoribus* *De quinque zonis*
De ortu et occasu signorum et diversitate dierum et noctium et de diversitate climatum. Capitulum tertium *De ortu et occasu signorum secundum astrologos* *De diversitate dierum et noctium quae sit habitantibus in diversis locis terrae* *Quorum zenith est inter tropicum cancri et circulum arcticum*[b] *Quorum zenith est in circulo arctico* *Quorum zenith est inter circulum arcticum et polum mundi* *Quorum zenith est in polo arctico* *De divisione climatum*	*Capitulum tertium De ortu et occasu signorum, de diversitate dierum et noctium, et de diversitate climatum* *De ortu et occasu signorum secundum astrologos* *De diversitate dierum et noctium quae sit habitantibus in diversis locis terrae* *Quorum zenith est inter aequinoctium et tropicum Cancri* *Quorum zenith est in tropico Cancri* *Quorum zenith est inter tropicum cancri et circulum arcticum* *Quorum zenith est in circulo arctico* *Quorum zenith est inter circulum arcticum et polum mundi* *Quorum zenith est in polo arctico* *De divisione climatum*
De circulis et motibus planetarum et de causis eclipsium solis et lunae. Capitulum quartum	*Capitulum quartum De circulis et motibus planetarum, et de causis eclypsium Solis et Lunae* *De statione, directione et retrogradatione* *De eclypsi Lunae*

[a]The subtitles marked with an asterisk correspond to a long addition on the visible constellations of the eighth sphere and their correspondence with those of the ninth and tenth spheres.
[b]The first two subtitles «Quorum zenith est» are in the flow of the text.

Fig. 9.2 *Quod Terra sit rotunda*. From (Sacrobosco 1526, A6r). State Library Regensburg--999/Philos.1325, urn:nbn:de:bvb:12-bsb11110162-1

However, the main improvement lay in the illustration. The diagrams of the Apianus edition bore an evident link of continuity to those of its Venetian and Leipzig predecessors in that they concerned the same topics, but they had all been redrawn. Several of these new diagrams were inspired by diagrams already present in the Venetian (Fig. 9.2) as well as the Leipzig editions (Table 9.5), but Apianus had found new visual solutions whenever the models at his disposal lacked clarity or relevance, or when they perpetuated out-of-date representations. An improved mastery of three-dimensional images provided better readability.

For instance, his figure of the elementary and celestial spheres (Table 9.1, No. 3) showed ten celestial spheres (instead of nine) according to the standard opinion of the astronomers at the beginning of the sixteenth century: it corresponded to the model adopted in the *Alphonsine Tables* and explained in Peuerbach's *Theoricae novae planetarum* (of which Apianus was to publish an edition in 1528). Thus, in Apianus's edition, the figure was more up to date than the text, which remained true to the original version and listed only nine celestial spheres.[17] Furthermore, in this diagram, like in that of the four elements (Table 9.1, No. 5), the sphere of earth was no longer included into the sphere of water, for earth and water formed one single

[17] The discrepancy in Apianus's work is the reverse of that in the Venetian editions, which modernized the text and kept the traditional figures.

sphere, according to the new representation of the terraqueous globe.[18] Such a representation was then still quite rare among *Sphaera* editions, for the printers reused the same woodcuts from edition to edition. It had first appeared in the edition supervised by Oronce Finé and printed in Paris by Vincent Quignon (active 1514–1557) for Regnault Chaudière (active 1509–1554) in 1516.[19]

More generally, the images were better distributed throughout all parts of the text (Table 9.5), and Apianus managed to replace diagrams that were not intelligible without the aid of the text with figures that made sense by themselves. To begin with, instead of the ambiguous diagram showing "The revolution of heaven" (Table 9.1, No. 6), there appeared the image of a celestial globe, the horizon of which separated the diurnal from the nocturnal sky (Sacrobosco 1526, A4v).

Indeed, Apianus created a coherent visual language with which his reader could become familiar, as the diagrams often constituted series. For instance, in chapter III, the figure of the "circles of natural days" (Table 9.1, No. 21) is followed by a series of eight diagrams that show what happens to the respective length of night and day (and to related phenomena) around the year under different latitudes (that is, different horizons).[20]

The figure of the "circles of natural days" (Apianus 1526, C1r), which inaugurates the series, shows the sphere of the sun, a small Earth at its center, the oblique track of the ecliptic, and the tight quasi-parallel (in fact spiral) circles that symbolize the path of the sun, day after day, as it follows the diurnal revolution of the heavens while progressing along the ecliptic of its own proper motion.

The next diagram (Apianus 1526, C1r) shows that in the "right sphere," that is at the equator, the horizon passes through the poles of the world and always divides the "circles of the days" into equal parts, so that nights are equal to days, whatever the position of the sun. The ecliptic and the earth are no longer visible, but we recognize the sphere and the band of solar spiral circles ("circles of the days"). These circles, drawn like lines, are vertically positioned, perpendicular to the line of the horizon (dividing night and day). Three of them are more heavily marked: the middle one (the path of the sun at the equinox that corresponds to the axis of the horizon) and those at both extremities (the solar path at the summer and winter solstices).

---

[18] The change concerning the representation of the spheres of water and earth was a consequence of the navigation of Amerigo Vespucci along the east coast of Brazil in 1501, which proved that the southern hemisphere was not immersed in water. Joachim Vadianus was probably the first to infer that the earth constituted one single sphere with the water, from which it "partly emerged," doing so in a letter to Rudolf Agricola that was published for the first time in Wien in 1515. See (Randles 1980, 44–48; Grant 1994, 635–37).

[19] In the (Chaudière 1516) edition (and in the 1519, 1524, and 1527 re-editions), the terraqueous globe is represented in the diagram of the four elements (a2v), but not in the diagram of the elementary and celestial spheres (a2r). In the *Sphaera* printed in Wien in 1518 by Johann Singriener for Lucas Alantsee, the terraqueous globe, with a map drawn on it, figures in the diagram of the elementary and celestial spheres (a3r, a6v).

[20] This part of chapter 3, entitled "on the variation of phenomena according to latitude," is not illustrated in the Venetian editions; in the Leipzig editions, it is illustrated with seven diagrams, which have only remote links to Apianus's diagrams (Fig. 9.3).

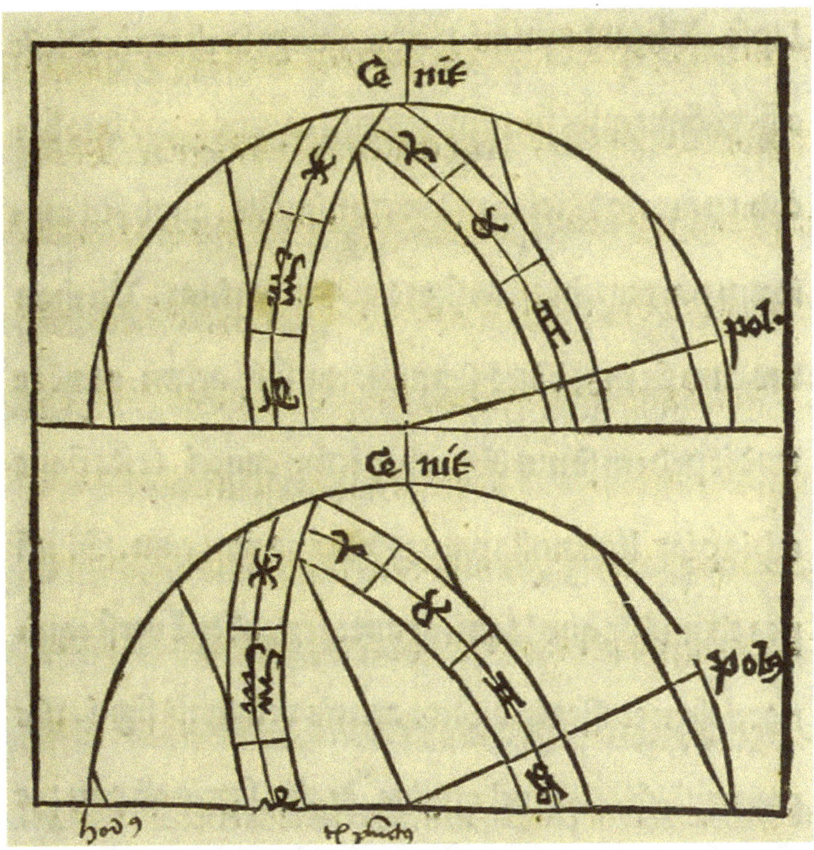

Fig. 9.3 *Opusculum spericum cum figuris optimis et novis*. When the zenith is between the Equator and the Tropic of Cancer (first diagram), and on the Tropic of Cancer (second diagram). From (Sacrobosco 1494). Courtesy of the John Carter Brown Library

At the right and left extremities of the horizon line, small stars symbolize the north and south poles: the plan of the horizon coincides with that of the axis of the world. The band of the "circles of the days" is divided by twelve lines (portions of meridians), parallel to the horizon and equally spaced. They are numbered and symbolize the perfect equality of all diurnal and nocturnal hours, whatever the season, in the right sphere (*sphaera recta*).

In the seven next diagrams, the basic elements are the same, but the respective positions of the horizon (with its axis), the axis of the world, and the "circles of the days" vary, as the text describes what happens to people "whose zenith is" at different latitudes (Fig. 9.4). Thanks to the consistency of the visual language, the principle of the variation of the duration of day and night (and of the direction and length of shadows) is easier to grasp. Apianus's effort towards transforming the traditional *Sphaera* into a modern manual capable of transmitting useful and applicable cosmological knowledge ended there, but this effort was to be pursued by followers and imitators.

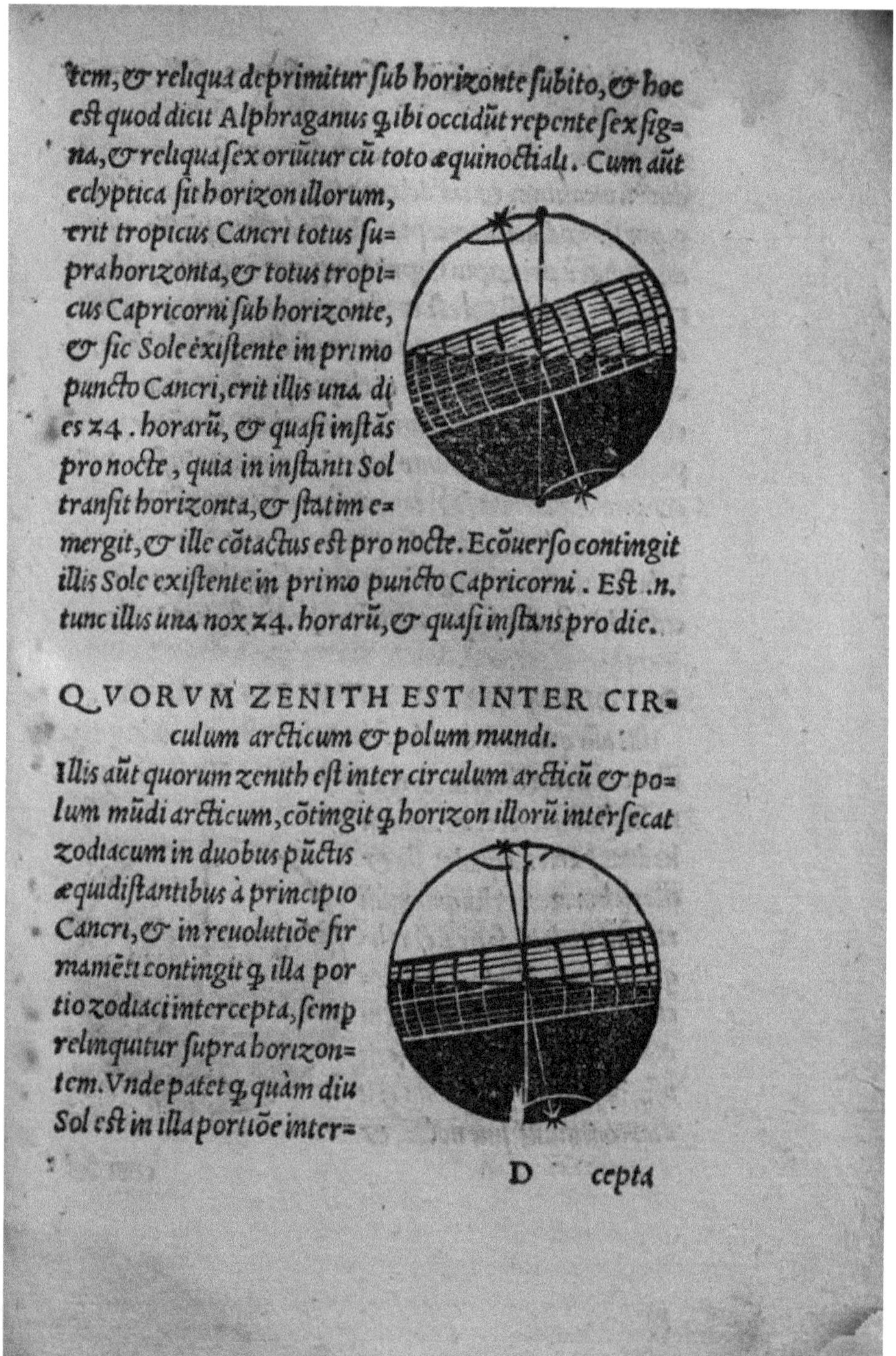

tem, & reliqua deprimitur sub horizonte subito, & hoc est quod dicit Alphraganus q̃ ibi occidũt repente sex signa, & reliqua sex oriũtur cũ toto æquinoctiali. Cum aũt eclyptica sit horizon illorum, erit tropicus Cancri totus supra horizonta, & totus tropicus Capricorni sub horizonte, & sic Sole existente in primo puncto Cancri, erit illis una dies 24. horarũ, & quasi instãs pro nocte, quia in instanti Sol transit horizonta, & statim emergit, & ille cõtactus est pro nocte. Ecõuerso contingit illis Sole existente in primo puncto Capricorni. Est .n. tunc illis una nox 24. horarũ, & quasi instans pro die.

QVORVM ZENITH EST INTER CIRculum arcticum & polum mundi.

Illis aũt quorum zenith est inter circulum arcticũ & polum mũdi arcticum, cõtingit q̃ horizon illorũ intersecat zodiacum in duobus pũctis æquidistantibus à principio Cancri, & in reuolutiõe firmamẽti contingit q̃ illa portio zodiaci intercepta, semp relinquitur supra horizontem. Vnde patet q̃ quàm diu Sol est in illa portiõe inter-

D cepta

Fig. 9.4 When the zenith is on the Arctic Circle (first diagram), and between the Arctic Circle and the North Pole (second diagram). From (Sacrobosco. 1526, D1r). State Library Regensburg--999/ Philos.1325, urn:nbn:de:bvb:12-bsb11110162-1

5 The Model Perfected: The Wittenberg Editions

Apianus's *Sphaera* was never exactly copied, but the model it provided was soon imitated and improved. The University of Wittenberg, founded in 1502, was then at a critical juncture in its reformation process. The disorders provoked by radical reformers like Thomas Müntzer (ca. 1489–1525) and the Zwickau Prophets, notably the 1525 Peasant War, had convinced Martin Luther (1483–1546) and Philipp Melanchthon (1497–1560) of the necessity of founding a strong educational system in order to spread the teaching of moral philosophy, physics, the arts of language, and mathematical disciplines (Kusukawa 1995). The elaboration of new programs at Wittenberg and the publication of manuals, meant to be circulated in Germany and even throughout Europe, were the focus of this scheme.[21]

The publication of a new edition of the *Sphaera* took place in a still more particular context. Melanchthon, who was the professor of Greek at Wittenberg, was a strong advocate of mathematics. He thought it provided key knowledge for reading ancient texts, history, and geography. He furthermore grew more and more convinced that the attentive study of celestial signs was a Christian duty, as it helped to pursue a better understanding of God's providential governance. He expressed this conviction in a well-argued apology of astronomy and astrology addressed to Simon Grynaeus (1493–1541) and appended as a preface to the new edition of Sacrobosco prepared for the beginning of the 1531–1532 academic year (Sacrobosco 1531).[22] This letter to Grynaeus was written in August 1531 while Melanchthon was observing the return of Halley's comet, which made a deep impression on him (Kusukawa 1995, 124–34).

This momentous preface was instrumental in the success of the book,[23] whose other attractive features were borrowed from the Apianus edition. The Wittenberg editor imitated Apianus's airy layout, his clear subtitles, and his use of paragraphs. He even added several new paragraphs and introduced many interlinear blank spaces.[24] He suppressed a few diagrams[25] and copied the rest of them, either closely

[21] On the amazing transformation of the modest town of Wittenberg into one of the largest book production centers in Germany, thanks to the publication of religious pamphlets, in the first place, but also of educational manuals, see (Pettegree 2010, 91–106).

[22] On Melanchthon's vision of astrology, see notably (Caroti 1986; Müller-Jahncke 1998).

[23] On the success of Melanchthon's letter, see (Pantin 1987).

[24] The description of the colures (see Chapter II of *Tractatus de Sphaera* by Sacrobosco) in the Apianus edition (B2v), for instance, does not use a paragraph to separate the general description from what concerns each colure; it uses two asterisks instead ("…** Colurus igitur distinguens solsticia…"). The Wittenberg 1531 edition (C3v) adds a paragraph and an interlinear blank, and it also keeps the asterisks (probably because their meaning was no longer understood). A second example: the beginning of the section on the natural days ("Ex praedictis etiam patet quod dies naturales. Est enim dies naturalis…") is only marked by an intralinear blank (before "Ex praedictis") in the Apianus edition (B8v). In the Wittenberg edition (D3v), a new paragraph, made more visible by an interlinear blank, begins at "Est enim dies." As a consequence of this airier layout, the text of the Wittenberg 1531 edition is printed on forty-three folios (instead of thirty-one in the Apianus edition).

[25] The diagrams on the effect of refraction, supposed to enlarge the image of the stars near the horizon, were suppressed.

(in almost all cases) or more freely, to improve their clarity.[26] In only one case, he followed another model.[27]

But the important thing was that the Wittenberg edition set a dynamic process in motion. The printers of the *Sphaera* had always worked for the universities that provided their first market. But in this case, the association and the cooperation were still closer. The regular reprinting of the Wittenberg *Sphaera* over decades shows that the printers had a monopoly on the production of textbooks for the University of Wittenberg and for other Lutheran universities, and that they were supposed to adapt this production to any change in the syllabus, with the aid and under the supervision of the professors. That was probably a good business model, and the large quantity of books produced year after year allowed the 'Wittenberg Sacrobosco' to be widely known and influential, even outside Germany: already in 1532, the Venetian bookseller Melchior Sessa copied the Wittenberg edition instead of reproducing one of the versions of the *Sphaera* that had been regularly printed in Venice since 1472 (Sacrobosco 1532).[28] Moreover, the low commercial risk and the easy access to teachers of mathematics created the favorable conditions for a smooth evolution. Moderate additions were progressively introduced, and two significant revisions took place before 1545.

In the 1531 edition, the additions were limited to four mnemonic Latin verses on the cosmic, acronychal, and heliacal risings and settings (*Mane vehit supra terram tibi cosmicus ortus*...), and to an extract of Regiomontanus's *Epitome* (III, 21) on "the double cause of the inequality of natural days" (*Dies naturales duplici causa inaequales esse*) (Sacrobosco 1531, F2r–F3r). The 1534 and 1536 Wittenberg editions (Sacrobosco 1534a, 1536) followed the 1531 edition line to line, but the next edition, in 1538, was carefully revised, with important changes in the layout and the subtitles, the addition of several notes,[29] and even some corrections to the text. Furthermore, at the end of the book, an extract from Alfraganus (Sacrobosco 1538, G6v–G8r)[30] followed the one from Regiomontanus. However,

[26] Examples of improved diagrams include two figures of the terrestrial zones in Apianus's edition: the first one is a crude geometrical diagram; in the second one, a kind of map covers all the space between the polar circles. In the Wittenberg 1531 edition, there is only one larger and more precise geometrical diagram, with the name of the zones inscribed.

[27] The image of the lathe that illustrates the definition of the sphere (Sacrobosco 1531, B1v) is closer to the diagram introduced by Jacques Lefevre d'Étaples in Parisian treatises on the *Sphaera* (Chap. 2).

[28] For a general view of the development of the in-octavo tradition, see the next sections.

[29] For instance: (Sacrobosco 1538, B1v): a note on Euclid's definition, which is quoted in Greek and translated; (D4v): a marginal note on 'CHRONICUS:' "Pro Chronico legendum [in Greek:] *akronuchos*...," with a quotation in Greek from Proclus's commentary on Hesiod; (F2v): a marginal note on the evaluation of the width of the first climate: "Unui gradui latitudinis, tribuuntur hoc loco 56. Milliaria et duae tertiae unius. Est enim hic locus omnino fere ex Alfragano desumptus."

[30] The text "De ortu et occasu Planetarum, et occultationibus eorum de sub radiis solis. Diff. xxiv" is taken from the edition printed in Nuremberg in 1537, with Regiomontanus's *Oratio habita Patavii in praelectione Alfragani* and a preface by Melanchthon: *Rudimenta astronomica Alfragani. Item Albategni...de motu stellarum.*

the more obvious additions concerned the diagrams, notably with the introduction of four volvelles.[31] This imitation of the device used in Apianus's 1524 *Cosmographia* signaled the moderate but significant tendency towards practicality (at least as far as pedagogical methods were concerned) that was to characterize the Wittenberg treatises on the *Sphaera*. The volvelles were all the more useful as pedagogical tools in that their movable parts, printed on a folded sheet at the end of the volume, had to be cut out and assembled by the reader. Besides, they were true paper instruments, meant to complement certain demonstrations or to show some of the elementary procedures of astronomical calculation. For instance, the second volvelle bears this caption:

> Instrument by which the roundness of Earth according to latitude [that is, from south to north] is proven, and by which all that the author says in the third chapter about the natural days is very easily understood.[32]

The underlying diagram (Fig. 9.7) shows a graduated circle (a meridian), the axis of the world, the southern star Canopus, "Helice" (that is *Ursa Major*), and the quasi-parallel circles (more exactly, spiral) that describe the daily motion of the sun as it progresses along the zodiac, from one tropic to the other. These circles are divided into equal portions by twelve numbered arcs (which are portions of meridians). The lines of the climes above the arctic circle are also drawn. The movable part of the volvelle is a half-disk, bearing the words "HORIZON" and "NULLA DIES SINE LINEA" ("Not a day without writing a line"), a proverb taken from Pliny (23–79) (*Natural history*, XXXV, 84), which is an admonition against laziness. This half-disk is meant to recover the part of the hemisphere under the horizon; a kind of paper alidade, showing the zenith, is attached to it (Fig. 9.8).

The next edition, in 1540, strictly followed the 1538 edition, with only one (probably accidental) change: the letter of Melanchthon was dated August 1540 (Sacrobosco 1540, A8v).[33] Then, in 1543, a new revision occurred leading to significant additions: in Chapter III of *Tractatus de Sphaera* by Sacrobosco, an "Annotatio" on the three causes of the inequalities of natural days (with a new diagram) (Sacrobosco 1543b, E2r), two more marginal notes on the climes (Sacrobosco 1543b, F4v, F5r), and a "Table of the maximal lengths of the natural days in every latitude between the arctic circle and the pole."[34] Aside from this, the extract from Alfraganus at the end of the book was replaced by a brief treatise on the "Poetical

---

[31] On the probable author of the revision of the 1538 Wittenberg edition, Georg Joachim Rheticus, and on the changes to the illustration and the subtitles, see *infra*.

[32] (Sacrobosco 1538, B8r): "Instrumentum quo et rotunditas terrae secondum latitudinem probatur, et facillime omnia ea., quae autor in tertio capite de diebus artificialibus tradit dijudicantur."

[33] The change to the date of the preface was probably a mistake that originated in the printer's shop (because of the habit of making the date of the preface match with that of the printing). In the next Wittenberg editions (1543, 1545a, and so on) the initial date (1531) was restored. However, the error was transmitted to Paris and Antwerp editions (Table 9.3).

[34] (Sacrobosco 1543b, F1v–F2r): "Tabula maximorum dierum naturalium ad singulas elevationes poli habitantium a circulo arctico usque ad polum arcticum."

risings of the stars" (*De ortu poetico*), which provided all the information needed to inventory and to interpret the astronomical passages in classical texts. The *De ortu poetico* was accompanied by two astronomical tables printed on two folded sheets (Sacrobosco 1543b, G3v–I7v).[35]

Thus, thanks to the Wittenberg editions, the series of small treatises that derived from the edition of Apianus fully became a tradition: the transmission of the model, always clearly recognizable, was kept alive by its ability to evolve, unjeopardized by changes and additions.

6 The Development of the In-Octavo Tradition

The success of the Wittenberg *Sphaera* launched a long-lasting movement. During the rest of the century, about seventy editions following the same model, more or less improved and with different sets of additions, were printed in several European towns. It will be clearest to present the progress and ramification of this tradition in tabular form (Table 9.3).

In the first column, the editions that were used as models are listed, for they introduced major innovations that were adopted by later editions. Each of these editions is identified by the initial letter of the town where it was printed, followed by a number. For instance, "W3" (for "Wittenberg 3") means the third model of edition (with a particular set of texts and of diagrams) that was published in Wittenberg. The relationship between this model and its predecessors is briefly indicated in underlined notes. If the model-edition in question was reproduced in the same town without significant changes, the dates of the editions are given

[35]The tables are referred to in G7v ("Huc pertinet Tabula continens ingressum Solis in XII signa Zodiaci") and G8r ("Huc referatur Tabula quae habet gradus eclipticae cum quibus stellae insigniores olim oriebantur et occidebant"). They have disappeared from most copies.

below (and the names of the booksellers if they were different from the editors of the preceding editions).

In the second column, the editions that copied a model are described according to the same principles. They are placed in front of the model they copied, and the variants they could introduce are briefly indicated in underlined notes. The editions that showed significant variants but were not copied, or that transmitted limited variants but not a whole model (notably Paris 2–5 and Cologne 2), are in this second column.

In the third column, publications are mentioned that were linked to the Sacrobosco editions, but were outside the "in-octavo tradition" proper.

Table 9.3 The in-octavo tradition 1526–1601[a]

Innovative editions used as models (Main variants and additions are underlined)	Following editions (Main variants are underlined)	Sideways publications
Ingolstadt 1 [I1] 1526. *Sphaera Iani de Sacrobusto*, ed. Petrus Apianus: P. Apianus.		
Wittenberg 1 [W1] 1531. *Liber Johannis de Sacro Busto de sphera. Addita est praefatio...*: Joseph Klug. = I1 revised + Melanchthon's preface + extract from Regiomontanus. New iconography (see above and Table 9.6). 1534a; 1536.	**Venice 1 [V1] 1532**: M. Sessa; 1534b; 1537a, b: F. Bindoni; 1541; 1541: M. Sessa; 1545c; 1553a: F. Bindoni. = **W1**. **Venice 2 [V2]** 1548: *Sphaera...Addita sunt quaedam ad explanationem eorum quae in Sphaera dicuntur facentia*: M. Sessa; 1550a; 1552; 1554; 1557a; 1561a; 1564c: F. Rampazeto; 1572a: Sessa; 1577a; 1580; 1587; 1594; 1601a. = W1, but Melanchthon's preface is replaced by a short treatise of elementary geometry (with diagrams).	

(continued)

Table 9.3 (continued)

Innovative editions used as models (Main variants and additions are underlined)	Following editions (Main variants are underlined)	Sideways publications
Wittenberg 2 [W2] 1538. *Joannis de Sacro Busto Libellus de Sphaera. Ejusdem autoris libellus, cujus titulus est Computus...Cum praefatione...et novis quibusdam typis, qui ortus indicant*: J. Klug. = W1 revised and augmented, with new diagrams (and four volvelles) ± extract from Alfarganus + Sacrobosco, *De anni ratione.* 1540 (= **W2**, but the date of the preface has been accidentally modified: "August 1540" instead of "August 1531").	**Paris 1 [P1]** 1542/1543.[b] *Joannis de Sacrobusto de Sphæra liber. Plurimis novis typis auctus et illustratus. Praemissa Philippi Melanchthone doctiss. praefatione...*: J. Loys and G. Richard. = **W2**, copied from the 1540 edition[c], without *De anni ratione*. **Antwerp [A1]** 1543a. *Joannis de Sacro Busto Libellus de Sphaera. Ejusdem authoris libellus, cujus titulus est Computus...Cum praefatione...* = **W2**, copied from the 1540 edition like P1. **Paris 2 [P2]** 1545b. *Sphaera Ioannis de Sacrobosco typis auctior quam antehac, atque ex diligenti manu scriptorum impressorumque codicum collatione castigatior, praemissa... praefatione...*: **J.** Loys & G. Richard. Follows W2 and P1 with significant changes in the layout (the text is divided into books and chapters), and numerous additions: some anonymous *scholia*, two new diagrams, several marginal references and a biographical note: *J. de Sacrobosco vita.* **Antwerp 2 & Louvain 1 [A2 & L1]** 1547: M. Nutius for J. Richard and J. Waen. = W2 without Melanchthon's preface and without Sacrobosco, *De anni ratione*. New dedication to Hieronymus Ruffaut. Imperial privilege (8th of March 1547). Antwerp: J. Richard: 1551; 1559; 1561b.	1546. *La Sphere*, transl. Martin de Perer: J. Loys.

(continued)

Table 9.3 (continued)

Innovative editions used as models (Main variants and additions are underlined)	Following editions (Main variants are underlined)	Sideways publications
Wittenberg 3 [W3] 1543b. *Joannis de Sacro Busto Libellus de Sphaera. Accessit ejusdem autoris Computus ecclesiasticus et alia quaedam in studiosorum gratiam edita*: P. Seitz. W2 revised and augmented, with one new diagram + *De ortu poetico.* 1545a: V. Creutzer; 1549: J. Krafft; 1550b; 1553b; 1558; 1561d; 1563; 1568; 1574a: P. Seitz; 1578a: heirs of P. Seitz; 1601b: Z. Schurer and J. Krafft.	**Paris 3 [P3] 1549/1550.** *Sphaera... typis auctior...*: T. Richard for G. Cavellat. Follows both W3 and P2. All additions are cumulated. **Paris 4 [P4]** 1550c. *Sphaera...typis auctior...cum annotationibus, quae locis aliquot obscuris magnam lucem afferent*: G. Cavellat. P3 + numerous new anonymous annotations + changes in the lay-out (the folios are numbered) + G. P. Valeriano: *Compendium in sphaeram.* Royal privilege (27th of March 1550 n.st.) **Paris 5 [P5] 1551.** *Sphaera...typis auctior...cum annotationibus, et scholiis doctissimi uiri Eliae Vineti, quæ locis aliquot obscuris magnam lucem afferent*: G. Cavellat, 1551. P4 + letter from Vinet to Tacitus (1550) + changes in the lay-out (the division into books and chapters is abandoned) + new *Scholia* by Vinet (which are not discernable from the older annotations that are still there). New layout: the division in chapter introduced in P2 disappears.	

(continued)

Table 9.3 (continued)

Innovative editions used as models (Main variants and additions are underlined)	Following editions (Main variants are underlined)	Sideways publications
Paris 6 [P6] 1555/1556.[d] *Sphaera...emendata. Eliae Vineti...scholia in eamdem sphaeram, ab ipso auctore restituta. Adiunximus...:* G. Cavellat. Follows P5, but Vinet's *Scholia* are revised and marked 'Scholium Vineti;' almost all the older notes are removed. + Pedro Nunez, *Annotatio in extrema verba capitis de climatibus*, transl. by Vinet. New privilege: 6th of February1555 n.st. **[P6b]** 1557b/**1558/1559**.[e] Paris: G. Cavellat. Follows P6, with some corrections, and the order of the texts has been slightly modified: Nuñes's *Annotatio* is printed first. 1561. **[P6c]** 1562a. Paris: G. Cavellat. Follows P6b, but the names of Melanchthon and Grynaeus have been suppressed. The letter becomes "Praefatio." 1564b.	**Venice 3 [V3]** 1562b. *Sphaera... emendata. Eliae Vineti Santonis scholia...*: G. Scoto. Follows P6, but the letter by Melanchthon has been suppressed. 1569; 1574; 1586. **Cologne 1 [C1]** 1566a. *Sphaera... emendata*: M. Cholinus. = P6c. **Paris 7 [P7]** 1569a. *Sphaera... emendata...*: J. de Marnef and G. Cavellat. Follows P6c, but Vinet's *scholia* are all printed together at the end of the *Sphaera.* 1572b; 1577b: J. de Marnef and widow Cavellat; 1584. **Cologne 2 [C2]** 1581. *Sphaera... emendata...*: M. Cholinus. Follows C1 and P6c + a dedication by Albertus Hero to Paulus Kuechovius (Cologne 04–03–1581) + *scholia* by Hero. 1591: G. Cholinus; 1594a.	1570. *La sphère*, transl. Guillaume des Bordes with some of Vinet's *Scholia*. Paris: J. de Marnef and G. Cavellat. 1576; 1584.
Lyons 1 [Ly1] 1564a/**1567**.[f] *Sphaera...emendata. Cum additionibus in margine, & indice rerum & locorum memorabilium, & familiarissimis scholijs, nunc recenter compertis, & collectis à Francisco Iunctino*: S. Barbier for the heirs of J. Giunta. P6 without *De ortu poetico.* + Dedication by Giuntini to Thomas de Guadagne (Lyons, 15.06.1564) + a large commentary by Giuntini (with some new diagrams) + new appendices + an Index. Privilege: 10th of January 1564 n.st.	**Antwerp 3 [A3]** 1566b. *Sphaera... emendata...*heirs of A. Birckman and J. Richard. Follows **Ly1**, without Valeriano's *Compendium* and without the index. **Antwerp 4 [A4]** 1573. *Sphaera Ioannis de Sacro Bosco emendata. In eandem Francisci Junctini Florentini et Eliæ Vineti Santonis Scholia*: J. Bellere. = Ly1 + Sacrobosco, *Algorismus*.	
	Antwerp 5 [A5] 1582. *Sphaera... emendata. In eandem...Iunctini... Vineti...et Alberti Heronis scholia*: J. and P. Bellere. = A4 + Hero's *scholia* from C2.	

(continued)

Table 9.3 (continued)

Innovative editions used as models (Main variants and additions are underlined)	Following editions (Main variants are underlined)	Sideways publications
	Lyons 2 [Ly2] 1578b. *Sphaera... emendata a Fr. Iunctino Theologo..., qui etiam in capite libri adjunxit Principia Geometrica...In calce libri habes Scholia Eliae Vineti*: F. Tinghi. With Vinet's *Scholia* printed as in **P7** + Giuntini, *Principia geometrica.* No other text.	Giuntini, Francesco. 1577–1578. *Commentaria in Sphaeram Joannis de Sacrobosco.* Lyons: F. Tinghi, 2 vols.
Cologne 3 [C3] 1601c. *Sphaera... emendata...*: G. Cholinus. Follows C2 + some notes taken from Clavius's commentary + some anonymous comments + a poem by Thomas Abel, physician and professor of mathematics, perhaps the editor of the volume.		

[a]In this table, only true editions are listed. In some towns, particularly in Paris, the practice of re-issuing earlier editions with a new date on the title was quite common. As the products of this practice ("separate issues" in English, "émissions" in French) had strictly no impact on tradition, they are not taken into account to prevent skewing the statistics. Editions posterior to 1601 are not listed here.
[b]1542 is the date at the colophon; 1543 is the date on the title page.
[c]The 1540 Wittenberg edition copies line for line W2, with one exception: all marginal notes are inserted into the text, and the preface is dated 1540 (instead of 1531).
[d]1555 on the "achevé d'imprimer;" 1556 on the title-page.
[e]1557. Two reissues with the date modified: 1558, 1559.
[f]1564. Reissue with the date modified: 1567.

This tabular presentation shows that the development of the tradition was not chronologically linear. For instance, the tracts published in Venice by the Sessa firm until 1601 (V1 and V2) reproduced (with nonessential variants) a model conceived in Wittenberg in 1531 (W1); whereas in the same town, the Scoto firm published, from 1569 to 1586, a *Sphaera* (V3) that followed a 1556 Parisian model (P6).

It also shows that the rhythm of the innovations was rather rapid up to 1556, at which point it slowed down, and that the initiative of the innovations passed from town to town.

Thirdly, the analysis of these progressively introduced changes suggests that it would be quite misleading to make a sharp distinction between innovations and borrowed material: in the in-octavo tradition, the innovation process was largely a braiding process in which old and new materials, either produced inside the tradition or borrowed from outside, were closely interlocked. All these indications will be examined further, as they bring us back to the initial question: what kinds of actors were mostly responsible for the changes?

7 Publishers Versus Mathematicians

My point is that the evolution of the in-octavo tradition was the result of close cooperation between the publishers and the mathematicians, but also that the former, with a few exceptions, kept control over the process, which accounts for the longevity of this tradition.

7.1 The Shift from Town to Town

As we have seen, in the in-octavo tradition, the power to innovate passed from town to town. Almost each time, the existence of close cooperation between mathematicians and publishers figures as an important factor. At first (between 1531 and 1543), the most creative publishers were in Wittenberg; afterward in Paris. Between 1564 and 1582, the main feature of the editorial landscape was a kind of rivalry between Lyons and Antwerp, while Cologne also entered the game.

In the cases of Wittenberg and Paris, the main causes of the shifts are apparent. Apianus, the founder of the tradition, was simultaneously a printer and a mathematician, but he did not work in association with a university, and his interest in astronomical textbooks (the *Sphaera* in 1526 and Peuerbach's *Theoricae novae* in 1528) soon dwindled, as he embarked on a career at the imperial court. Then, around 1530, the reform of the university of Wittenberg experienced a significant turning point, as we have seen. Melanchthon supervised an editorial program that produced several innovative textbooks, notably in astronomy, written or edited by the professors and printed at the university print shop. The Wittenberg innovations, which did so much to give the in-octavo tradition its identity, were thus an exception in that their principal initiator was neither a mathematician nor a bookseller, but the reformer of the university syllabuses. However, when his efforts in this field had borne fruit, Melanchthon turned his attention to other priorities, such as the reform of the teaching of philosophy (the first version of his *Initia doctrinae physicae* was to appear in 1549). Moreover, the Wittenberg mathematicians, notably Kaspar Peucer (1525–1602), went on to publish alternative models of astronomical textbooks (Peucer 1550), and the publication of successive editions of the *Sphaera*, which remained in use, became simple routine work.

Paris had become one of the main centers of the European book trade during the first decades of the sixteenth century. It had benefited from the importance of its university and from the decline of Venice, which had been the center of the book world in the fifteenth century, but was now handicapped by a succession of political and economic crises. Major proponents of mathematical humanism were then teaching in Paris, notably Jacques Lefèvre d'Étaples (ca. 1450–1536) (Chap. 2) and his disciples, and in the next generation Oronce Finé (Chap. 8). These scholars were convinced of the necessity of spreading their ideas through the publication of innovative books, and could rely on their relationship with different interconnected milieus to do so: the Art Faculty and the Parisian colleges, the humanists engaged in pedagogical reform, the courtly circles that supported the foundation of the *Collège royal*, and prestigious printers able to assert themselves in the European market, namely Henri I Estienne (active 1502–1520) and Simon de Colines (active 1520–1546). Then, towards the end of the 1540s, less distinguished booksellers decided to widen the range of mathematical books offered on the market in order to attract new customers: Jean Loys, Guillaume Richard, and above all Guillaume Cavellat.[36] Their first targets were students, and they eagerly copied the pocket-sized textbook published in Wittenberg. They could furthermore count upon the cooperation of teachers who were willing to revise the annotation while preparing their own lectures: from 1545 to 1556, five editions of the *Sphaera* (P2 to P6) were published in Paris with significant additions and changes.

The move to Lyons (Ly1) was short-lived and probably mainly due to the initiative of the astrologer Francesco Giuntini, who wished, as we shall see, to affirm his mastery of all astronomical knowledge, starting with the *Sphaera*. At first, the in-octavo tradition probably appealed to him as an easy way to enter the field. But he abandoned it as soon as he saw that it offered too narrow a framework and did not permit him to expand his ambitions.

Antwerp was a "commercial megapolis" (Pettegree 2010, 250), and each time a kind of book proved successful, Antwerp booksellers often tried to enter its market. Until 1561, Jean Richard had on several occasions published two variants of the Wittenberg editions (A1 and A2). In 1566, in association with the heirs of Arnold I Birckmann (died 1542), he published, with some suppressions, a copy of Ly1 (A3). Then, in 1573, another Antwerp bookseller, Jean Bellere (1526–1595), issued a new edition (A4) that combined A3 and the Paris 1556 edition (P6). Finally, in 1582 (A5), the same Bellere still added Albertus Hero's (1549–1589) notes that had been printed in Cologne the preceding year (C2). This practice of combining borrowed material, which needed little mathematical expertise, can barely be called innovation. In any case, these attempts were not successful, as the in-octavo tradition was already declining.

The Cholinus firm in Cologne resorted to similar practices: it first followed a Parisian model (C1); then it freshened it up by adding some modest *scholia* written by Albertus Hero (C2). The masterpiece was achieved by Goswin Cholinus (active 1588–1610) in 1601 (C3): he published a new version of the *Sphaera emendata* that looked like a patchwork of diverse material, for an anonymous mathematician[37] had added to Elie Vinet's (1509–1587) and Hero's *scholia* several extracts from Clavius's commentary, some new notes (under the heading "Commentarius"), and even one *scholium* of Giuntini, one extract from Peuerbach's *Theoricae*, and two new (but not original) diagrams.

---

[36] On Cavellat's career, see (Pantin and Renouard 1986).

[37] This anonymous editor of the Cholinus edition was perhaps Thomas Abel, a physician and professor of mathematics, who wrote a poem "De hoc libro sphaerico emendato," printed between Hero's 1581 dedication and the beginning of the *Sphaera*.

7.2 *Editorial Policies and Clustered Publications*

For printers and booksellers, the publication of the *Sphaera* was as a rule part of a larger editorial program that included printing several mathematical and cosmological textbooks. Thus, as we have seen, Peter Apian began by editing a *Cosmographicus liber* in 1524, even before he settled as a printer in Ingolstadt. In 1529, he published a modified version of the *Cosmographiae introductio* of Mathias Ringmann (1482–1511) and Joannes Waldseemüller (ca. 1470–ca. 1521) (Ringmann and Waldseemüller 1529). He published the *Sphaera* in 1526, then Peuerbach's *Theoricae novae planetarum* in 1528, according to the same editorial principles (Peuerbach 1528). The printers of the university of Wittenberg imitated first Apianus's *Sphaera* (1531), then Apianus's *Theoricae novae* (1535); both editions were copied by Melchior Sessa the elder in Venice, in 1532 (the *Sphaera*) and 1534 (the *Theoricae*).

In 1538, the Wittenberg editors adopted the coupling of the *Sphaera* and Sacrobosco's *De anni ratione*, alias *Computus ecclesiasticus*, which was largely imitated by printers outside Germany, either in one volume or two separate editions. Other works, partly complementary to the *Sphaera*, partly redundant with it, were published in parallel by the same firms, such as Hartmann Beyer's (1516–1577) *Quaestiones novae in sphaeram*, an adaption of the *Sphaera*, first printed in Frankfurt (1549), then reproduced in Wittenberg (1550) and in Paris (1551) (Beyer 1549, 1550, 1551). From around 1550 on, as we have seen, the Wittenberg publishers continued to print regularly the 1543 model of the *Sphaera*, but they minimized the impact of obsolescence by printing, in parallel, the works of the actual professors in mathematics, Kaspar Peucer and Sebastianus Theodoricus (Sebastian Dietrich) (died in 1574) (Theodoricus 1563, 1564).[38]

The Parisian bookseller Guillaume Cavellat, active from 1549 to 1575, had in his catalogue a remarkable set of mathematical and astronomical textbooks (Pantin and Renouard 1986; Pantin 1988). He seems to have been keen to find new titles in this field. He was also responsible for important changes in the Sacrobosco in-octavo tradition, notably the adjunction of Pierio Valeriano's (1477–1558) *Compendium in Sphaeram*. Numerous similar examples—too many to enumerate—show that the publication of the *Sphaera* has to be evaluated in the context of booksellers's editorial policies.

7.3 *Anonymity Versus Signature*

Until the first publications of Elie Vinet's *Scholia* (Paris, 1551, P5), the changes and additions to the text and the illustration were always anonymous, with the noticeable exception of Apianus's founding work. For instance, there were some *scholia* in W2 and W3, and P1, P2, P3, and P4 added new ones, but their authors did not sign their names.

[38] Theodoricus's adaptions of the *Sphaera* were reprinted in Wittenberg in 1567, 1570, 1573, 1578, and 1583.

The case of the Wittenberg editions is interesting. They were published under the authority of Melanchthon, who wrote the preface, but certainly did not supervise the editorial work. For W2 (1538) and W3 (1543), this work may probably—at least partially—be ascribed to Georg Joachim Rheticus (1514–1574), then titular of the chair of lesser mathematics in Wittenberg, before and after his momentous stay with Nicolaus Copernicus (1473–1543) in Frombork.[39] In any case, Melanchthon remained the inspiration for the evolution of the Wittenberg treatises on the *Sphaera*. The addition, in W3, of the small anonymous treatise *De ortu poetico*, which gives technical explanations on heliacal, acronycal, and cosmical risings and settings followed by a number of poetical quotations with commentary, corresponded exactly to Melanchthon's conception of what ought to be the *lectio poetarum* (Ben-Tov 2009). Thus, in Wittenberg, the anonymity of the editors and commentators responded to a conscious policy: all publications displayed the pedagogical views of the university, under the authority of Melanchthon. However, the adaptions later published in Wittenberg by Kaspar Peucer and Sebastianus Theodoricus (Sebastian Dietrich) bore the name of their authors.

Indeed, in the second part of the century there came a change. As a probable result of a more competitive situation, the publishers now obviously recognized that signatures added value to their editions. A signature, often strengthened by a dedication,[40] highlighted the novelty of an edition and could warrant the request for a privilege. In March 1550, Guillaume Cavellat obtained a royal privilege for the printing of the *Sphaera*, which expired in March 1555. In France, as a rule, it was impossible to obtain a second privilege for the same work without demonstrating that the edition had been significantly improved, as the criterion of newness had been, since the origins of the system, essential in the granting of privileges (Armstrong 1990, 92–99).[41] Cavellat successfully applied for a privilege for the *Sphaera Joan. De Sacro Bosco emendata cum scholiis Eliae Vineti.*[42] The heirs of Giacomo Giunta (1487–1546) in Lyons also obtained a royal privilege for the

[39] The 1538 edition (W2) adds an edition of the *De anni ratione*, with a preface by Melanchthon addressed to A.P. Gasser. This preface states that the editorial work for the *De anni ratione* was done by Caspar Borner, but that the idea for the addition was Rheticus's. The intervention of Rheticus in W2 is thus documented. In a letter to Paul Eber (1 March 1562), Rheticus affirmed that he had "prepared the *Sphaera* and *Computus* for the press" in 1550 and complained of having been miserly paid (then, the W3 model was firmly installed in Wittenberg and the supervisor's work was routine). From the summer of 1542 on, Rheticus no longer taught in Wittenberg, but before that date the supervision of the printing of mathematical textbooks had been part of his job (Rosen 1970, 1974).

[40] Vinet wrote a dedication to Johannes Tacitus (P4), and Francesco Giuntini one to Thomas de Guadagne (L1). Even Albertus Hero wrote a dedication to Paulus Kuechovius (C2), though the few *scholia* he added to those of his predecessors were quite small indeed.

[41] A great change occurred in the French book-privilege system in February 1566, when the ordinance of Moulins (art. 78) stipulated that it was henceforth mandatory to publish under royal privilege. However, this still concerned "nouveaux livres." See (Saugrain 1744, 357–58).

[42] This privilege was first used in P6 (1555/1556). It is worth noting that Vinet's *scholia* had first been printed by Cavellat in 1551 (P5) under the first privilege. But P6 was *emendata* and added, for the first time, Pedro Nuñes's *Annotatio in extrema verba capitis de climatibus* translated by Vinet.

treatise of the sphere "corrigé et augmenté par maistre François Junctini professeur en Mathematique."[43]

When no author was available, a bookseller could write the dedication himself and obtain a privilege. That was the case with Jan Waen (active 1545–ca. 1565), who dedicated to Hieronymus Ruffaut (died 1563), abbot of Saint-Vaast in Arras, a reprint of W2 (and A1), but without Melanchthon's preface, under an imperial privilege (A2 & L1, 1547).[44] Thus, the decision for signature or for anonymity responded to editorial policies and commercial strategies. It was for the bookseller to decide which signatures were worth displaying. On the title page of the Cologne 1601 edition (C3), there is only one new name (in addition to those of Sacrobosco, Vinet, Hero,[45] Pierio Valeriano, and Pedro Nunez): that of Clavius. The Cholinus were the printers of the Jesuits in Cologne and used their emblem as their mark. The names of Giuntini and Peuerbach appear inside the volume, but not that of the author of the new "commentarii," who probably also supervised the organization of the new textual patchwork—unless we identify him as Thomas Abel, who discreetly signed his name under a poem at the end of the introductory part of the book.

7.4 The Large Range of Innovations and the Importance of the mise en texte

Innovation in the in-octavo tradition concerned a large range of elements. The additions, substitutions, and corrections in the *scholia* were viewed as fundamental—that is why the names of Elie Vinet, Francesco Giuntini, and even the modest Albertus Hero were displayed on the title-pages. So were the improvements in the illustration (as we shall see below).

The text itself was frequently submitted to revision. The corrections were superficial and only meant to improve the clarity of the style. At first, they were mainly due to the fact that nobody held the original author in particular respect. But, with the Parisian editions, a 'humanist turn' progressively occurred. The title of the 1545 edition printed by Jean Loys (P2) already boasted that the text was "more correct, thanks to the diligent collation of manuscript and printed exemplars" (*ex diligenti manuscriptorum impressorumque codicum collatione castigatior*), and this evolution lead to the production in 1555/1556 of a *Sphaera emendata* (P6) that enjoyed great success: until the end of the century, all the new editions (and their imitations) retained this phrase at the beginning of their titles.

---

[43] Privilege of Ly1.

[44] Imperial privileges could be granted to "republication as well as new publication, although in most cases even republication was claimed to be an improved and enlarged version of the first edition" (Maclean 2012, 140). On imperial privileges and their use in the Netherlands, see (Gompel 2011, 61–64). There, as in France, in the first part of the sixteenth century, book-privileges were mainly meant to protect the commercial interests of the beneficiaries; afterwards, they became mainly instruments of censorship.

[45] Vinet's and Hero's dedications are printed at the beginning of the volume, in addition to Melanchthon's letter, now an anonymous preface.

Indeed, the titles played a role in the innovation process: there never was a new model of edition without a new title—in turn faithfully copied in the imitations of this model. The new titles always advertised the innovations: the *scholia* and the improvement in the text, as we have seen,[46] but also the changes in the illustration[47] and other additions: "addita est praefatio...Philippi Mel. ad Simonem Grynaeum" (W1), later "Cum praefatione..." (W2) and "Praemissa praefatione..." (P1); "with a small treatise of the same author entitled the Comput" ("Ejusdem autoris libellus, cujus titulus est Computus...," W2); "and other materials printed to please the students" ("et alia quaedam in studiosorum gratiam edita," W3). On V2's title-page appear the words: "with some additions that help to explain what is told in the *Sphaera*" ("Addita sunt quaedam ad explanationem eorum quae in Sphaera dicuntur facentia"), which means that the preface of Melanchthon has been replaced by a short treatise of elementary geometry. The addition, in P4, of Valeriano's *Compendium in Sphaeram* is also advertised, as is that of Pedro Nunez's *Annotatio* in P6. Even the new *marginalia* and the index introduced in the 1564 Giuntini edition (Ly1) are signaled: "Cum annotationibus in margine, et indice rerum et locorum memorabilium."

This highlighting of the index and the *marginalia* shows that what the French historians of the book analyze as the elements of the "mise en texte" and "mise en livre" (Martin and Vezin 1990; Martin 2000) was viewed as an integral part of the innovation process. For instance, the Apianus and the Wittenberg editions had no running titles, and their pages were unnumbered. Guillaume Cavellat, in 1550 (P4), was the first to number the folios and add running titles.

As we have seen, the editors of the in-octavo editions, from the beginning, improved the clarity of the divisions of the text, using more subtitles, interlinear blanks, and other typographical devices. In this respect, W1 had already made progress in comparison to I1. Then W2 changed the wording of the subtitles (some of them became short summaries) and introduced a hierarchy between them to mark more clearly the logical structure of the text. For instance, in the last part of Chap. 2 (Table 9.4).

Table 9.4 The subtitles of the last part of chap. I in W1 and W2

W1	W2
QUOD TERRA SIT ROTUNDA QUOD AQUA SIT ROTUNDA	DE TERRA I Terram cum aqua globum constituere
QUOD TERRA SIT CENTRUM MUNDI DE IMMOBILITATE TERRAE	II Terram esse centrum mundi, hoc est, in medio universi sitam et velut punctum respectu firmamenti esse, immobilemque consistere.
DE QUANTITATE ABSOLUTA TERRAE	III Ambitum terræ, et ex eo Diametrum invenire.

[46]The anonymous *scholia* of P3 were mentioned in the title too: "with notes that throw much light on somewhat obscure passages" ("cum annotationibus, quae locis aliquot obscuris magnam lucem afferunt").

[47]W2: "with new figures that explain the risings [of the zodiacal signs]" ("et novis quibusdam typis, qui ortus indicant"), echoed in P1 by "with the addition and illustration of numerous new figures" ("Plurimis novis typis auctus et illustratus").

From 1545 (P2) on, the Parisian editions adopted an important innovation: the four chapters became four books, each divided into numbered chapters. However, this tendency to modernize the *Sphaera* was checked by the humanist turn operated by Elie Vinet. In 1551 (P5), the *Sphaera* returned to the original division into four chapters.

7.5 The Illustration: Scientific and Commercial Issues

From 1526 to 1538, the principal innovations concerned the illustration (Tables 9.5 and 9.6). Apianus and the anonymous editor or editors of Wittenberg did all the work, and the result was deemed so satisfactory that their successors simply copied their set of diagrams. Only Franco Burgersdijk, in 1626, was to undertake a complete reworking of the illustration (Chap. 11).

The creation of the complete set of diagrams that illustrated the in-octavo *Sphaeras* from 1538 to 1620 was progressive. Its authors borrowed from their predecessors (the Venetian incunabula, the Leipzig quartos, and above all the Apianus edition), but also from books outside this strict tradition. Thus, the editor of W2 (1538) borrowed fifteen diagrams from Oronce Finé's *Protomathesis*, published in Paris in 1532. As a result, some figures of a type that had previously been excluded from the tradition were introduced into it: four true geometrical diagrams meant to explain the ascensions of the signs (in Chapter III of *Tractatus de Sphaera* by Sacrobosco). However, as we have seen, the more striking innovation of W2 was the arrival of four volvelles, of which the different parts were printed on folded sheets bound at the end of the volume—the reader had simply to cut the parts and assemble them.

If we set aside a few additions that were not retained for long in the tradition, the iconography of the in-octavo editions was thus conceived between 1526 and 1538, with a strong dynamic of innovation. It was the main contribution of the German mathematicians to this tradition, and their work in that field was so well received by the public that a certain commercial logic cemented this contribution's place in future publication. The typical set of diagrams and volvelles of the in-octavo tradition was such a success that to modify it was out of the question. It could even be used as an identifying trademark.

The main reason why iconographical innovation ceased at an early stage in the in-octavo tradition was thus probably commercial. Such immobility could have represented a detrimental shortcoming, as it was indicative of a lack of serious mathematical work on the text: the invention of new diagrams was bound to happen as soon as mathematicians worked on a text and made commentary. However, this did not happen in this case, thanks to the particular character of Sacrobosco's short introduction to cosmology, whose large and enduring success was due to its completeness, the clarity of its style, but also its avoidance of all mathematical complexities. The *Sphaera* was more descriptive—even narrative—than demonstrative. As Thorndike has observed, it achieved a happy compromise between the literary tradition, derived notably from Macrobius's *Commentarii in somnium Scipionis*, and a then (in 1220) recent and more scientific approach to cosmology, permitted by the first translations of Arabic astronomers. Rival manuals composed at around the

Table 9.5 The diagrams from the Venetian incunabula to the Apianus edition[a]

Venetian incunabula[b] V	Leipzig 1486 Le1	Leipzig 1494– & Faber von Budweis (1495–) Le2	Apianus 1526 I1
Preliminaries			
Armillary sphere: 1	1 different	1 different	1 different (title)
Geometry: 24 diagrams	0	2 (copied from V?)	0
Chap. 1	**Chap. 1**	**Chap. 1**	**Chap. 1**
Diffinitiones 0	1 new	0	3 new
4 elementary and 9 celestial spheres: 1	1 different (1 elementary and 9 celestial spheres)	1, between V and Le1 (1 elementary, 9 celestial spheres)	1 new (terraqueous globe, air, fire, and 10 celestial spheres)
Sphaera recta/obliqua: 1	1 new	= Le1, but simplified	2 new
Elemental spheres: 1	1 new	1 different from Le1	1 new
De caeli revolutione: 1	1 similar to V	2 new	1 new
De caeli rotunditate: 1	3 new	3 similar to Le1	4 new
Effects of refraction: 2	0	0	2 (linked to V)
Quod terra sit rotunda: 2 (Fig. 9.1)	2 new	2: 1 similar to V; 1 similar to Le1	2 new (linked to V) (Fig. 9.2)
Quod aqua sit rotunda: 1	1 new	1 similar to V	1 new
Quod terra sit centrum mundi: 0	1 new	1 similar to Le1 (clearer)	1 new
De immobilitate terrae: 0	0	0	0
De quantitate terrae: 0.	0	0	0
Chap. 2	**Chap. 2**	**Chap. 2**	**Chap. 2**
0	Celestial circles	1 new	0
The equinoctial: 1	0	0	1 new (linked to V)
De zodiaco circulo: 1	0	0	1 new (linked to V)
Zodiacal signs: 2	4 (2 similar, 2 new)	4 similar to Le1	4 new (linked to Le1/2)
De duobus coluris:1	0	0	1 new
Meridian and horizon: 1	2 new	2 similar to Le1	1 new
Celest. zones & circles: 1	0	0	0
Terrestrial zones (with landscape): 1	1 new	1 new (with a map)	2 new (1 with landscape and map)
Chap. 3	**Chap. 3**	**Chap. 3**	**Chap. 3**
Poetical risings and settings: 1	2 new	2 similar to Le1	1 new (unclear)
The circles to measure astronomical risings: 1	2 new	2 similar to Le1	2 new
Natural day: 1	0	0	0

(continued)

Table 9.5 (continued)

Venetian incunabula[b] V	Leipzig 1486 Le1	Leipzig 1494– & Faber von Budweis (1495–) Le2	Apianus 1526 I1
Circles of natural days: 1	1 new	1 much clearer	1 (= L2 reduced)
Circles of the sphere: 1	0	0	0
12 celestial houses: 3	0	0	0
Shadows at the equator: 1	7 new (*diversitas dierum*)	7 similar to Le1 but clearer (Fig. 9.3)	8 new (*diversitas dierum*) (Fig. 9.4)
Terrestrial zones and climes: 2	1 new	1 new, with a map	2 new (only climes, 1 linked to V) (Figs. 9.5 and 9.6)
Chap. 4	**Chap. 4**	**Chap. 4**	**Chap. 4**
Circles of the Sun: 1	1 new	1 similar to Le1	2 new
Moon's circles, *caput & cauda draconis*: 1	1 new	1 similar to Le1	2 new
Circles of other planets 0	0	0	1 new
Station, retrogradation: 1	1 new	1 similar to Le1	1 new
The lunar eclipse (1)	2 new	2 similar to Le1	2 new (+1: phases)
The solar eclipse (1)	2 new	2 similar to Le1 (with variants)	1 new

[a]In this table all diagrams indicating an imitation or an influence are underlined. They are described with "similar to" if the imitation is obvious or with "linked to" if the imitation is less complete
[b]For more details, see (Table 9.1)

same time, such as those of Robert Grosseteste and John Peckham, were far less successful, although they were more up-to-date, probably because they were too dryly technical and mathematical and less elegantly written, and because they suppressed the quotations of classical poets (Thorndike 1949, 21). Indeed, technicity was not banned altogether, as it subsisted in largely diffused commentaries very soon to be associated with Sacrobosco's *Sphaera*, like the one attributed to Michael Scot that was composed in the first half of the thirteenth century.

Thus, from the beginning up to the Renaissance, the successful formula was the association of a short, clear, and elegant treatise with commentaries that could afford further information. In the in-octavo tradition, as soon as the *Sphaera* was provided with a set of clear, precise, pedagogical, and (if possible) self-explanatory diagrams, the effort turned elsewhere. The editors made improvements of a different kind. They wrote pedagogical or erudite notes—notably to rectify some blunders owing to Sacrobosco's lack of humanist training, and to explain the meaning of a few Greek words.[48] And they also resorted to additions: the addition of a few modest

[48]The Paris editors were particularly keen on this point. For instance, the Wittenberg 1538 editor (W2) already tells, in a sober marginal note (D4v), that in Sacrobosco's text "chronicus ortus" means "achronichos," referring to Proclus's commentary on Hesiod, whereas the 1545 Paris editor

Table 9.6 From the Apianus edition to the Wittenberg editions[a]

Topics	I1: Ingolstadt 1526 Cf Table 9.5.	W1: Wittenberg 1531–1536 Crudely copied in V1 (Venice, 1532–1541, 1553), and, with some additions, in V2 (1548–1601)	W2: Wittenberg, 1538–1540 P1–P2: Paris 1542/1543, 1545[b] A1–L1: Antwerp and Louvain, 1547; Antwerp 1551, 1559, 1561b With one more diagram: W3: Wittenberg 1543 Copied with a few eventual additions in: P3–P7: Paris 1549–seventeenth cent. V3: Venice 1562b–1586 Ly1–2: Lyons, 1564a, 1578b A2–A4: Antwerp 1566b–1582 C1–C3: Cologne 1566a–1601c
Preliminaries			
Armillary sphere	1new (title)	1 copied from I1 (title)	Id.
Geometrical diagrams	0	0	Id.
Chap. 1	**Chap. 1**	**Chap. 1**	**Chap. 1**
Diffinitiones	3 new	2 inspired by I1	Id. + 2 new
Elementary and celestial spheres	1 new: Elements, 10 celestial spheres	1 copied from I1	Id.
Sphaera recta / obliqua	2 new	2 copied from I1	Id.
Elemental spheres	1 new	0	0
De caeli revolutione	1 new	0	1 new +2 copied from Finé
De caeli rotunditate	4 new	4 copied from I1	Id. + 1 inspired by V + 1 inspired by Le2
Effects of refraction	2	0	2 copied from I1 ± 1 copied from Finé
Quod terra sit rotunda	2 (Fig. 9.2)	2 copied from I1	2 new **(= volvelles)**(Figs. 9.7 and 9.8)
Quod aqua sit rotunda	1 new	1 copied from I1	Id.
Quod terra sit centrum mundi	1 new	1 copied from I1	Id. ± 1 copied from Finé

(continued)

Table 9.6 (continued)

De immobilitate terrae	0	0	0
De quantitate terrae	0	0	2 new
Chap. 2	**Chap. 2**	**Chap. 2**	**Chap. 2**
Armillary sphere	0	1 (= title)	Id.
The equinoctial	1 new	0	1 copied from Finé
De zodiaco circulo	1 new	0	1 new = **volvelle** + 1 copied from Finé
Zodiacal signs	4	4 copied from I1	Id. + 1 copied from Finé
De duobus coluris	1 new	1 copied from I1	Id.
De meridiano et horizonte	1 new	1 copied from I1	Id.
Terrestrial zones (with landscape)	2 new (1 with landscape and map)	1 new	Id. + 1 copied from Finé: terrestrial globe and celestial circles
Chap. 3	**Chap. 3**	**Chap. 3**	**Chap. 3**
Poetical risings	1 new (unclear)	1 copied from I1	**1 new = volvelle**
Circles to measure astronomical risings	2 new	2 copied from I1	Id.
De ascensionibus	0	0	4 copied from Finé
Dies naturalis	0	0	1 copied from Finé
Circles of natural days	1	1 copied from I1	Id.
De diversitate dierum...in diversis locis terrae	8 new (Fig. 9.4)	8 copied from I1	Id.
Terrestrial zones and climes	2 new (climes) (Fig. 9.5)	1 copied from I1	1 (with map) copied from Finé. (Fig. 9.6)
Chap. 4	**Chap. 4**	**Chap. 4**	**Chap. 4**
Circles of the Moon, nodes of the dragon	4 new	4 copied from I1	Id.
Circles of other planets	1 new	1 copied from I1	Id.
De statione, directione...:	1	1 copied from I1	Id.
The lunar eclipse	3 new	3 copied from I1	Id.
The solar eclipse	1 new	1 copied from I1	Id.

[a]In this table as in (Table 9.5) the diagrams copied from other editions (or inspired by them) are underlined. The new diagrams in W2 are in bold

[b]P2 (Paris, 1545b) adds two crude and completely redundant diagrams. The first one (in a note on Euclid's definition of the sphere) remains in the Paris editions until P5 (1551), the second one (on the solar eclipse) until P6 (1555/1556)

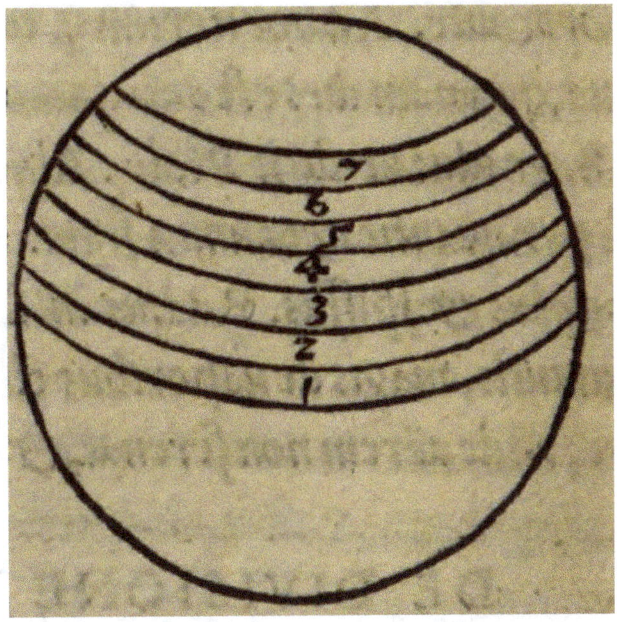

Fig. 9.5 *De divisione climatum*. From (Sacrobosco 1526, D2v). State Library Regensburg--999/Philos.1325, urn:nbn:de:bvb:12-bsb11110162-1

tables inside the commentary,[49] and above all the addition of other texts to complement the *Sphaera*.[50]

8 The Limits of the Tradition: The Case of Francesco Giuntini

This cumulative process had its limits. The control of the publishers over the evolution of this tradition fostered innovation to a certain degree; but it could also hinder it. The iconography, as we have seen, consisted of a fixed set of diagrams; the additions were welcome, but only if they could fit into the handy format of the in-octavo textbook; the original contributions of skilled commentators were sought-after, provided that they let the model remain recognizable. The ideal new edition in the in-octavo tradition was an improved item that continued on the lines of previous

(P2) expands on this topic in a long note ("Corrupta scriptura…" D5r). Then Elie Vinet, in 1551 (P5), writes a new version of the note with more erudite details (34r). The 1550 Paris editor (P4) adds some notes on difficult words, like "Archetypus" (B7r) and "Diaphanum" (B8r).

[49] In W3, the table of the longest natural days for all degrees of latitude between the arctic circle and the pole appeared (Sacrobosco. 1543b, F1v–F2r), and in P2 a table of the climes (Sacrobosco. 1545b, F3v). The first massive addition of tables occurred thanks to Giuntini, but never to appear again in further editions (see *infra*).

[50] For a general view of additions, see (Table 9.3).

titudine ſecundi, & ſic deinceps. Longitudo autem climatis, poteſt appellari linea ducta ab oriente in oc=

cidentem, æquidiſtans æquinoctiali. Vnde longitudo primi climatis eſt maior longitudine ſecundi, & ſic deinceps, quod contingit propter anguſtiam ſphæræ.

Capitulum

Fig. 9.6 *De divisione climatum*. From (Sacrobosco 1538, F4r). Bavarian State Library Munich—Astr.u. 154, urn:nbn:de:bvb:12-bsb00020992-1

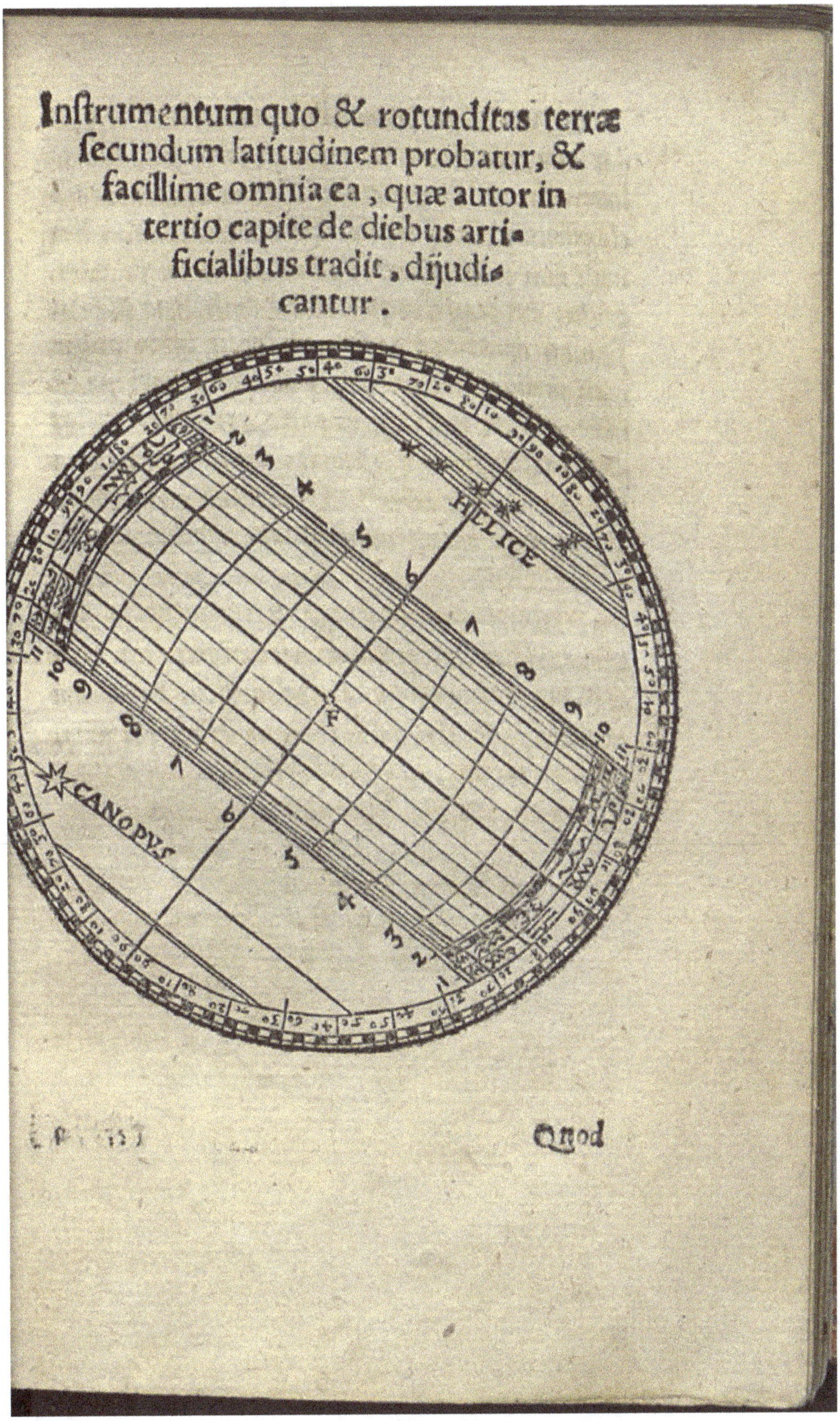

Inſtrumentum quo & rotunditas terræ
ſecundum latitudinem probatur, &
facillime omnia ea, quæ autor in
tertio capite de diebus arti-
ficialibus tradit, dijudi-
cantur.

Quod

Fig. 9.7 Unmounted volvelle. From (Sacrobosco 1538, B8r). Bavarian State Library Munich—Astr.u. 154, urn:nbn:de:bvb:12-bsb00020992-1

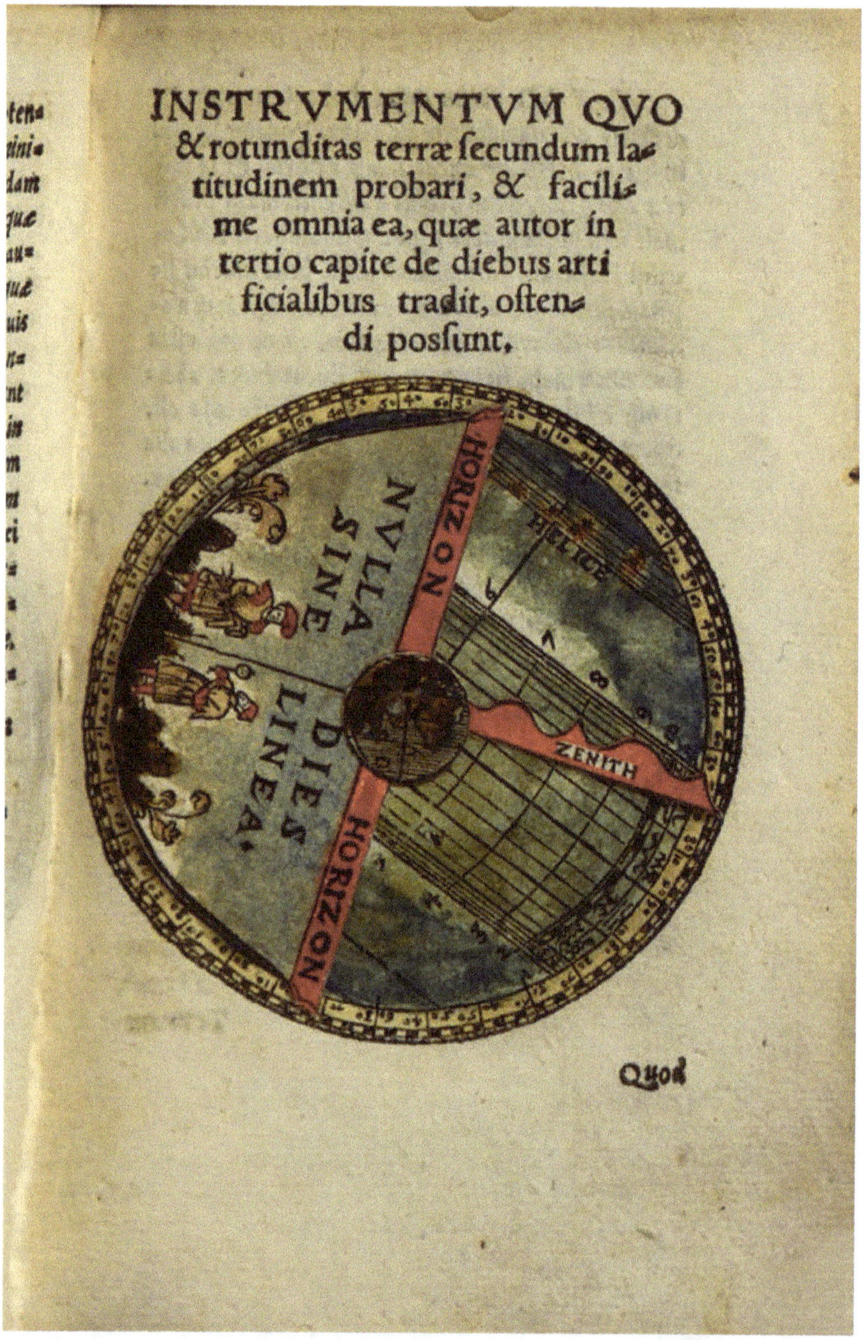

Fig. 9.8 Mounted volvelle. From (Sacrobosco 1543b, B8r). Staats- und Stadtbibliothek Augsburg—Math 745, urn:nbn:de:bvb:12-bsb11267683-6

editions. The case of Francesco Giuntini shows that more ambitious projects were bound to leave this framework.

Francesco Giuntini, a Florentine by birth, had been a Carmelite priest and a doctor of theology. His assiduous practice of astrology and, above all, his Protestant sympathies led him into trouble. In 1561, he went to Lyons as a religious exile, publicly renounced his heresy, and entered a new career as a mathematician and astrologer with the support of the Italian colony in Lyons. After a time, he won the patronage of royal officers and even began to aim higher.[51] His first commentary on Sacrobosco was published in 1564, when he was still at the beginning of this successful second career (Sacrobosco 1564a). It fit into the framework of the in-octavo tradition, which likely appeared to him a good medium to widen his fame outside the Lyons circles. The presence of Melanchthon's preface in praise of astrology (which had to be printed without the name of its author) probably appealed to him as well (Sacrobosco 1564a, 3r–7v).

In the dedication of the work to Thomas de Gadagne (ca. 1539–1594), lord of Bellegarde, Giuntini states that Filippo Tinghi (died 1580), a Florentine printer and bookseller settled in Lyons, has asked him to emendate the *Sphaera* and add some "very brief notes" ("brevissimis notis") to explain the difficult passages, a demand which he has eagerly answered out of his zeal for promoting such a universally useful science as astronomy (Sacrobosco 1564a, 2r–v). Indeed, the "very brief notes" were to consist of innumerable *marginalia*, abundant *scholia* (even lengthier than Vinet's, and printed in bigger type), and long technical appendices, which concerned the method of determining the polar altitude at any location in the northern hemisphere,[52] the method of determining longitudes,[53] the method of determining the duration of natural days at different latitudes,[54] and instructions for the calendar (Sacrobosco 1564a, 143–46). The Giuntini edition thus brought into the in-octavo tradition a noticeable amount of material borrowed from other types of treatises on the *Sphaera*—adaptions or huge commentaries in which the original text was buried. Whereas the standard in-octavo model included only two modest tables, as we have seen, Giuntini added a series of tables (none of them original), which transformed the original textbook into an introduction to astronomical practice and calculation.[55]

[51] On Giuntini's career, see (Ernst 2001; Pantin 2013).

[52] (Sacrobosco 1564a, 117–27): "Quomodo altitudo poli Aquilonaris et aequatoris sit investiganda."

[53] (Sacrobosco 1564a, 128–29): "Quomodo Astrologi invenerunt locorum longitudines."

[54] (Sacrobosco 1564a, 136–42): "De quantitate diei et noctis, ortu et occasu Solis."

[55] The tendency to transform the original textbook into an introduction to astronomical practice and calculation, already present in Lefevre's and Finé's work on the *Sphaera* (Chaps. 2 and 8), had been reinforced in many adaptions and 'hypercommentaries' of Sacrobosco, notably those of (Schreckenfuchs 1569) and (Clavius 1570).

A table of the measures of the earth (perimeter, diameter etc.) in different units of measurement (leagues, miles, etc.).[56]
A table to compare the measures of the obliquity of the ecliptic by Ptolemy, Albategnius, etc. until Johannes Werner. (Sacrobosco 1564a, 41–42)
A table of the cosmic and chronic risings and settings of the twelve zodiacal signs. (59)
Two tables of the astronomical risings and settings of the zodiacal signs under the right sphere, measured in arcs of the equator, and in hours and minutes. (66–67)
A table of the astronomical risings and settings of the zodiacal signs under the oblique sphere, measured in arcs of the equator and calculated for each of the eight climes. (70–71)
For the determination of latitudes (first appendix), a table of the true position (*verus locus*) of the sun at midday for each day of the year according to the Prutenic Tables[57] (119–20), a table of the equation of the sun[58] (123), and a table of the declination of the sun. (125)
A table of the longitudes and latitudes of the principal towns of the world (130–36)
For the determination of the length of natural days, a table of the semi-diurnal arcs and of the latitudes of the sun at midday under different latitudes. (139–42)
Two calendar tables. (243, 246)

Giuntini was not satisfied with this first work on the *Sphaera* and he planned more ambitious publications, both to attract the attention of more important patrons and to advance the project he took most to heart: asserting the complete legitimacy of astrology as an integral part of the science of the stars, and as the science of Providence, perfectly compatible with the Catholic faith, even under the new Tridentine rules.[59] In 1573, he dedicated to Catherine de'Medici (1519–1589), the queen mother, the first version of his *Speculum astrologiae* (Lyons: Filippo Tinghi), which contained a series of astrological treatises and new astronomical tables (*Tabulae resolutae*). Then he published a two-volume commentary on the *Sphaera* in 1577–1578, which rivaled that of Clavius, though it was far less methodical and much more digressive (Giuntini 1577–1578).[60] Its most exceptional feature was the space it devoted to astrological and theological questions. The content of the former commentary was immersed into this new one, which still retained the typical diagrams conceived in Wittenberg (mixed with new ones), and even the in-octavo format, but, all the same, no longer belonged to the tradition founded by Apianus. The

[56] (Sacrobosco 1564a, 30): "Tabula quantitatis terrae secundum Ptolemaeum et experientiam."

[57] On Giuntini's reliance on Erasmus Reinhold's "Copernican tables," though he had otherwise no Copernican sympathies, see (Omodeo 2014, 136–39; Proverbio 1997).

[58] The equation of the sun (*aequatio Solis*) is the arc of the ecliptic between the true position (*verus locus*) of the sun, determined by a line issued from the center of the world and passing through the center of the body of the planet, and its mean position (*medius motus*) determined by a line issued from the center of the world and parallel to a line issued from the center of the eccentric deferent of the Sun and passing through the center of the body of the planet. All these notions are part of the knowledge taught in the *Theoricae planetarum*.

[59] Although the Council of Trent had hardened the control of the practice of astrology, Giuntini felt confident that he could promote this practice by using the same arguments as those developed by Melanchthon. On Giuntini's astrology, see (Thorndike 1941, 129–33; Pantin 2013).

[60] One volume of Giuntini's work is dated 1577, the other one 1578, but both were issued together. This commentary was reprinted in 1583 as part as the second in-folio volume of the much enlarged new version of the *Speculum astrologiae* (Lyons: Filippo Tinghi).

same year (1578), Filippo Tinghi published a new edition of the *Sphaera emendata* (Ly2). The *scholia* added in 1564 had been removed from it, replaced by a short treatise of elementary geometry. Giuntini's commentary and the in-octavo tradition had parted company for good.

9 Conclusion

In this paper, I focused on a kind of collective authorship, which was responsible for the development of a long-running tradition that developed from 1526 until the seventeenth century. I chose the privileged example of the in-octavo tradition to provide an interesting point of view on the means, the ways, and even the rhythm of innovation in the Sacrobosco field. This tradition was a successful attempt at standardization, producing a kind of manual that was able to retain, for over half a century, an essential core of original features even as it continued to evolve. This success proves the efficiency of a commercial model. The never-relenting interest in this particular model of *Sphaera* among a succession of booksellers in several European countries was certainly a motor for innovation: it would not have been possible to perpetually reprint the same book without drying up the market. It was necessary to innovate, at least to a certain extent, to gain enough new customers. But on the other hand, the constraints of the tradition, the obligation to retain the model, and the weight of trade imperatives imposed a limit on innovation, which, at the end, made decline inescapable. Between Giuntini's intervention and Burgersdijk's late attempt at reviving the tradition in 1626 (Chap. 11), no significant innovation occurred: the Wittenberg 1538 (W2) and the Paris 1555/1556 (P6) models were still printed, either in a repetitive manner (in Paris and Venice), or with different attempts at refreshment (in Antwerp and Cologne). In the second part of the sixteenth century, the concurrence of numerous new types of tracts on the *Sphaera* had marginalized the type studied in this paper.

In spite of the importance of the German mathematicians who conceived a remarkable set of cosmological pedagogical diagrams (following Melanchthon's educational plan), and in spite of the patient work of annotation achieved by a succession of professors (notably Elie Vinet), the actors who did most to support first the dynamic of the tradition, then its longevity, were probably the printers and booksellers: they did not write commentaries, but in many cases[61] they acted as intermediaries between the authors and the public, and they managed to keep control over the process. Innovation, as they conceived it, was a braiding process: it was often achieved through the artful combination of diverse borrowed material. In this way and in this specific instance, the printers and booksellers were largely responsible for blurring the distinction between borrowers and innovators.

[61] Not in all cases printers acted as intermediaries: as we have seen, in Wittenberg, the printers followed the instructions of the university.

References

Editions of Sacrobosco's Sphere Anterior to the In-Octavo Tradition

Sacrobosco. ca. 1472a. *Tractatum de spera*. Venice: Florentius de Argentina. http://hdl.handle.net/21.11103/sphaera.100685.

———. 1472b. *Spaera mondi*, ed. Pietro Buono Avogaro. Ferrara: Andreas Belfortis. http://hdl.handle.net/21.11103/sphaera.101122.

———. 1478a. *Spera mundi*. Venice: Franciscus Renner from Heilbronn. http://hdl.handle.net/21.11103/sphaera.100686.

———. 1478b. *Spera*. Venice: Adam Burkardt from Rottweil. http://hdl.handle.net/21.11103/sphaera.100687.

———. 1482. *Sphericum opusculum*. Venice: Erhard Ratdolt. http://hdl.handle.net/21.11103/sphaera.100692.

———. 1485. *Sphericum opusculum*. Venice: Erhard Ratdolt. http://hdl.handle.net/21.11103/sphaera.101123.

———. 1486. *Spericum opusculum*. Leipzig: Drucker des Capotius.

———. ca. 1487. *Spericum opusculum*. Leipzig: Martin Landsberg. http://hdl.handle.net/21.11103/sphaera.101126.

———. 1488. *Spaerae mundi compendium*. Venice: Johannes Santritter. http://hdl.handle.net/21.11103/sphaera.100822.

———. 1489. *Spericum opusculum*. Leipzig: Conrad Kachelofen. http://hdl.handle.net/21.11103/sphaera.100266.

———. 1490. *Sphaera mundi*. Venice: Bonetus Locatellus for Octavianus Scotus. http://hdl.handle.net/21.11103/sphaera.101124.

———. 1494. *Opusculum spericum cum figuris optimis et novis*. Leipzig: Martin Landsberg. http://hdl.handle.net/21.11103/sphaera.100271.

———. ca. 1495a. *Opusculum spericum cum notabili commento*, comm. Wenceslas Faber von Budweis. Leipzig: Martin Landsberg. http://hdl.handle.net/21.11103/sphaera.100886.

———. 1495b n.st. *Textus de Sphera*, comm. Jacques Lefèvre d'Etaples. Paris: Wolfgang Hopyl. http://hdl.handle.net/21.11103/sphaera.101126.

———. 1498. *Uberrimum sphere mundi comentum*, comm. Pierre d'Ailly and Pedro Sanchez Ciruelo. Paris: Jean Petit. http://hdl.handle.net/21.11103/sphaera.100038.

———. 1499a. *Opusculum spericum cum notabili commento*. Leipzig: Wolfgang Stöckel. http://hdl.handle.net/21.11103/sphaera.100888.

———. 1499b. *Sphaera mundi*, comm. Cecco d'Ascoli, Franciscus Capuanus de Manfredonia, Jacques Lefèvre d'Etaples, with Peuerbach, *Theoricae novae planetarum*, comm. Franciscus Capuanus. Venice: Simone Bevilaqua. http://hdl.handle.net/21.11103/sphaera.100265.

———. 1500. *Opus sphericum...figuris et notabili commento illustratum*, comm. Wenceslas Faber von Budweis. Cologne: Heinrich Quentell. http://hdl.handle.net/21.11103/sphaera.100891.

———. 1501a. *Opus sphericum...figuris et perutili commento illustratum*, comm. Wenceslas Faber von Budweis. Cologne: Heinrich Quentell. http://hdl.handle.net/21.11103/sphaera.100023.

———. 1501b. *Sphaera mundi*. Venice: Melchior Sessa. http://hdl.handle.net/21.11103/sphaera.100892.

———. 1503a. *Opus sphericum...figuris verissime exculptis et interpretatione familiari...illustratum*, comm. Wenceslas Faber von Budweis. Cologne: Heinrich Quentell. http://hdl.handle.net/21.11103/sphaera.100024.

———. 1503b. *Textus Spere materialis...cum lectura Magistri Conradi Norici*, comm. Conrad Tockler. Leipzig: Martin Landsberg. http://hdl.handle.net/21.11103/sphaera.100931.

———. 1505. *Opus sphericum...figuris verissime exculptis et interpretatione familiari...illustratum*, comm. Wenceslas Faber von Budweis. Cologne: Heinrich Quentell. http://hdl.handle.net/21.11103/sphaera.100028.

———. 1508. *Opus sphericum...figuris verissime exculptis et interpretatione familiari...illustratum*, comm. Wenceslas Faber von Budweis. Cologne: Heirs of Heinrich Quentell. http://hdl.handle.net/21.11103/sphaera.100183.
———. 1509. *Textus Spere materialis...cum lectura Magistri Conradi Norici*, comm. Conrad Tockler. Leipzig: Martin Landsberg. http://hdl.handle.net/21.11103/sphaera.101042.
———. 1513. *Sphaera mundi*. Venice: Melchior Sessa. http://hdl.handle.net/21.11103/sphaera.101023.
———. 1516. *Mundialis sphere opusculum*. Paris: Vincent Quignon for Regnault Chaudière. http://hdl.handle.net/21.11103/sphaera.100991.
———. 1518. *Opusculum de sphaera*. Wien: Johann Singriener for Lucas Alantsee. http://hdl.handle.net/21.11103/sphaera.100883.
———. 1519. *Sphaera mundi*. Venice: Jacobus Pentius for Melchior Sessa. http://hdl.handle.net/21.11103/sphaera.100992.

Editions in the In-Octavo Tradition (1526–1601)

———. 1526 **[I1]**. *Sphaera...per Petrum Apianum accuratissima diligentia denuo recognita ac emendata*. Ingolstadt: Petrus Apianus. http://hdl.handle.net/21.11103/sphaera.100070.
———. 1531 **[W1]**. *Liber de sphaera. Addita est praefatio in eundem librum Phlippi Melanchthonis ad Simonem Gryneum*. Wittenberg: Joseph Klug. http://hdl.handle.net/21.11103/sphaera.100138.
———. 1532 **[V1]**. *Liber de sphaera*. Venise: G.A. Nicolini da Sabio for Melchior Sessa. http://hdl.handle.net/21.11103/sphaera.100118.
———. 1534a. *Liber de sphera*. Wittenberg: Joseph Klug. http://hdl.handle.net/21.11103/sphaera.100085.
———. 1534b. *Liber de sphaera*. Venice: G.A. Nicolini da Sabbio for Melchior Sessa. http://hdl.handle.net/21.11103/sphaera.100179.
———. 1536. *Liber de sphaera*. Wittenberg: Joseph Klug. http://hdl.handle.net/21.11103/sphaera.101008.
———. 1537a. *Liber de sphaera*. Venice: Maffeo Pasini and Francesco Bindoni. http://hdl.handle.net/21.11103/sphaera.101059.
———. 1537b. *Liber de sphaera*. Venice: Giovanni Antonio Nicolini da Sabbio for Melchior Sessa. http://hdl.handle.net/21.11103/sphaera.101060.
———. 1538 **[W2]**. *Libellus de sphaera*. Wittenberg: Joseph Klug. http://hdl.handle.net/21.11103/sphaera.101106.
———. 1540. *Libellus de sphaera*. Wittenberg: Joseph Klug. http://hdl.handle.net/21.11103/sphaera.101025.
———. 1541a. *Liber...de sphaera. Addita est praefation in eundem librum Philippi Mel. Ad Simonem Grineum*. Venice: Francesco Bindoni and Maffeo Pasini. http://hdl.handle.net/21.11103/sphaera.100140.
———. 1541b. *Liber...de sphaera. Addita est praefatio in eundem librum Philippi Mel. Ad Simonem Grineum*. Venice: Giovanni Antonio Nicolini da Sabbio for Melchior Sessa. http://hdl.handle.net/21.11103/sphaera.100144.
———. 1542/1543 **[P1]**. *De sphaera liber. Plurimis novis typis auctus*. Paris: Jean Loys. http://hdl.handle.net/21.11103/sphaera.101027.
———. 1543a **[A1]**. *Libellus de sphaera*. Antwerp: Jean Richard. http://hdl.handle.net/21.11103/sphaera.100180.
———. 1543b **[W3]**. *Libellus de sphaera*. Wittenberg: Peter Seitz. http://hdl.handle.net/21.11103/sphaera.101029.
———. 1545a. *Libellus de sphaera*. Wittenberg: Veit Kreutzer. http://hdl.handle.net/21.11103/sphaera.100818.
———. 1545b **[P2]**. *Sphaera typis auctior, quam antehac, atque ex diligenti manu scriptorum impressorumque codicum collatione castigatior*. Paris: Jean Loys and Guillaume Richard. http://hdl.handle.net/21.11103/sphaera.101054.

———. 1545c. *Liber de sphaera*. Venice: Giovanni Antonio Nicolini da Sabbio for Melchior Sessa. http://hdl.handle.net/21.11103/sphaera.100145.

———. 1546. *La Sphere…traduicte de latin en langue francoyse*, transl. Martin de Perer. Paris: Jean Loys. http://hdl.handle.net/21.11103/sphaera.101030.

———. 1547 **[A2 & L1]**. *Sphaera*. Antwerp and Louvain: Martinus Nutius for Jean Richard and Jan Waen. http://hdl.handle.net/21.11103/sphaera.100917 [A2]. http://hdl.handle.net/21.11103/sphaera.100125 [L1].

———. 1548 **[V2]**. *Liber de sphaera. Addita sunt quaedam ad explanationem eorum, quae in sphaera dicuntur facientia*. Venice: Pietro Nicolini da Sabbio for Melchior Sessa. http://hdl.handle.net/21.11103/sphaera.101031.

———. 1549a. *Libellus de sphaera*. Wittenberg: Veit Kreutzer. http://hdl.handle.net/21.11103/sphaera.100136.

———. 1549b **[P3]**. *Sphaera typis auctior, quam antehac…atque ex diligenti manu…castigatior*. Paris: Thomas Richard and Guillaume Cavellat. http://hdl.handle.net/21.11103/sphaera.100137.

———. 1550a. *Sphaera. Addita sunt quaedam…*Venice: Pietro Nicolini da Sabbio for Melchior Sessa. http://hdl.handle.net/21.11103/sphaera.100163.

———. 1550b. *Libellus de sphaera*. Wittenberg: Johann Krafft. http://hdl.handle.net/21.11103/sphaera.100157.

———. 1550c **[P4]**. *Sphaera typis auctior, quam antehac…cum annotationibus*. Paris: Guillaume Cavellat. http://hdl.handle.net/21.11103/sphaera.100952.

———. 1551a. *Sphaera*. Antwerp: Jean Richard. http://hdl.handle.net/21.11103/sphaera.100161.

———. 1551b. *Sphaera typis auctior, quam antehac…cum annotationibus, et scholiis… Eliae Vineti*, comm. Elie Vinet. Paris: Guillaume Cavellat. http://hdl.handle.net/21.11103/sphaera.100181.

———. 1552. *Sphaera. Addita sunt quaedam…*Venice: Cornelio Nicolini da Sabbio for Melchior Sessa. http://hdl.handle.net/21.11103/sphaera.101045.

———. 1553a. *Liber de sphaera*. Venice: Francesco Bindoni. http://hdl.handle.net/21.11103/sphaera.100193.

———. 1553b. *Libellus de sphaera*. Wittenberg: Johann Krafft. http://hdl.handle.net/21.11103/sphaera.101102.

———. 1554. *Sphaera. Addita sunt quaedam…*. Venice: Francesco Rampazetto for Melchior Sessa. http://hdl.handle.net/21.11103/sphaera.100164.

———. 1555/1556 **[P6]**. *Sphaera emendata. Eliae Vineti Santonis scholia…ab ipso autore restituta*, comm. Elie Vinet. Paris: Guillaume Cavellat. http://hdl.handle.net/21.11103/sphaera.100198.

———. 1557a. *Sphaera. Addita sunt quaedam…*. Venice: heirs of Giovanni Padovani for Melchior Sessa. http://hdl.handle.net/21.11103/sphaera.101048.

———. 1557b **[P6b]**. *Sphaera emendata*, comm. Elie Vinet. Paris: Guillaume Cavellat. http://hdl.handle.net/21.11103/sphaera.100207. Reissues in 1558: http://hdl.handle.net/21.11103/sphaera.100210, and in 1559. http://hdl.handle.net/21.11103/sphaera.100819.

———. 1558. *Libellus de sphaera*. Wittenberg: Johann Krafft. http://hdl.handle.net/21.11103/sphaera.101049.

———. 1559. *Sphaera*. Antwerp: Jean Richard. http://hdl.handle.net/21.11103/sphaera.101107.

———. 1561a. *Sphaera. Addita sunt quaedam…*. Venice: heirs of Melchior Sessa. http://hdl.handle.net/21.11103/sphaera.101066.

———. 1561b. *Sphaera*. Antwerp: Jean Richard. http://hdl.handle.net/21.11103/sphaera.101058.

———. 1561c. *Sphaera emendata*, comm. Elie Vinet. Paris: Guillaume Cavellat. http://hdl.handle.net/21.11103/sphaera.100211.

———. 1561d. *Libellus de sphaera*. Wittenberg: Johann Krafft. http://hdl.handle.net/21.11103/sphaera.101109.

———. 1562a **[P6c]**. *Sphaera emendata*, comm. Elie Vinet. Paris: Cavellat. http://hdl.handle.net/21.11103/sphaera.101067.

———. 1562b **[V3]**. *Sphaera emendata…ex postrema impressione Lutetiae*, comm. Elie Vinet. Venice: Girolamo Scoto. http://hdl.handle.net/21.11103/sphaera.101094.

———. 1563. *Libellus de sphaera*. Wittenberg: Johann Krafft. http://hdl.handle.net/21.11103/sphaera.101070.

———. 1564a [Ly1]. *Sphaera...emendata cum additionibus in margine...nunc recenter compertis et collectis a Francisco Junctino Florentino sacrae theologiae doctore*, comm. Elie Vinet and Francesco Giuntini. Lyons: Symphorien Barbier for Filippo Tinghi and the heirs of Giunta. http://hdl.handle.net/21.11103/sphaera.101071. Reissue with the date 1567. http://hdl.handle.net/21.11103/sphaera.101111.

———. 1564b. *Sphaera emendata*, comm. Elie Vinet. Paris: Guillaume Cavellat. http://hdl.handle.net/21.11103/sphaera.100110.

———. 1564c. *Sphaera. Addita sunt quaedam....* Venice: Francesco Rampazetto. http://hdl.handle.net/21.11103/sphaera.100165.

———. 1566a **[C1]**. *Sphaera emendata*, comm. Elie Vinet. Cologne: Maternus Cholinus. http://hdl.handle.net/21.11103/sphaera.101076.

———. 1566b **[A3]**. *Sphaera emendata cum additionibus in margine*, comm. Elie Vinet and Francesco Giuntini. Antwerp: Jean Richard and the heirs of Arnold Birckmann. http://hdl.handle.net/21.11103/sphaera.101114. http://hdl.handle.net/21.11103/sphaera.101115.

———. 1568. *Libellus de sphaera*. Wittenberg: Johann Krafft. http://hdl.handle.net/21.11103/sphaera.101112.

———. 1569a **[P7]**. *Sphaera emendata*, comm. Elie Vinet. Paris: Guillaume Cavellat and Jérôme de Marnef. http://hdl.handle.net/21.11103/sphaera.100240.

———. 1569b. *Sphaera emendata...ex postrema impressione Lutetiae*, comm. Elie Vinet. Venice: Girolamo Scoto. http://hdl.handle.net/21.11103/sphaera.101040.

———. 1572a. *Sphaera. Addita sunt quaedam....* Venice: heirs of Melchior Sessa. http://hdl.handle.net/21.11103/sphaera.101086.

———. 1572b. *Sphaera emendata*, comm. Elie Vinet. Paris: Guillaume Cavellat and Jérôme de Marnef. http://hdl.handle.net/21.11103/sphaera.100268.

———. 1573 **[A4]**. *Sphaera emendata*, comm. Elie Vinet and Francesco Giuntini. Antwerp: Gilles van der Rade for Jean Bellere. http://hdl.handle.net/21.11103/sphaera.100460.

———. 1574a. *Libellus de sphaera*. Wittenberg: Peter Seitz. http://hdl.handle.net/21.11103/sphaera.101056.

———. 1574b. *Sphaera emendata...ex postrema impressione Lutetiae*, comm. Elie Vinet. Venice: heirs of Girolamo Scoto. http://hdl.handle.net/21.11103/sphaera.100309.

———. 1577a. *Sphaera. Addita sunt quaedam....* Venice: heirs of Melchior Sessa. http://hdl.handle.net/21.11103/sphaera.100478.

———. 1577b. *Sphaera emendata*, comm. Elie Vinet. Paris: widow of Guillaume Cavellat and Jérôme de Marnef. http://hdl.handle.net/21.11103/sphaera.100269.

———. 1578a. *Libellus de sphaera*. Wittenberg: heirs of Peter Seitz. http://hdl.handle.net/21.11103/sphaera.101116.

———. 1578b **[Ly2]**. *Sphaera emendata a Fr. Junctino...qui etiam in capite libri adjunxit Principia geometrica*, comm. Elie Vinet and Francesco Giuntini. Lyons: Filippo Tinghi. http://hdl.handle.net/21.11103/sphaera.100311.

———. 1580. *Sphaera. Addita sunt quaedam....* Venice: heirs of Melchior Sessa. http://hdl.handle.net/21.11103/sphaera.100299.

———. 1581. *Sphaera emendata*, comm. Elie Vinet and Albertus Hero. Cologne: Maternus Cholinus. http://hdl.handle.net/21.11103/sphaera.100655.

———. 1582 **[A5]**. *Sphaera emendata*, comm. Elie Vinet, Francesco Giuntini and Albertus Hero. Antwerp: Pierre Bellere. http://hdl.handle.net/21.11103/sphaera.100898.

———. 1584. *Sphaera emendata*, comm. Elie Vinet. Paris: widow of Guillaume Cavellat and Jérôme de Marnef. http://hdl.handle.net/21.11103/sphaera.100324.

———. 1586. *Sphaera emendata...ex postrema impressione Lutetiae*, comm. Elie Vinet. Venice: heirs of Girolamo Scoto. http://hdl.handle.net/21.11103/sphaera.100325.

———. 1587. *Sphaera. Addita sunt quaedam....* Venice: heirs of Melchior Sessa. http://hdl.handle.net/21.11103/sphaera.100300.

———. 1591. *Sphaera emendata*, comm. Elie Vinet and Albertus Hero. Cologne: Goswin Cholinus. http://hdl.handle.net/21.11103/sphaera.100326.

———. 1594a. *Sphaera emendata*, comm. Elie Vinet and Albertus Hero. Cologne: Goswin Cholinus. http://hdl.handle.net/21.11103/sphaera.100656.
———. 1594b. *Sphaera. Addita sunt quaedam....* Venice: heirs of Melchior Sessa. http://hdl.handle.net/21.11103/sphaera.100301.
———. 1601a. *Sphaera. Addita sunt quaedam....* Venice: heirs of Melchior Sessa. http://hdl.handle.net/21.11103/sphaera.100302.
———. 1601b. *Libellus de sphaera*. Wittenberg: Johann Krafft for Zacharias Schurer. http://hdl.handle.net/21.11103/sphaera.100646.
———. 1601c **[C3]**. *Sphaera emendata, aucta, et illustrata...nunc accessere R.P. Christoph. Clavii breves commentarii*, comm. Elie Vinet, Albertus Hero and Christoph Clavius, ed. Thomas Abel [?]. Cologne: Goswin Cholinus. http://hdl.handle.net/21.11103/sphaera.100327.

Large Commentaries on the Sphere

Clavius, Christopher. 1570. In *Sphaeram Ioannis de Sacro Bosco commentarius*. Rome: Vittorio Eliano. http://hdl.handle.net/21.11103/sphaera.100365.
Giuntini, Francesco. 1577–1578. *Commentaria in Sphaeram Joannis de Sacrobosco,* 2 vols. Lyons: Jean de Tournes for Filippo Tinghi, 1577–1578. http://hdl.handle.net/21.11103/sphaera.100921; http://hdl.handle.net/21.11103:sphaera.100393.
Idem. 1583. *Speculum astrologiae*, ed. Giuntini, Francesco, vol. II. Lyons: Filippo Tinghi. http://hdl.handle.net/21.11103/sphaera.101101.
Schreckenfuchs, Erasmus. 1569. *Commentaria in sphaeram*. Basel: Henricpetri. http://hdl.handle.net/21.11103/sphaera.101080.

'Adaptions' of the Sphere and Other Cosmological and Astronomical Books

Apianus, Petrus. 1524a. *Cosmographicus liber*. Landshut: Johann Weissenburger for Petrus Apianus.
———. 1524b. *Ein Kunstlich Instrument oder Sonnen Ur*. Landshut: Johann Weissenburger for Petrus Apianus.
Beham, Lazarus. 1476. *Buch von der Astronomie*. Cologne: Nicolaus Götz.
Beyer, Hartmann. 1549. *Quaestiones novae in sphaeram*. Frankfurt: Peter Braubach. http://hdl.handle.net/21.11103/sphaera.100645.
———. 1550. *Quaestiones novae in sphaeram*. Wittenberg: Heirs of Peter I Seitz. http://hdl.handle.net/21.11103/sphaera.101121.
———. 1551. *Quaestiones novae in sphaeram*. Paris: Guillaume Cavellat. http://hdl.handle.net/21.11103/sphaera.101038.
Dietrich, Sebastian, see Theodoricus.
Maurolico, Francesco. 1543. *Cosmographia in tres dialogos distincta*. Venice: heirs of Luca Antonio Giunta. http://hdl.handle.net/10.3931/e-rara-2055.
Peucer, Kaspar. 1550. *De dimensione terrae, et fontibus doctrinae longitudinis et latitudinis locorum*. Wittenberg: heirs of Peter I Seitz.
———. 1551. *Elementa doctrinae de circulis coelestibus, et primo motu*. Wittenberg: Johann Krafft. http://hdl.handle.net/21.11103/sphaera.100251.
Peuerbach, Johannes. 1528. *Novae theoricae planetarum*, ed. Petrus Apianus. Ingolstadt: Petrus Apianus.
Regiomontanus, Johannes. [1472–1474?]. *Kalender*. Nüremberg: Johannes Regiomontanus.
Ringmann, Mathias, and Joannes Waldseemüller. 1529. *Cosmographiae introductio cum quibusdam Geometriae ac Astronomiae principiis ad eam rem necessariis*. Ingolstadt: Petrus Apianus.
Stoeffler, Johann. 1514. *Tabulae astronomicae*. Tubingen: Thomas Anshelm.
Theodoricus, Sebastianus (Sebastian Dietrich). 1563. *Breve, perspicuum et facile compendium logisticae astronomicae*. Wittenberg: Hans Lufft.

——— (Sebastian Dietrich). 1564. *Novae quaestiones sphaerae, hoc est, de circulis coelestibus, et primo mobile*. Wittenberg: Johann Krafft. http://hdl.handle.net/21.11103/sphaera.101072.

Secondary Literature

Armstrong, Elizabeth. 1990. *Before copyright. The French book-privilege system 1498–1526*. Cambridge: Cambridge University Press.

Bennett, James A. 1998. Practical geometry and operative knowledge. *Configurations* 6 (2): 195–222.

Ben-Tov, Asaph. 2009. *Lutheran humanists and Greek antiquity. Melanchthonian scholarship between universal history and pedagogy*. Leiden: Brill.

Caroti, Stefano. 1986. Melanchthon's astrology. In *Astrologi hallucinati*, ed. Paola Zambelli, 115–121. Berlin: De Gruyter.

Clagett, Marshall. 1974. The works of Francesco Maurolico. *Physis* 16: 148–198.

Ernst, Germana. 2001. Giuntini, Francesco. In *Dizionario biografico degli Italiani*, vol. 57, 104–108. Rome: Treccani.

Gingerich, Owen. 1993. Astronomical instruments with moving parts. In *Making instruments count. Essays on historical scientific instrument*, ed. R.J.W. Anderson, J.A. Bennett, and W.F. Ryan, 63–74. Aldershot: Variorum.

van Gompel, Stef. 2011. *Formalities in copyright law. An analysis of their history, rationales, and possible future*. New York: Wolters Kluwer.

Grant, Edward. 1994. *Planets, stars, and orbs: The Medieval cosmos, 1200–1687*. Cambridge: Cambridge University Press.

Kremer, Richard. 2011. Experimenting with paper instruments in fifteenth- and sixteenth-century astronomy: Computing syzygies with isotemporal lines and salt dishes. *Journal for the History of Astronomy* 42: 223–258.

Kusukawa, Sachiko. 1995. *The transformation of natural philosophy. The case of Philip Melanchthon*. Cambridge: Cambridge University Press.

Lattis, James. 1994. *Between Copernicus and Galileo. Christopher Clavius and the collapse of Ptolemaic cosmology*. Chicago: The University of Chicago Press.

Maclean, Jan. 2012. *Scholarship, commerce, religion. The learned book in the age of confessions, 1560–1630*. Cambridge, MA: Harvard University Press.

Martin, Henri-Jean. 2000. *Mise en page et mise en texte du livre français. La naissance du livre moderne (XIV*[e]*–XVII*[e] *siècles)*. Paris: Cercle de la Librairie.

Martin, Henri-Jean, and Jean Vezin. 1990. *Mise en page et mise en texte du livre manuscrit*. Paris: Promodis.

Moscheo, Rosario. 1998. L'insegnamento delle matematiche nell'antico *Studium Messinese*: Francesco Maurolico. *Annali di Storia delle Università italiane* 2: 123–145.

Mosley, Adam. 2004. Tycho Brahe, Erasmus Oswald Schreckenfuchs and the mysterious planetary motion machine. In *Instruments of mystery*, ed. Patrick Boner, 14–26. Cambridge: Cambridge Whipple Museum of the History of Science.

Müller, Kathrin. 2008. *Visuelle Weltaneignung. Astronomische und kosmologische Diagramme in Handschriften des Mittelalters*. Göttingen: Vandenhoeck & Ruprecht.

Müller-Jahncke, Wolf-Dieter. 1998. Melanchthon und die Astrologie: Theoretisches und Mantisches. In *Melanchthon und die Naturwissenschaften seiner Zeit*, ed. Günther Frank and Stefan Rhein, 105–136. Sigmaringen: Thorbecke.

Obrist, Barbara. 2004. *La cosmologie médiévale. I. Les fondements antiques*. Florence: Sismel.

Omodeo, Pietro Daniel. 2014. *Copernicus in the cultural debates of the Renaissance*. Leiden: Brill.

Pantin, Isabelle. 1987. La lettre de Melanchthon à Simon Grynaeus: les avatars d'une défense de l'astrologie. In *Divination et controverse religieuse au XVI*[e] *siècle*, 85–101. Paris: Cahiers V.L Saulnier.

———. 1988. Les problèmes spécifiques de l'édition des livres scientifiques à la Renaissance: l'exemple de Guillaume Cavellat. In *Le Livre dans l'Europe de la Renaissance*, ed. Pierre Aquilon et al., 240–252. Paris: Promodis.

———. 1999. New philosophy and old prejudices: Aspects of the reception of copernicanism in a divided Europe. *Studies in History and Philosophy of Science. Part A* 30–32: 237–262.

———. 2013. Francesco Giuntini et les nouveautés célestes. In *Celestial novelties, science and politics on the eve of the scientific revolution*, ed. Patrick Boner and Dario Tessicini, 85–104. Florence: Olschki.

Pantin, Isabelle, and Philippe Renouard. 1986. *Imprimeurs et libraires parisiens du XVIe siècle: Cavellat – Marnef et Cavellat*. Paris: Bibliothèque nationale.

Pettegree, Andrew. 2010. *The book in the Renaissance*. New Haven: Yale University Press.

Proverbio, Edoardo. 1997. Francesco Giuntini e l'utilizzo delle tavole copernicane in Italia nel XVI secolo. In *La diffusione del copernicanesimo in Italia, 1543–1610*, ed. Massimo Bucciantini and Maurizio Torrini, 37–55. Florence: Olschki.

Randles, W.G.L. 1980. *De la terre plate au globe terrestre. Une mutation épistémologique rapide 1480–1520*. Paris: Armand Colin.

Rhodes, Neil, and Jonathan Sawday. 2001. Paperworlds: Imagining the Renaissance computer. In *The Renaissance computer: Knowledge technology in the first age of print*, 1–17. London: Routledge.

Rosen, Edward. 1957. Maurolico's Attitude toward Copernicus. *Proceedings of the American Philosophical Society* 101 (2): 177–194.

———. 1970. Rheticus' earliest extant letter to Paul Eber. *Isis* 61: 384.

———. 1974. Rheticus as editor of Sacrobosco. In *For Dirk Struik*, ed. R.S. Cohen et al., 245–248. Dordrecht: Reidel.

Röttel, Karl, ed. 1995. *Peter Apian. Astronomie, Kosmographie und Mathematik am Beginn der Neuzeit*. Buxheim-Eichstätt: Polygon-Verlag.

Saugrain, Claude-Marin. 1744. *Code de la librairie...avec les anciennes ordonnances*. Paris: Communauté des libraires.

Schmidt, Suzanne Karr. 2011. Printed instruments. In *Prints and the pursuit of knowledge in early modern Europe*, ed. Suzan Dackerman, 267–315. New Haven: Yale University.

Stijnman, Ad, and Elizabeth Upper. 2014. Color prints before Erhard Ratdolt. Engraved paper instruments in Lazarus Beham's *Buch von der Astronomie*. *Gutenberg-Jahrbuch* 89: 86–105.

Thorndike, Lynn. 1941. *A history of magic and experimental sciences*. Vol. VI. New York: Columbia University Press.

———. 1949. *The Sphere of Sacrobosco and its commentators*. Chicago: University of Chicago Press.

Valleriani, Matteo. 2017. The tracts on the *Sphere*. Knowledge restructured over a network. In *The structures of practical knowledge*, ed. Matteo Valleriani, 421–473. Dordrecht: Springer.

Vanden Broecke, Steven. 2000. The use of visual media in Renaissance cosmography: The cosmography of Peter Apian and Gemma Frisius. *Paedagogica Historica* 36: 130–150.

6

A Lathe and the Material *Sphaera*: Astronomical Technique at the Origins of the Cosmographical Handbook

Richard J. Oosterhoff

Abstract Even though cosmographers loved to drape their discipline in the ancient dignity of Ptolemy, actual manuals of cosmography often depended on Johannes de Sacrobosco's medieval introduction to spherical astronomy. In fact, certain strains of cosmography shared organization, principles, and even visual apparatus with Sacrobosco's *Sphaera* and its growing commentary tradition, to the extent that these cosmographies can be seen as themselves commenting on the *Sphaera*. This paper traces the origins of certain Renaissance cosmographical handbooks to the commentary on the *Sphaera* by Lefèvre d'Étaples and his colleagues at Paris around 1495. By focusing on the visual elements of this commentary, its instructions for calculating techniques, and the emergence of the "lathe" model of the "material sphere," this chapter argues that one of the mixed mathematical genres now seen as most characteristic of the Renaissance—cosmography—in fact was based on a medieval textbook.

1 Introduction[1]

The first printed commentary on the *Sphaera* of Johannes de Sacrobosco (died ca. 1256) was published at Paris in February of 1495. Early in the book the Paris arts master Jacques Lefèvre d'Étaples (ca. 1455–1536) turns to the artisan's workshop to explain the sphere. There one would find a "description of wonderful efficacy,"

[1] I wish to thank Matteo Valleriani for helpful comments, the other contributors to this volume for stimulating conversation, and an anonymous reviewer for a useful suggestion. This research has been supported in part by the European Research Council under the European Union's Seventh Framework Programme (FP7/2007-2013)/ERC grant agreement no 617391.

> which clearly teaches (insofar as sensible matter can take it) how to make an artificial sphere. Artisans in our time who wish to fashion figures with a lathe in metal or another material should find its use worth its weight in gold. So, having taken a compass of thin steel or iron, a semicircle is inscribed on some line which is then cut out from the arc to the diameter, and moreover the diameter in between as well; then it is fit for cutting and dividing, and you have a tool very much suited for turning a sphere, just as a compass is for turning circles.[2]

The example of turning a sphere on a lathe makes a point about the status of astronomy, a classic example of a mixed science. The compass, a paradigmatic instrument of geometry, embeds the description firmly within the objects of mathematical study: lines and curves. At the same time, Lefèvre sets before his readers a concrete example, a lathe with a semi-circular blade (Fig. 2.1). A compass inscribes the edge of a tool, cutting the curve into the surface of a metallic chunk.

As we shall see, the lathe is a visual analogy that pervades later versions of the *Sphaera*; but this tradition originates in Lefèvre's commentary. As the first printed commentary within the genre, the *Textus de sphera* shows the changes the printed medium and new modes of commentary could bring to the *Sphaera*.[3] Lefèvre's version was widely available in the first part of the sixteenth century, through the two great exporting centers of print, Paris and Venice. In 1499 it was republished in Venice, together with commentaries by Cecco d'Ascoli (ca. 1269–1327) and Capuano de Manfredonia (died ca. 1490)[4]; Lefèvre's *Textus de sphera* thereafter formed the core of omnibus editions (1508, 1518, 1519, 1531), which included cosmographical writing such as travel narratives of Amerigo Vespucci (1454–1512) and medieval alternatives to Sacrobosco such as the textbook of spherical astronomy by Robert Grosseteste (ca. 1175–1253) (Barker 2011). On its own, the *Textus de sphera* remained a venerable standard in the printing repertoire of Henri Estienne the Elder (1460–1520) (printed in 1500, 1503, 1507, 1511, 1516) and his successor Simon de Colines (ca. 1480–1546) (1521, 1527, 1532, 1534, 1538).[5] Colines framed the work with new frontispieces and marginal notes by the next generation's most eminent mathematical practitioner, Oronce Fine (1494–1555),[6] but retained the diagram woodblocks and layout of the commentary's original edition of 1495.

---

[2] (Lefèvre 1495, a iiiir): "Et hec profecto mire efficacie descriptio est, que aperte docet (quantum sensibilis materia recipere valet) artificialem constituere spheram, cuius utilem commodamque intelligentiam nostre tempestatis artifices multi auri pondo comparare deberent, qui metalo, ligno, aut alia materia figuras torno exprimere volunt. Si itaque in levi calybe aut ferro, sumpto circino supra quancunque lineam semicirculus educatur qui ab arcu ad diametrum usque excavetur, quin immo et medium diametri interstitium, et mox ad arcum circumferentiamque excavatur ut ea. ex parte ad scindendum secandumque fiat aptus, exurget instrumentum tornandis spheris (haud secus quam circinus circulis) aptissimum."

[3] The scholarly framework here includes (Johnson 1953; Gingerich 1988, 1999; Hamel 2004, 2006, 2014; Pantin 2001, 2012; Crowther and Barker 2013; Oosterhoff 2015; Valleriani 2017).

[4] On Francesco Capuano de Manfredonia's work concerning Sacrobosco's *Sphaera* see (Chap. 4).

[5] The relationship of Lefèvre to these printers is addressed by (Armstrong 1952; Veyrin-Forrer 1995). (Rice 1972) edited Lefèvre's prefatory letters to these works. For further context, see also (Bedouelle 1976).

[6] On Oronce Fine's work concerning Sacrobosco's *Sphaera* see (Chap. 8).

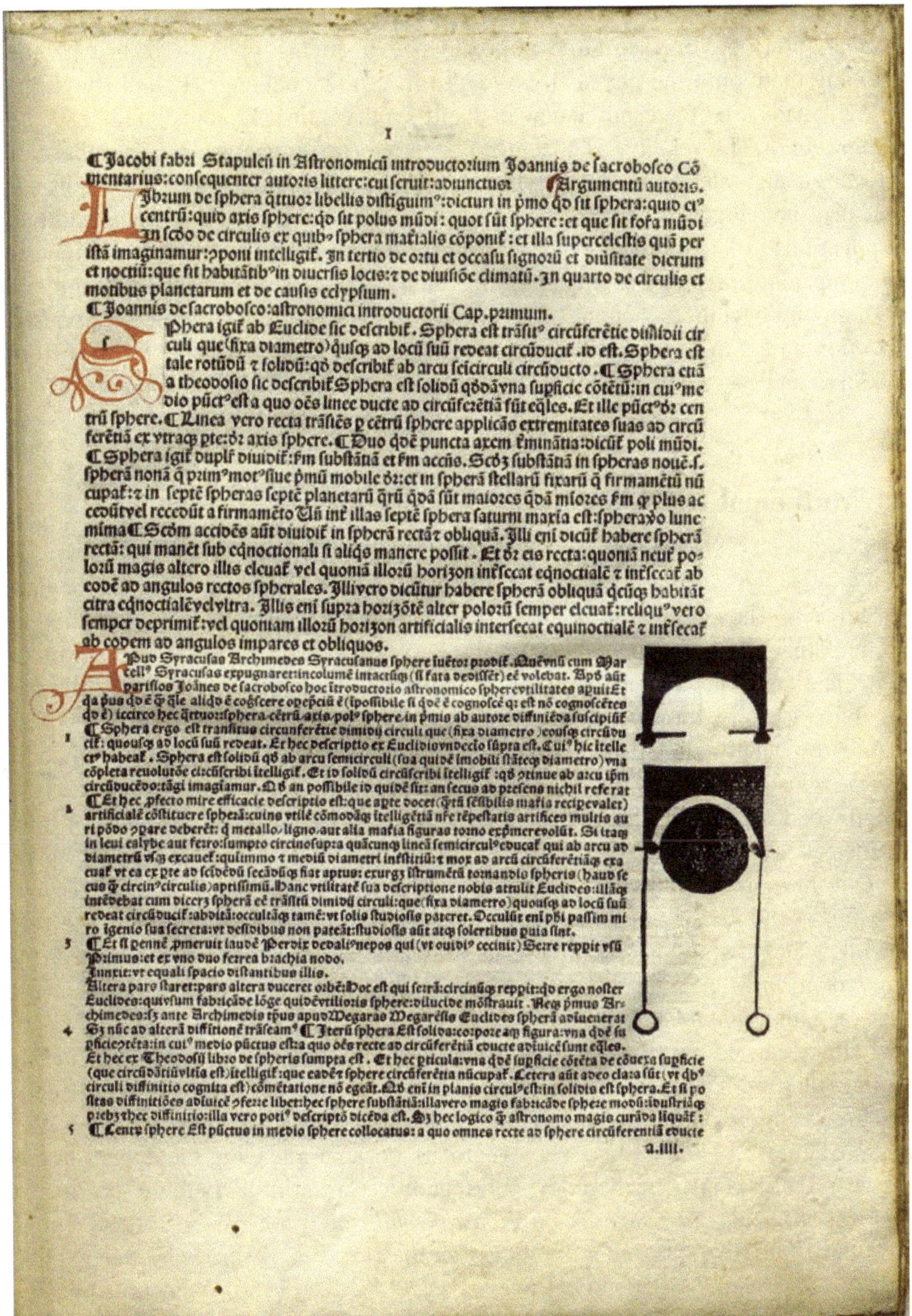

I

Jacobi fabri Stapulensis in Astronomicum introductorium Joannis de sacrobosco Commentarius: consequenter autoris littere: cui servit: adiunctus. Argumentum autoris.

Librum de sphera quattuor libellis distinguimus: dicturi in primo quid sit sphera: quid eius centrum: quid axis sphere: quid sit polus mundi: quot sunt sphere: et que sit forma mundi. In secundo de circulis ex quibus sphera materialis componitur: et illa supercelestis quam per istam imaginamur componi intelligitur. In tertio de ortu et occasu signorum et diversitate dierum et noctium: que fit habitantibus in diversis locis: et de divisione climatum. In quarto de circulis et motibus planetarum et de causis eclypsium.

Joannis de sacrobosco: astronomici introductorii Cap. primum.

Sphera igitur ab Euclide sic describitur. Sphera est transitus circumferentie dimidii circuli que (fixa diametro) quousque ad locum suum redeat circumducitur. id est. Sphera est tale rotundum et solidum: quod describitur ab arcu semicirculi circumducto. Sphera etiam a theodosio sic describitur Sphera est solidum quoddam una superficie contentum: in cuius medio punctus est a quo omnes linee ducte ad circumferentiam sunt equales. Et ille punctus dicitur centrum sphere. Linea vero recta transiens per centrum sphere applicans extremitates suas ad circumferentiam ex utraque parte: dicitur axis sphere. Duo quidem puncta axem terminantia: dicuntur poli mundi. Sphera igitur dupliciter dividitur: secundum substantiam et secundum accidens. Secundum substantiam in spheras novem. scilicet spheram nonam que primus motus sive primum mobile dicitur: et in spheram stellarum fixarum que firmamentum nuncupatur: et in septem spheras septem planetarum quarum quedam sunt maiores quedam minores secundum quod plus accedunt vel recedunt a firmamento. Unde inter illas septem sphera saturni maxima est: sphera vero lune minima. Secundum accidens autem dividitur in spheram rectam et obliquam. Illi enim dicuntur habere spheram rectam: qui manent sub equinoctionali si aliquis manere possit. Et dicitur eis recta: quoniam neuter polorum magis altero illis elevatur vel quoniam illorum horizon intersecat equinoctialem et intersecatur ab eodem ad angulos rectos sphericales. Illi vero dicuntur habere spheram obliquam quicumque habitant citra equinoctialem vel ultra. Illis enim supra horizontem alter polorum semper elevatur: reliquus vero semper deprimitur: vel quoniam illorum horizon artificialis intersecat equinoctialem et intersecatur ab eodem ad angulos impares et obliquos.

Fig. 2.1 Lefèvre opens his commentary on Sacrobosco's *Sphaera* with the illustration of a semi-circular blade cutting a sphere on a lathe

The case of the lathe evokes a central problem of early modern mathematical learning: how does one learn to manipulate mathematical tools with integrity, while mapping them onto the physical world? To meet this challenge, I shall show that Lefèvre made use of a resource new in the fifteenth century, Ptolemy's rediscovered *Geographia.*[7] The result was that the first printed commentary on Sacrobosco's *Sphaera* inaugurated a particular tradition of cosmographical handbooks. I will suggest that the merger of astronomy with new handbooks for making maps had two implications for the ramifying genres of the *Sphaera*. The first is a higher emphasis on competent *practice* within astronomy, raising the epistemic value of technique even among beginners. Second, later versions of Sacrobosco's *Sphaera* refocused attention on the metaphorical nature of astronomical reasoning—the transfer from model to reality—a shift embodied in Lefèvre's material metaphor of the lathe.

2 Mathematical Reform

Lefèvre's *Textus de sphera* of 1495 was the first public result of a turn to mathematics at Paris.[8] The goal was to renovate the University of Paris, Lefèvre explained in his prefatory letter to a bureaucrat in the Paris Parliament. After all, he reported, Plato had said that mathematics "is of the greatest importance not only for the republic of letters, but also for the civil republic—Plato thinks those with the best natures especially should be taught in it."[9] Lefèvre named George of Trebizond (1395–1486)—an arch-Aristotelian in Florence and Rome who had written a lengthy controversial commentary on Ptolemy's *Almagest*—as an example of how mathematics might benefit learning (Monfasani 1976, 105–8; Shank 2002, 2007a). Lefèvre himself was already deep into his project, having prepared an extensive revision of the *Elementa arithmetica* of Jordanus (fl. thirteenth century). The following year, he would have this edition printed together with his own introductory study of arithmetic, an innovative study of Pythagorean music theory, and the arithmetical game *Rithmimachia* (Lefèvre 1496).

Lefèvre carried out this project during a 17-year tenure as regent master of the Collège du Cardinal Lemoine, one of the older and smaller colleges of the University of Paris, with the help of an expanding circle of students. His right hand man was Josse Clichtove (1472–1543), who eventually became a leading theologian in the powerful Paris faculty of theology, but apprenticed in print as Lefèvre's corrector and editor in his *Paraphrases on Natural Philosophy* of 1492 (Lefèvre 1492; Massaut 1968). In the summer of 1495, after the *Textus de sphera* was already published, Lefèvre met Charles de Bovelles (ca. 1475–ca. 1566), who immediately

[7] For the subject of the rediscovery of Ptolemy's *Geographia,* I build on an argument sketched in (Oosterhoff 2018, 133–50).

[8] On this mathematical turn more generally, see (Oosterhoff 2018) and (Chap. 3).

[9] (Lefèvre 1495, a iv; Rice 1972, 27): "Mathemata, inquit, que (si Platoni septimo de republica credimus) non modo reipublice litterarie, sed et civili momentum habent maximum, et in his (ut sentit Plato) precipue erudiendi sunt qui naturis sunt optimis."

joined him at Cardinal Lemoine. Bovelles later identified this moment as the origin point of his life-long preoccupation with mathematical figures and numbers in philosophy, responding to Ramon Lull (1232–1316) and Nicholas of Cusa (1401–1464).[10] The extent to which Lefèvre depended on such close students in making his books, especially his mathematical books, can be seen in the *Textus de sphera*, where in the prefatory note Lefèvre recorded a debt to his *familiaris* Jean Grietan for doing many calculations; the colophon also thanked three correctors besides Grietan. Lefèvre often worked with students: Clichtove and Bovelles published a suite of textbooks on number theory, practical arithmetic, geometry, optics in 1503, which took its starting point in an extended commentary on Lefèvre's epitome of arithmetic, and closed with Lefèvre's *Astronomicon*, a contribution to the genre of theorics that was usually read after the *Sphaera*.[11] This mathematical project was marked by intensive habits of collaboration (Oosterhoff 2019b).

These habits matched a language of friendship and harmony. The preface to the *Textus de sphera* closed on the *benevolentia* or goodwill Lefèvre shared with his patron. The idiom of shared goodwill as a bond between two people built on Aristotle's (384–322 BCE) account of friendship.[12] It also echoed the intuition that the world's deep, hidden structures include patterns of sympathy and repulsion, friendship and repulsion. Lefèvre hinted as much in his manuscript treatise *De magia naturali* of the mid-1490s—just when he published the *Textus de sphera*—which offered a learned astrology organised around these forces and included a book on "Pythagorean magic."[13] Lefèvre's interest in arithmetic and its subaltern science of music revolved around the Pythagorean notion of concord. Mathematically, the idea addresses the relation of quantities, ratios, or proportions: number theory or arithmetic offers a classification scheme for understanding different ratios, while music theory offers a way to use ratios of small numbers to divide the scale. This could have much larger implications. Lefèvre's edition of Euclid's (fifth century to fourth century BCE) *Elements* set a new humanist translation alongside the classic medieval translation of Campanus of Novara (1220–1296), who introduced Book 6 with a long excursus on proportion as the *habitudo* or relation of any one thing to another, citing Aristotle's *Categories* as support (Lefèvre 1517, 57r). Although the immediate question was how to relate arithmetical numbers to geometrical points and lines, the discussion opened larger questions. Campanus cited Boethius (early sixth century) on music, noting that the question related to sound. He also cited Plato's (ca. 428–ca. 348 BCE) *Timaeus*, which suggested that weights and powers relate in mathematical proportion. The fact that arithmetic theory seemed to get at

---

[10] (Bovelles 1511, 168v): "Parhisiis, quod anno 1495 peste affecti sunt…te ruri illustrem disciplinarum solem ostenderi. Tu nempe per introductiones numerorum, per arithmetice discipline preludia, Pythagorico more, totius mei philosophici profectus ac litterarii studii extitisti causa." On Bovelles more generally, see (Cassirer 1927; Faye 1998; Klinger-Dollé 2016).

[11] The *Astronomicon* was republished under the editorship of Oronce Fine in 1515, and again in a deluxe edition with Clichtove's commentary in 1517.

[12] This culture of friendship and its relation to mathematics is described in (Oosterhoff 2016).

[13] Book 2 is titled "de Pythagorica philosophia que ad Magiam introducit" (Olomouc, University Library, MS M.I.119, 198r). See (Mandosio 2013 and 2018).

the forces governing the physical experience of sound, weight, and natural powers offered a suggestive hint that mathematics could explain and even control physical causes too.

These intuitions sparkle under the surface of Lefèvre's commentary on the *Sphaera*. At the beginning of Book 2, he added to his explanation of the great circles (equator, zodiac, colures, etc.) a report from the *magi*. Magicians, he said, divide the heavens into four points: the eastern one is that of God; the midday point is that of "the intelligences;" the western point belongs to the fallen dead (*caducorum*), while the midnight point to "evil powers" (Lefèvre 1538, 9r–v). Elsewhere, on the topic of "crespuscular risings" he alerted readers that those who concern themselves with such matters are not the good sort of *mathematicus*, engaged as they ought to be with arithmetical, musical, geometrical, or astronomical profit. Rather, the *mathematici* who indulge in such topics are "those we call vain and poisonous, such as we read poisonous wise-women were, especially that notorious Thessalian, or Circe, or Medea…."[14]

Despite evident familiarity with operative magic, Lefèvre quickly steered readers clear of such dangerous waters. The account of the four points of heaven turns into an anodyne analogy for contemplative theology; the movement of the heavens from first light, to midday sun, shadow, and then darkness "unfolds the movement of contemplation" (*contemplationis motus explicatur*). Lefèvre urged readers to read this in light of Romans 1:20, which proclaimed that "through those visible things that were made, we sense an understanding of the invisible things of God."[15] Although he recognised that "the magicians foretell great and hidden things through these four points," Lefèvre offered a deflationist reading: readers should learn to use such insights into the movements of the heavens in order to gain an understanding (*idea cognita*) of everything, to feed a cycle of higher intellectual insight, not to manipulate the occult forces of the natural world.

In hindsight, we can distinguish both Platonist and Aristotelian sources on mathematics flowing into Lefèvre's approach to the *Sphaera*. The broadly Platonist assumption that ideas are the most powerful objects that exist also nourished the Pythagorean preoccupation with numbers as the most fundamental category of being, stretching even to the possibility that numbers might underly magic. These ideas were filtered through a Christian modulation of Plato, especially through late antique authorities such as Pseudo-Dionysius (early sixth century) and Boethius, both reimagined by the fifteenth-century cardinal Nicholas of Cusa (1401–1464)—whose *Opera omnia* Lefèvre and colleagues edited in 1514. Using such resources, Lefèvre offered a middle way between realism and nominalism: number exists out there in the world, but is only accessed mentally through human creative conjectures. In fact, the human mind *makes up* number and so, in the act of measuring, imitates God's creation of number in the world. Since Lefèvre thought that late

[14] (Lefèvre 1538, 16r): "Eo enim tempore quod a crepusculo vespertino principium sumit, mathematici utuntur, non qui Arithmeticen, Musicen, Geometricam, Astronomicamque dignitatem profitentur, sed quos vanos veneficosque nuncupamus, qualibus veneficis sagisque mulieribus, maxime infamis Thessalia fuisse legitur, qualem fuisse Circen, qualemque Medeam…."

[15] (Lefèvre 1538, 9v): "ex iis quae visibilia facta sunt, invisibilia dei comprehensa cernamus."

antique Platonists had actually stolen the insights of New Testament thinkers, he believed this roughly Platonist framework to be distinctively Christian. But like the ancient Platonists themselves, within this general framework, he had no trouble using Aristotle as a reliable guide in specific domains of scholarly study such as natural philosophy, meteorology, or astronomy.[16]

By making mathematics foundational for inculcating Aristotelian intellectual virtue, Lefèvre emphasised an educational program of soul-craft slightly different from Platonist mentors such as Marsilio Ficino (1433–1499) (Chap. 5). This can be seen clearly in Lefèvre's account of the "fatal number" that Plato gave in the *Republic* as the enigmatic number that encoded the shifts between political regimes. Aristotle had commented on the passage in his *Politics*, in a passage that elicited Lefèvre's longest scholium in his entire study of the work. Ficino had offered a magical explanation of the passage: souls were constructed as a geometrical set of relations, so to change political systems or groups of souls was at root a matter of manipulating mathematics.[17] For Ficino, contemplation was valuable because it had pragmatic uses. Lefèvre reversed this approach. He agreed with Ficino that the number was 1728, and the bulk of his comments coached the reader through the mathematical theory and techniques necessary to arrive at this conclusion. But he constantly warned his reader to treat this solely as a matter of mathematical exegesis, not vatic soothsaying.[18] The point of moving one's mind along the paths of a mathematical problem *was just the exercise of the soul*, not for technological power, but for the pedagogical acquisition of *habitus*. Although the goal was theoretical, motivated by a contemplative vision of exercising the soul's virtues, this metaphysical end was only achieved through an operative emphasis on practical technique.

3 Vision and Technique

What techniques then did a reader exercise in the *Sphaera*? Since nearly every university student in Renaissance Europe encountered Sacrobosco's *Sphaera* at some point in their studies, its opening pages give an especially good snapshot of what standard mathematical knowledge could look like. After all, astronomy was a classic example of a "mixed" science, taking the pure principles of geometry and applying them to the moving heavens.[19] Therefore a student opening the *Sphaera* needed

[16] On Lefèvre's theory of numbers, see (Oosterhoff 2018, 199–204); on his self-positioning vis-à-vis Plato and Aristotle, see (Oosterhoff 2019a).

[17] The relevant passages are Plato, *Republic,* 8.546a1-d3; Aristotle, *Politics*, 5.1316a1b26. See (Ficino 1491, 225r; 1496, unnumbered final quire). On Ficino's metaphysics here, see (Allen 1994, 1999).

[18] (Lefèvre 1506, 87r): "verum vaticinari ex illis querere, et futili coniectura divinum scrutari velle secretum, vanum."

[19] Most students could have traced this classic account to the quadrivium of Boethius: e.g. *De arithmetica* I.1.

to recall the language of geometry (the exotic vocabulary of points, lines, surfaces, and solids) in order to reimagine the heavens as an orderly *machina mundi*, which was what Sacrobosco named the cosmos early in Book 1. With these tools in hand, the risings of stars, the conjunctions of planets, and the shadows of eclipses might be measured, explained, and—with the help of Lefèvre's commentary—calculated.

Before 1495, a student gained mostly qualitative and general geometrical knowledge from the *Sphaera*. As other chapters in this book reveal, the genre expanded and diversified in many ways between the first printed editions of 1472 and the late sixteenth century. The first editions were slim octavos, visually sparce, with no prefatory material, and although they left space for diagrams, it appears that readers were expected to pen them in (de Sacrobosco 1472a, b).[20] Erhard Ratdolt (ca. 1447–ca. 1527) introduced only three woodcuts in 1478 and 1482; these grew to 24 in his edition of 1485.[21] Many of these appear to be available in the manuscript tradition. The next major change occurred in a Venetian edition of 1488, which added a lengthy prefatory note as well as a "definition of the sphere and of certain presupposed geometrical principles."[22] These five pages gave the terminology of circumferences, poles, lines, curves, and surfaces, etc., accompanying a regime of diagrams; this can be seen as the first step towards including commentaries with the printed *Sphaera*.

This expansion of the *Sphaera*'s visual apparatus between 1485 and 1495 suggests an important function for Sacrobosco in this period. Certainly, the competition of printers partly explains the proliferation of diagrams. But those printers seem to have spotted a need in late medieval learning of astronomical phenomena. Students needed to develop their capacity to image mentally—to imagine—the mathematical structure of heavenly movements. Imagination could be unreliable, of course, and since Pierre Duhem a classic strand in the history of science has focused on this language of *imaginatio* as a problem in "saving the phenomena."[23] Recent scholarship has noticed that this language was ambivalent—the faculty of imagination also mediated trustworthy knowledge (Crowther and Barker 2013) (Chap. 9).[24] For example, mental imaging allowed one to identify the mathematical shape of physical bodies. A culminating proposition in Sacrobosco observes how "Euclid imagined that a sphere is caused by the revolution of a semicircle firmly set on a chord, returned around to the place it started from."[25] The geometrical rudiments added to fifteenth-century manuscript and printed editions of the *Sphaera* suggest the effort

[20] For a more detailed overview of the developments described in this paragraph, see (Baldasso 2007), and especially (Crowther and Barker 2013).

[21] This edition, and many others from Venice and Paris up to 1500 also included the *Theorica* of Georg Peuerbach and the *Contra Cremonensem* of Johannes Regiomontanus.

[22] (de Sacrobosco 1488, a4r): "Diffinitio sphaerae et de quibusdam principiis geometricis supponendis."

[23] For an overview of this historiography, see (Shank 2002; Barker 2011).

[24] A wideranging study of vision's ambiguous place in early modern culture is (Clark 2007).

[25] (de Sacrobosco 1490, a6r): "Imaginatus est Euclides quod sphaera causetur ex revolutione semicirculi super chordam suam firmiter permanentis donec revertitur ad locum a quo caepit circumduci."

to foster a certain kind of spatial imagination as part of the everyday cultural and mental furniture of Renaissance intellectual life.

The same observation might be made of many diagrams throughout these early printed editions: the concentric orbs of the planets; the images of small stick figures walking around the globe as they observe the stars rising over the horizon (an argument for the sphericity of the world and heavens); the ubiquitous ship with observers at its mast and on its deck, in which the lower sailor's line of sight is blocked by the earth's bulge; the smiling sun as the moon eclipses it for a viewer from earth.[26] These figures supply the toolkit needed to reframe the heavenly bodies as mathematical objects within the mind. This diagrammatic tradition performs an important function—but it is sharply limited. It does help the mind gain certain geometrical intuitions and abstractly apply them to a heavenly model. But these diagrams do little to help one calculate anything. Before 1495, Sacrobosco did little to teach students the specialist skills of quantification.

This is by no means a presentist observation. Contemporaries would have recognised this point. The main purpose of early modern astronomy, Robert Westman has shown in abundant detail, was prognostication (Westman 2011). Even those interested primarily in the theoretical description of heavenly models—as Lefèvre claimed to be—were acutely aware of their utility for medicine or courtly advice. Physicians and courtly astrologers therefore depended on techniques of calculation to cast a horoscope: they had to understand the locations of the heavens in the past, and to decipher where they would be in the future. A would-be prognosticator required several things beyond some geometrical intuitions about how the heavens move.[27] First, a set of tables locating heavenly bodies, usually based on the thirteenth-century tables associated with King Alphonse of Spain, or (increasingly) the newly calculated tables of Johannes Regiomontanus (1436–1476). Second, a set of canons, which supplied the protocols for calculating from those tables. Third, skill in sexagesimal arithmetic, since all calculations were in degrees, minutes, and seconds. Fourth, and optionally, a prognosticator might take a shortcut with a calculation device, such as an equatorium or an astrolabe—these allowed one to read values off the instrument, rather than perform laborious calculations. None of these elements were part of Sacrobosco's introduction to the sphere.

All of these, however, were integrated into Lefèvre's commentary on the *Sphaera*, to one degree or another. The edition of 1500 even added Bonet de Lattes's (fifteenth century to sixteenth century) short treatise on a miniature astrolabe, which Lefèvre had found in his travels to Italy.[28] In the shape of a ring intended to fit a finger, it was too small for meaningful observations or for precise calculations—but it gestured towards the company Lefèvre expected his treatise to keep.

[26] (Crowther and Barker 2013, 442–63) have categorized these into three kinds of images: abstract geometrical diagrams; hybrid images that are diagrammatic with naturalistic features (such as the ship); and cosmic section, i.e. the cross-sections of the planetary orbs.

[27] An overview of recent work in this area can be found in (Chabás and Goldstein 2014; Kremer and Husson 2012).

[28] The first edition is (de Lattes 1493).

Throughout the commentary itself, Lefèvre took his reader by the hand, explaining the techniques needed to perform calculations. This starts with the opening *Introductoria additio*, which offers a brief overview of the same definitions of points, lines, and circles that we have already seen in versions of Sacrobosco from the previous decade. Lefèvre concluded this note with a telling primer on sexagesimal arithmetic. Although it is clear from annotated copies that readers did in fact sharpen their skills on the two specimen calculations he included, nevertheless Lefèvre sent readers to other sources if they wished to become proficient: "these things are added about the physical mode of calculation, not because this is enough for abacus or astronomical calculation, but so that those educated in this astronomical instruction might consult calculations and experts in calculation."[29]

Beyond his introduction to sexagesimal arithmetic, Lefèvre offered a series of tables for astronomy and cosmography. Elsewhere I have discussed in some detail the series of astronomical tables, but a few words about their function will be significant here (Oosterhoff 2020). In Book 3, Sacrobosco sets the topic of "risings," namely the times when the stars in the various signs of the zodiac rise over the eastern horizon. This is a fundamental task for spherical astronomy: it allows one to determine the speed of the heavenly sphere's movement. Therefore not only are these tables the basis for calculating the location of any other star at a given time, but they also set the starry backdrop against which to calculate the movement of the lower planets. The task is fairly straightforward when an observer stands at the equator ("right sphere"), so that is where Lefèvre begins. His commentary constructs six rules based on Sacrobosco's text, and joins them to worked up specimen examples. Using short, simplified tables, he connects the movement of the heavens with the passing of time: a quarter of the zodiac has passed when a quarter of a day has passed, and so on. The matter is simplified since, even though the zodiac wobbles with the sun's annual movement through the ecliptic, this wobble is symmetrical at the equator. The basics set, Lefèvre complicates the picture, moving the observer north from the equator, to the latitude of Paris. From this perspective ("oblique sphere"), the zodiac speeds up and slows down irregularly from the observer's perspective. Here Lefèvre's rules help the reader to respond and recalculate from a realistic point.

By this point in the text, the reader has worked through enough techniques to read a table of risings or ascensions. Lefèvre includes, therefore, two full tables of ascensions that synthesize the information taught in the earlier rules: one for the equator and one for the latitude of Paris. These are not intended to replace the much fuller tables used by professionals. In fact, Lefèvre offered recommendations to readers who wished to work with actual tables. They should avoid the older Alphonsine tables, "for they are not precise. Instead, it is from the tables of ascensions of Johannes Regiomontanus…that they should compute."[30] Lefèvre

[29] (Lefèvre 1538, 3r): "Haec de abaci physica ratione adiecta sunt, non quia ad abacum astronomicumque calculum sufficienter introducant, sed ut calculum calculique peritos consulant, qui hoc astronomico instituto sunt informandi."

[30] (Lefèvre 1538, 20r): "Caveant tamen abacistae…per ascensiones tabulis Alphonsinis adiectas numerando perquirere, nam praecisae non sunt; sed potius per tabulas ascensionum Ioannis Nurembergi ubilibet…computent."

did not claim that his commentary on Sacrobosco was everything needed for the competent astrologer. Nevertheless, the student who had mastered it could approach the canons and tables of professionals with some degree of confidence, now able to visualise the movements of the heavens and also to relate those movements to tables of measurements.

4 Introducing Cosmographical Practice

The other set of tables in Lefèvre's commentary on the *Sphaera* addressed cosmography. The word "cosmography" was a moving target throughout the period. The discipline itself was new to the Renaissance. Medieval authors had provided narrative accounts of cities, rivers, and lands, but the word gained new associations after ca. 1410, when one of Manuel Chrysoloras's (1353–1415) students Jacopo d'Angelo (ca. 1360–1411) finished the translation of a text his master had brought from Greece, Ptolemy's *Geographia*.[31] By titling his Latin version *Cosmographia*, d'Angelo linked the term to Ptolemy's practice of mapping: projecting the grid of stars on the celestial sphere onto the terrestrial globe—then projecting this globe, in turn, onto the flat surface of a map. Using this insight, the bulk of the book comprised lists of longitudes and latitudes for around 8000 cities, along with three modes of projection that would allow users to construct maps from these coordinates. By 1533, when Erasmus of Rotterdam (1466–1536) wrote a preface for the Basel *editio princeps* of the Greek text, the book had been published in dozens of editions and had spawned a whole subgenre of introductory handbooks.[32]

We have often identified the earliest cosmographical handbooks as those by Matthias Ringmann (1482–1511) (1507) and Heinrich Glarean (1488–1563) (1513).[33] Some have loosely linked these handbooks to Sacrobosco. Benjamin Weiss suggested that the reading patterns of Sacrobosco's *Sphaera*—heavily annotated witnesses to underlying manuscript notes and circles of readers—are analogous to the ways readers of Ptolemy's work labored over practices of recalculating and remapping the longitudes and latitudes in his *Cosmographia*. Such specialist readers existed, as we know from their noisy complaints that Ptolemy's data was incorrect, and by the 1530s Ptolemaic techniques of mapmaking were common across Europe. I would claim this connection between the reading practices of Ptolemy and Sacrobosco is more than an analogy; in fact, beginning in 1495 with Lefèvre's publication of the *Textus de sphera*, commentaries on the *Sphaera* actually introduced students to the techniques of Ptolemaic cartography.

---

[31] A range of useful recent studies includes (Tessicini 2011; Mosley 2009; Dalché 2009; Hankins 2003).

[32] On expansion of Ptolemaic handbooks and cartography, see (Buisseret 2003, 49–70; Weiss 2011). Erasmus's involvement in the edition of Ptolemy was minimal, as convincingly argued by (Reedijk 1989).

[33] Glarean's work is in manuscript (Glarean 1513) and was first printed in 1527.

We can see this in Lefèvre's commentary on Book 2 of the *Sphaera*, which introduces the major and minor circles of spherical geometry. Major circles are fundamental reference points for cartography, as they divide the sphere's surface into equal halves or hemispheres: e.g. the equator, the meridian, and the horizon. The equator runs east-west, while a meridian evenly splits the earth by running north-south. When projected up onto the heavenly sphere of the stars, these circles define the celestial equator, meridians, and celestial horizons—the latter is the major circle that defines the hemisphere of stars visible to any observer on earth. Minor circles are smaller circles on the surface of the globe, which divide the sphere unequally: e.g. the colures such as the lines of Cancer and Capricorn that lie above and below the equator; the smallest of these minor circles are the arctic and antarctic circles. Taken together, these various circles slice the sphere into five zones.

Lefèvre's commentary draws out the key assumption necessary for Ptolemaic cartography, namely that "the sphere" is *both the earth and the heavenly system of concentric orbs*. The zodiac is therefore the belt of stars embedded on the inner surface of the outer starry sphere, a belt that follows the celestial equator; but the zodiac also includes the space below it, projected ever more narrowly onto the earthly sphere. "And in this sense we may properly say that the planets are *in* signs" (Thorndike 1949, 125). To illustrate this point, previous versions of the *Sphaera* included an image of what Sacrobosco called a "pyramid" based on the zodiac, with its apex at the center of the earth (Fig. 2.2 left). Lefèvre's version of the pyramid is clearer; but his biggest change is to present a sphere in which the major circles of longitude are assembled, hinting at the gores of a globe (Fig. 2.2 right). Around the sphere's middle is the belt of the zodiac giving a physical sense of the pyramid bases linked together as a cross section of the celestial globe. To clarify the arrangement, a third diagram presents this cross section from above, like an orange sliced in half,

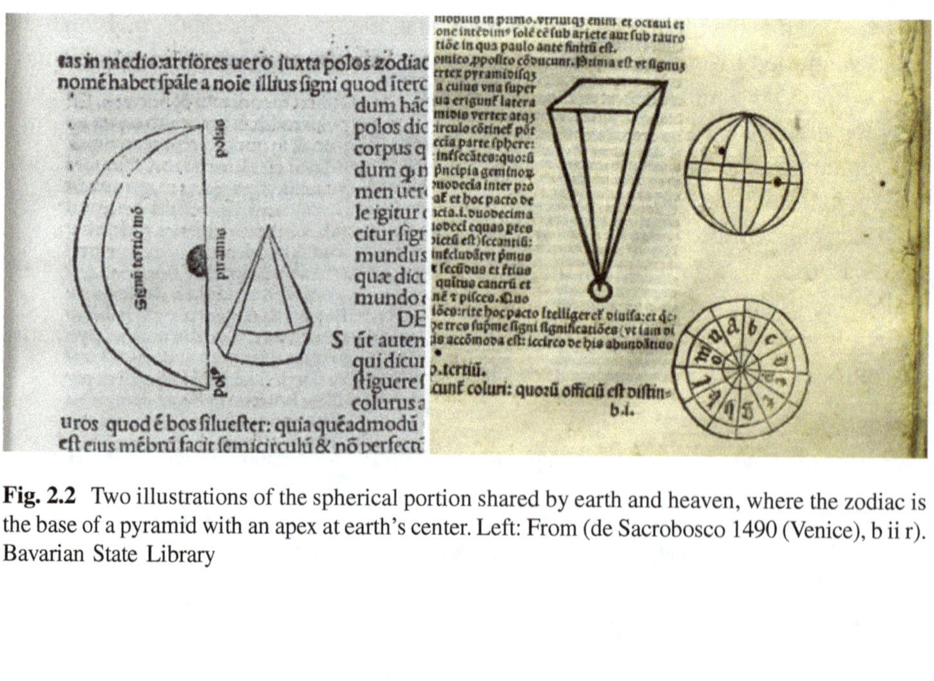

Fig. 2.2 Two illustrations of the spherical portion shared by earth and heaven, where the zodiac is the base of a pyramid with an apex at earth's center. Left: From (de Sacrobosco 1490 (Venice), b ii r). Bavarian State Library

revealing the pyramids fanning out from the center of the earth where their points meet. The image crystallises the assumption that the geometry of the heavenly spheres maps directly onto earth.

From this point, Lefèvre offers tools for deploying spherical astronomy to map locations on earth. In passing, Sacrobosco mentions that the meridian line is known as the longitude of a city. Lefèvre specifies precisely how to calculate the difference in degrees and minutes of two cities, and how to convert a difference of coordinates into a difference of time (useful for calculating horoscopes). A specimen example compared the time in Paris and Jerusalem, where the sun rises 2 hours and 47 minutes earlier. This gave the reader a simple task in order to begin using the central exhibit of the text: a four-page table of longitudes and latitudes, "taken from Ptolemy."[34]

Longitude—physically measured by travellers and shared in such tables—would have posed students no great challenge, since there was little to be done but read it off the table.[35] The more interesting technical challenge was to measure latitude, based on the altitude of the sun. Of course, this required additional information. The sun's altitude changes over the course of the year, between its maximum height at the summer solstice, and maximum depth in winter. As Sacrobosco pointed out, these solstitial points determine the place of the minor circles at the tropics of Cancer and Capricorn (since the sun would be in those signs during June and December). But if one can account for the sun's elevation, the geometry of greater and minor circles will allow a novice astronomer to calculate the latitude of a given location. More generally, the same techniques allow one to set a given location in relation to the various circles drawn on a map. Lefèvre presented these techniques as a set of seven rules, adding a small table of worked examples for the reader's benefit (Fig. 2.3).

Is this astronomy or cosmography? Lefèvre's specimen examples emphasised that the techniques were useful for both disciplines: they allow one to calculate distances on a star map or a on terrestrial map.[36] But both the tables and the techniques were explicitly taken from Ptolemy's *Cosmographia*, and Lefèvre clearly had in mind that his reader would be prepared to read geographical works. He concluded his list of seven techniques with the claim that "quickness in thinking through these intervals and distances will have great value for the *Cosmography* of Ptolemy and the *Geography* of Strabo."[37] The goal was cosmographical literacy.

The overlap of genre between the *Sphaera* and the new handbooks of cosmography has been noted several times before.[38] I would emphasise that this overlap

---

[34] (Lefèvre 1495, b iir): "…ex Ptolemaeo deprompta."

[35] Determination of longitude from astronomical principles required time-keeping devices sufficiently reliable over long distances to give a time measurement independent from the sun—a practical impossibility before the eighteenth century.

[36] (Lefèvre 1538, 14r): "Ex his quoque et determinatis in praecedente commento, distantias tum in caelo, tum in terra cognoscere promptum est."

[37] (Lefèvre 1538, 14v): "Et horum intervallorum distantiarumque cognoscendarum promptitudo non parvum ad Cosmographiam Ptolemaei et Geographiam Strabonis habet momentum."

[38] (Johnson 1953, 296–99) already offered some suggestive comments; see now (Weiss 2011; Mosley 2009).

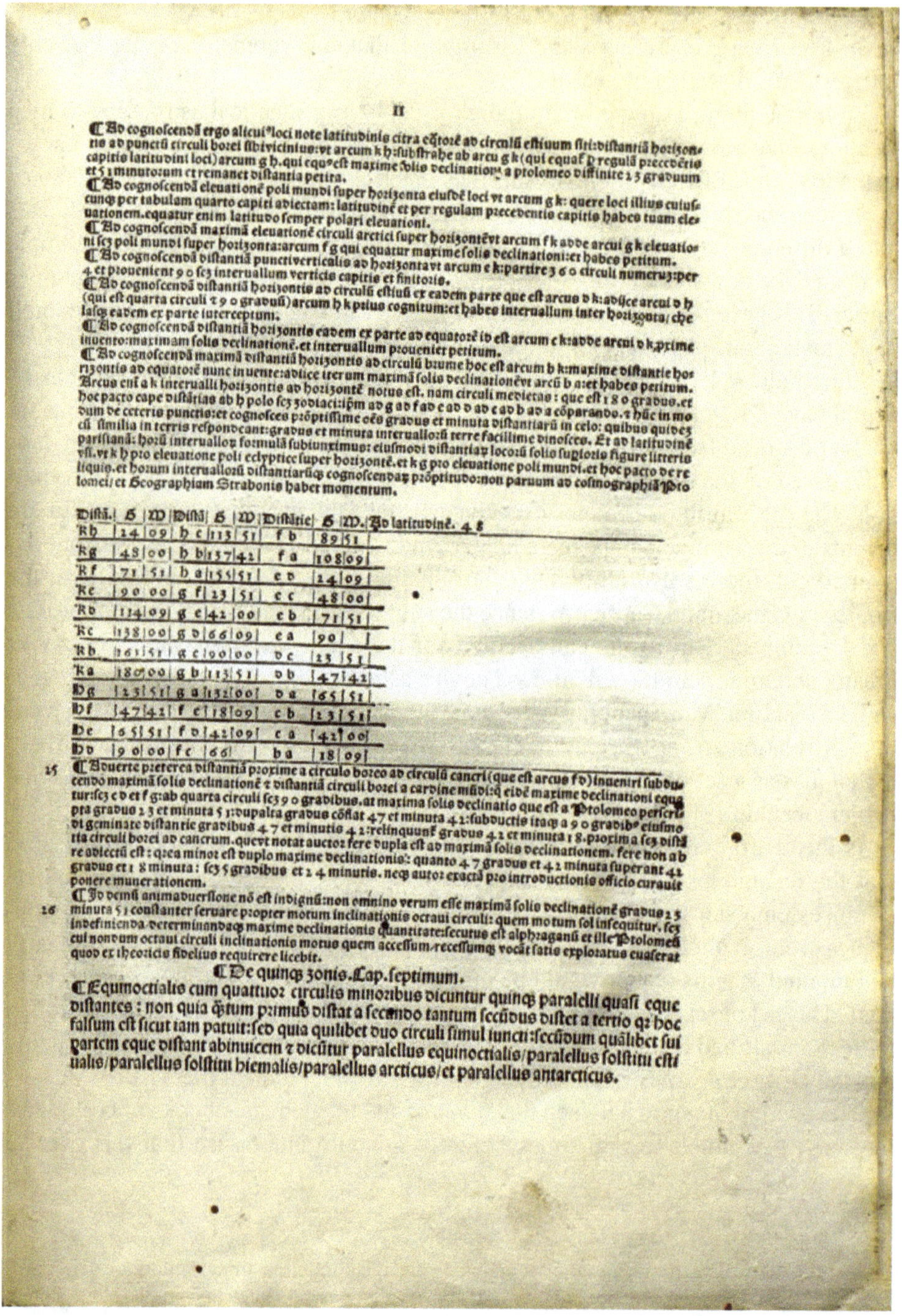

Fig. 2.3 Seven rules, with a specimen table, for calculating various problems of latitude.

begins quite deeply already within the Fabrist commentary on the *Sphaera*, well before the cosmography manuals just mentioned (see Appendix). It would be simplistic to claim a linear influence from Lefèvre's *Textus de sphera* to those early manuals, but there exist some suggestive links. Prosopographically, three of the earliest writers on the topic were connected to Lefèvre and his circle. The first is Matthias Ringmann, who wrote his *Cosmographiae introducto* (1507) to accompany the gores for a Ptolemaic map by Martin Waldseemüller (ca. 1472–1520).[39] Ringmann quite likely had been Lefèvre's student in Paris; by most accounts he had gone to the grammar school at Sélestat. Certainly he was close to the circle of Alsatian students who first studied in Sélestat and then went to Paris to study for the MA with Lefèvre, before returning to participate in the Rhineland community of humanists that would attract Erasmus to Basel: the sons of Johann Amerbach (1440–1513), Johann Sapidus (1490–1561), Michael Hummelberg (1487–1527), and Beatus Rhenanus (1485–1547).[40] All of these were Lefèvre's students; it is not surprising then that Ringmann identified Lefèvre's own textbooks as a model for his *Grammatica figurata* (1509) (Ringmann 1509, 2r).[41]

The second case is Henricus Glarean, whose manuscript handbook from 1513 is also among the earliest examples of the genre. The book seems to have been composed near the end of Glarean's stay at the University of Cologne, probably as part of his teaching at the Bursa Montis just before he moved to teach at Basel in 1514. Although it was not until 1517 that Glarean spent time in Paris with Lefèvre, there are suggestions that he was already familiar with the older humanist's commentary on Sacrobosco. Glarean's own teaching copy of the *Sphaera* from this period is a 1493 edition of Sacrobosco. Glarean had transformed the book into a compendium of annotations from a wide range of other texts, including other versions of the *Sphaera*. Two sets of annotations are enough to make the point. One is a table of the various climates, together with a diagram that is directly copied from Ringmann's *Cosmographiae introductio* (Fig. 2.4 right). The second telling note is the pyramids we have already seen from Lefèvre's 1495 commentary on the *Sphaera*, complete with a cross-section of the heavenly sphere showing the circle of pyramids fanning from earth to zodiac (Fig. 2.4 left). Evidently, already during his teaching at Cologne from 1507 to 1513 Glarean had set Ringmann's cosmography alongside Lefèvre's *Textus de sphera*, even as he was compiling his own *Geography* (with its own set of maps expanding on Waldseemüller's charts).[42] In this trend-setting early stage—Glarean's handbook would inform Peter Apian's (ca. 1495–1552) cosmographical handbooks (Chap. 9)—Ptolemaic cartography is impossible to separate from the genre of the *Sphaera*.

[39] The relation between Waldseemüller and Ringmann is discussed in (Laubenberger 1982; Johnson 2006).

[40] On this circle, see (Bietenholz 1971; Oosterhoff 2014).

[41] More generally on the relationship between Ringmann and Lefèvre, see (Schmidt 1879, 90–91, 121–23; Margolin 1972).

[42] Glarean's associated maps are extant. See the book list of (Fenlon and Groote 2013, nos. 57, 119). A description can be found in (Heawood 1905). More generally, on Glarean's geographical teaching, see (Johnson 2013), who presents Glarean's astronomy as part of his geography; the argument could equally go the other way.

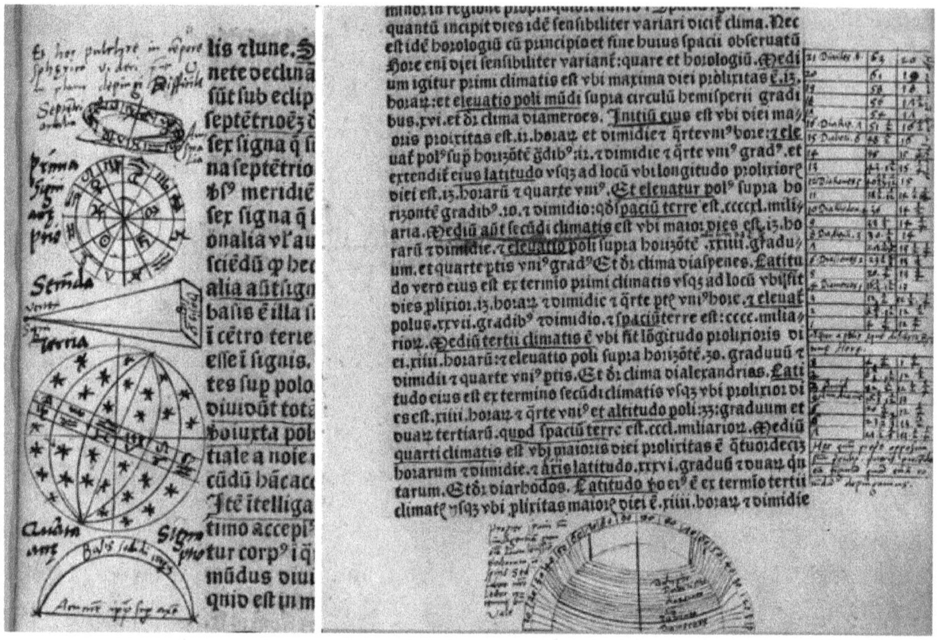

Fig. 2.4 On the left, the diagrams closely follow those given in Lefèvre's *Textus de sphera* (see also Fig. 2.2 right). On the right, the bottom figure has been hand-copied directly from (Ringmann 1507)

A third example is taken up by Angela Axworthy in much closer detail: Oronce Fine's *De mundi sphaera, sive Cosmographia* (first edition 1532) (Chap. 8). Here I simply wish to highlight the book's debt to Lefèvre. While it is not clear how closely Fine and Lefèvre associated in Paris, their intellectual filiation is not in doubt: Fine's first significant contribution to the mathematical writing and illustration that made him famous was his frontispiece for Simon de Colines' 1521 edition of Lefèvre's *Textus de sphera*.[43] The title of Fine's own *De mundi sphaera, sive Cosmographia* already betrayed its origins in Sacrobosco, and a glance at the contents confirms that its first five books quite closely follow the *Sphaera* (see Appendix). Moreover, the text itself depends considerably on Lefèvre's commentary, even augmenting Lefèvre's use of tables. Book 2 adds star charts, giving not only the locations of cities on the earthly grid, but the longitudes and latitudes of stars on the heavenly grid. Book 3 has at its core the same task as Lefèvre's *Textus de sphera*, using a selection of small charts to prime the reader for interpreting larger tables of right and oblique ascensions. Building on Lefèvre's work, Fine's *Cosmographia* binds astronomy and cosmography even closer together by underscoring the fact that the astronomer and cosmographer share the same techniques (Besse 2009).

[43] On this relationship, see (Axworthy 2016, 28; Pantin 2009a, b).

As Henrique Leitão has argued, Fine's work was open to critique from many angles; but this was precisely because Fine, as the prestigious royal professor of mathematics in Francis I's *Collège Royal*, was the pre-eminent mathematical practitioner of mid-sixteenth-century Europe (Leitão 2009). The cosmographies of Ringmann, Glarean, and especially Fine defined the shape of cosmography in the Renaissance—and they cemented Lefèvre's version of the *Sphaera* into the foundations of cosmography.

5 The Material *Sphaera*

The techniques of actual measurement and calculation discussed above bring us back to lathes and the question of how manipulation and models relate to the physical world. By the seventeenth century, the art of turning stood in for the manual use of mathematics to rival and control nature's untiring motions. Princes from Maximilian I (1459–1519) and Rudolf II (ca. 1401–1495) to Peter the Great sought recreation in the art of turning. Courtly collections of wonders from Dresden to the Palazzo Vecchio of Florence included fine pieces of ornamental turnery (Klaus 1985, 2004; Connors 1990). In 1565 Samuel Quiccheberg (1529–1567) recommended that such collections devote an entire room to such tools and their artefacts.[44] Turned ivories among surviving examples in the Kunstkammer of the Electors of Saxony, in Dresden, are material ruminations on the five platonic solids, examples of how turning became a metaphor for mechanical control of nature, or what Horst Bredekamp has called the "cult of the machine" (Korey 2007; Dupré and Korey 2009, 417; Bredekamp 1995).[45] Such lathework quoted in matter the *Timaeus*, where Plato set gave his analogy of the cosmos as spinning bowls formed by a divine Craftsman, setting up his account of the five solids as the building blocks of the universe. Over the course of the sixteenth century, the image of a lathe permeated versions of the *Sphaera*, part of the construction of turning as a material topos.

In 1495, the lathe was not yet a commonplace depiction of the heavenly sphere. When he introduced it in his *Textus de sphera*, Lefèvre likely had in mind the Platonic image of God as artifex; in his *Astronomicon* (1503) he offered planetary astronomy as a way for human souls to imitate the circular, productive motions of the first Artisan (Lefèvre et al. 1503, xcviir). With the lathe, Lefèvre set readers the problem of reasoning about how mathematical forms regulate matter, and especially the movement of that matter. Although axial motion was a common question in late medieval physics, the metaphor of the lathe was problematic for other reasons.[46]

[44] (Quiccheberg 2013, 72): "A workshop of turner's equipment and turning and joining tools, such as those considered among most princes and patricians to belong to the domain of the more congenial arts." More generally, see (Maurice 1985, 2004).

[45] On the metaphor *machina mundi* in this period, see (Popplow 2007).

[46] Late medieval reflections on revolving objects include Gaetano da Thiene, commenting on the Merton calculators: see (Wallace 1981, 55–56; Shank 2007b, 2009).

Even Plato had observed that mathematics was only fitted to matter by means of "bastard kind of reasoning" (Plato, *Timaeus*, 52b2).

How much could the metaphor describe what the *Sphaera* called the material sphere? Not all of Lefèvre's own readers were happy with the lathe. In 1498, Pedro Ciruelo's (1470–1554) own commentary on the *Sphaera* provides evidence that Lefèvre's commentary was already being read more widely at the University of Paris (Chap. 3). His ambivalence, however, shows that the lathe example touched on unresolved issues within the philosophy of mathematics. First he paraphrased Lefèvre's example at length, anonymously as the "account of others." Then he noted that "although this seems to be a beautiful and ingenious case, nevertheless this was not what Euclid had in mind."[47] The primary reason explicitly draws on Aristotle's misgivings about mathematical abstraction in *On the Soul* and the *Metaphysics*: "since sensible matter cannot take those forms or shapes that mathematicians think up."[48] Ciruelo was a talented mathematician, who regularly expressed his appreciation of Lefèvre, Clichtove, and Bovelles, yet he was worried that Lefèvre's lathe example was overoptimistic about how well mathematics can define a physical object.

Still, the visual echoes of Lefèvre's lathe suggest that the example was too powerful to let go. Readers of early books often, like Glarean, took images that they found useful from other versions of the *Sphaera* and copied them into the margins of their own copies. One 1508 version from Cologne reveals just this: at the top of the page the reader has drawn in the metal blade, defined as the "curve of the circumference;" a little ways down the page is drawn the lathe, which "is an example of the first definition [i.e. of Euclid's definition]" (Fig. 2.5). To someone whose mathematical literacy is set by twentieth- or twenty-first-century visual culture, the series of geometrical "principles" that introduced most editions of the *Sphaera*—and also cosmography manuals—may seem somewhat superfluous. But such visual cues required cultivation in the sixteenth century.

An important turning point in this visual topos comes in the 1530s, in the stripped-down versions of the *Sphaera* published at Wittemberg, often with Philip Melanchthon's oration on astronomy. The first chapter addresses the *definitio sphaerae*, and for the first several editions, the only commentary added to Sacrobosco's words is the diagram of a lathe, before moving on to the next chapter on the division of the sphere (Fig. 2.6 left). After 1538, a further brief scholium was added, citing the Greek edition of Euclid. This note did not eclipse the old lathe, however, but complemented it with a new image of a pseudo-lathe (Fig. 2.6 right). Where the first lathe shows how to create a solid sphere from a semi-circular hollow, the second image does something slightly less intuitive. It takes a semicircular surface, and asks the viewer to imagine it spun around an axis—the space it sweeps out exemplifies a *hollow* sphere. The place of this example between physical object and

---

[47] (Ciruelo 1498, b ir): "Sed quamvis hec pulchra et ingeniosa videantur, hanc tamen non fuisse Euclydis mentem."

[48] (Ciruelo 1498, b ir): "Cum materia sensibilis non tales formas seu figuras recipere possit quales mathematici concipiunt ut satis probatur in primo de anima et in tertio methaphysice."

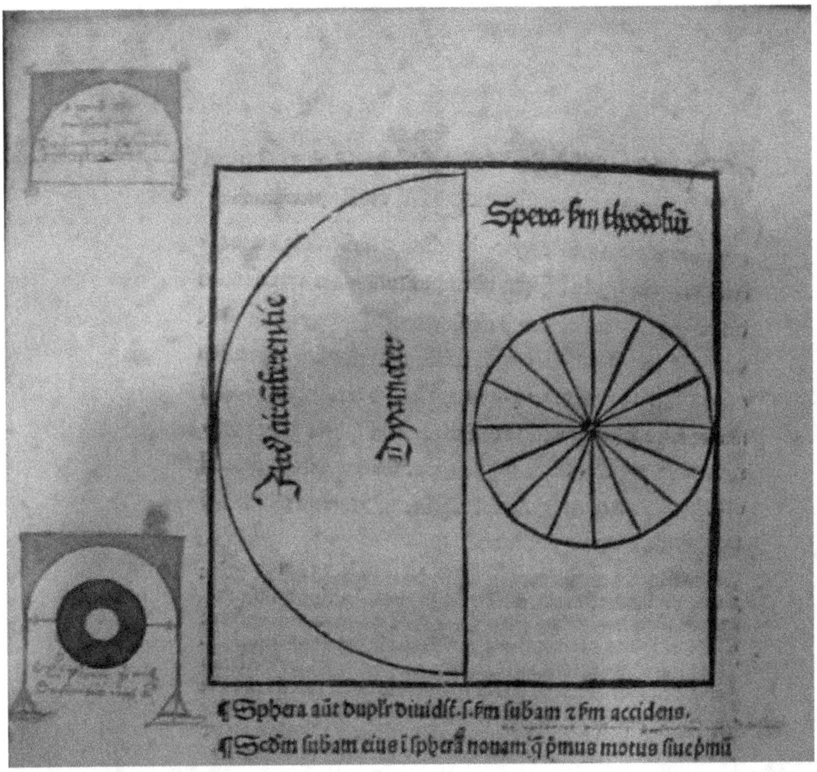

Fig. 2.5 A reader adds diagrams of the material lathe, inscribing above "arcus circumferentie," and below, "exemplum prime definitionis est."

imagined concept is underscored by two tiny figures standing on the ground below the sphere, allowing the reader to imagine themselves standing below an enormous space. By the second half of the sixteenth century, this second, inverted lathe was widely common in the genre of the *Sphaera*, including versions by Giuntino Giunta 1477–1521, Elie Vinet (1509–1587), and Franco Burgersdijk (1590–1635).[49]

To what extent was this object on a lathe identified with the physical cosmos? Typically, the figure introduced a discussion of the "material sphere." One might suppose this meant the actual nested orbs that were the subject of Sacrobosco's planetary astronomy (Barker 2011). This seems to have been the case for Lefèvre, who took no trouble to distinguish the "machina mundi materialis" from the actual cosmos. But over the course of the sixteenth century, the material metaphor seems to have become more troubling. One example is the *Epitome astronomiae* of Michael Maestlin (1550–1631), which gives the inverse, Wittenberg version of the lathe to exemplify the Euclidean definition of the sphere (Maestlin 1597, 10). Maestlin qualified the "material sphere" differently than Lefèvre had done: "We give the name 'material sphere' to the instrument that represents the outermost, convex sur-

[49] See also the versions of the *Sphaera* by Vinet and Burgersdijk (Chap. 11).

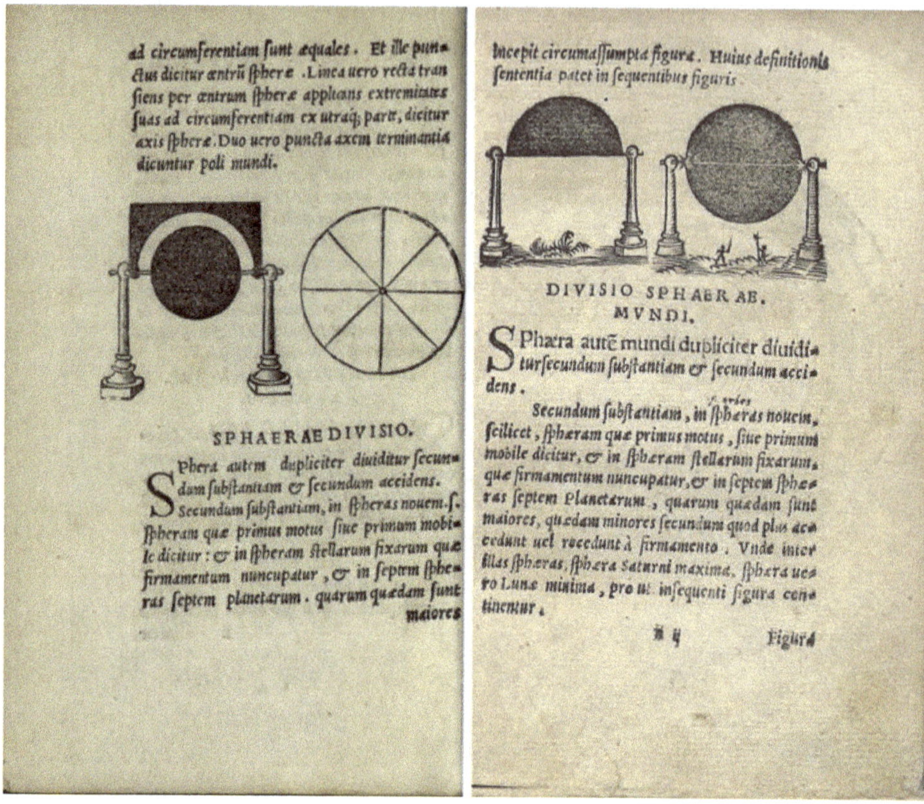

Fig. 2.6 Lefèvre's lathe example became mainstream in the enormously popular Wittenberg edition of 1531 (left), which was then augmented in 1538 with an additional outsized example of a revolving semi-circle (right). Left: From (de Sacrobosco 1531, B i v). Bavarian State Library, H.g.hum. 18 l, http://mdz-nbn-resolving.de/urn:nbn:de:bvb:12-bsb00039827-2. Right: From (de Sacrobosco 1538, B ii r). Bavarian State Library

face, or the circles that define it, of the highest heaven (which we call the 'natural sphere')."[50] In other words, Maestlin applied the term "material sphere" to the instrument or model (perhaps thinking of an armillary sphere) rather than to the heavens themselves.

The lathe had become a commonplace, within the genre of the *Sphaera* as well as at courts. Moreover, the function of that commonplace seems to have shifted ever so slightly. Only a much larger study could adequately consider the range of concerns implicit in such a commonplace.[51] But this brief sketch suggests that the lathe

[50] (Maestlin 1597, 13): "Ut sphaeram Materialem appellamus illud instrumentum, quod ultimi coeli extremam seu convexam superficiem (quam vocamus, Sphaeram Naturalem) vel Circulos in ea. conceptos, repraesentat."

[51] A study considering the range of concerns implicit in the lathe as a commonplace would include reflection on movement within mathematical argument, e.g. (Axworthy 2017, 2018). On the epistemic status of armillary spheres and globes, see (Mosley 2006a, b).

pushed to the foreground the question of what constitutes an adequate model. To what extent, that is, can a tool serve as a metaphor for the universe; to what extent do the qualities of one object or domain transfer over to another? Maestlin (and others who implicitly agreed by reusing the image) seems to have shifted away from accounts that seamlessly elided the lathe model with the materiality of the heavenly referent, opening a space for alternative theories of the heavenly sphere's composition. At the same time, he paid closer attention to the visual work of such machines *as* machines, suggesting that the model helped to think about physics.[52] Perhaps this is reading too closely—but, if not, then the paradigmatically Ptolemaic *Sphaera* helped to stage some of problems that a new mathematization of motion would set. At all events, for Maestlin as well as for Ciruelo, the example of the lathe prompted reflection on the relationship of model to original, of mathematics to matter.

6 Conclusion

Taken together with the tables and cosmographical tools discussed earlier in this chapter, the example of the lathe suggests the multiple trajectories that could meet within a capacious and growing genre such as the *Sphaera*. First, the *Textus de sphera*, first printed in 1495, presented techniques for calculation that became more widespread in books on the *Sphaera* during the course of the sixteenth century. Lefèvre offered commentary on literary and terminological questions, offering the kind of qualitative mastery of the science of the stars that any university educated man was expected to have in the Renaissance. As I show elsewhere, this skill set can be traced through later versions of the *Sphaera*, and constitutes an important shift in the wider cultural expectation of early modern Europe that educated people should be literate in the arts of number as well as the alphabet (Oosterhoff 2020).

Second, I have argued that Lefèvre's *Textus de sphera*, first published in 1495, brought together the techniques of Ptolemy's *Geography* with the genre of the *Sphaera*. Therefore, Lefèvre's book could be read as a cosmographical handbook; and I have suggested that it did set a precedent for Ringmann, Glarean, and Fine. Lefèvre's expectation that readers of the *Sphaera* should be interested and able to calculate for themselves locations on a map fed the new genre of cosmography.

A third trajectory is particularly clear with the lathe, a visual topos that originates in Lefèvre's commentary. The reformulation of this topos offers a chance to consider the accruing visual culture of the *Sphaera* in print, and the way that a visual topos that could travel independently of verbal commentary. While Lefèvre's verbal account was eventually made redundant by new versions of the *Sphaera* that incorporated new tables and rules for their use, the visual power of the lathe seems to have helped it outlast Lefèvre's own text. As a material metaphor it reprised the

[52] E.g. (Lefèvre 2004). These twin shifts would produce a paradox: mathematics is separated from the heavens, while at the same time the machine's motion is conceived of *more* mathematically. For one discussion of the tensions at play, see (Gal and Chen-Morris 2013, 117–160).

Timaean account of the universe as a crafted mathematical object; as a visual object, particularly in the Wittenberg tradition where it came to represent a *hollow* or absence, it kept the viewer in mind of the difference between model and reality—a crucial cognitive habit in the later sixteenth century as alternative world models became public.

Appendix. Contents of the Early Handbooks of Cosmography

Sacrobosco, *Sphaera*	Matthias Ringmann, *Cosmographiae* (1507).	Heinrich Glarean, *De geographia liber unus* (Basel: Jo. Faber. Emmeus, 1527) [MS c. 1513]	Peter Apian, *Cosmographia* (1529).	Oronce Fine, *Cosmographia* (1st ed. 1532)
Lib. I				Lib. I
– Definition of the sphere – Four elements – Revolutions of the heavens – Rotundity of the earth – Diameter of the earth	1. Basics of geometry 2. Sphere, axis, poles	1. Principles of geometry and spherical astronomy 2. Celestial and planetary spheres 3. Motions of the spheres	1. Definition of cosmographia 2. Motion of the sphere	– The world, its parts and motions
– Lib. II				Lib. II
– Celestial circles – North, south poles – Division of the zodiac – Ecliptic – Colures, meridian, horizon, – Tropics – Five zones [parallels]	3. Celestial circles 4. Various degrees of the sphere 5. The five heavenly zones 6. Parallels	4. Axis, poles 5. Right, oblique spheres 6. Greater and smaller circles (ecliptic, colures, meridian, etc.) 7. Composition of the material sphere 8. Elevation of the poles 9. The five zones 10. Parallels	3. Definition of a sphere, its circles, signs of zodiac 4. The five zones 5. The parallels	– Equator, poles, zodiac – Longitude and latitude – Meridian – Minor circles – Five zones

(continued)

Sacrobosco, *Sphaera*	Matthias Ringmann, *Cosmographiae* (1507).	Heinrich Glarean, *De geographia liber unus* (Basel: Jo. Faber. Emmeus, 1527) [MS c. 1513]	Peter Apian, *Cosmographia* (1529).	Oronce Fine, *Cosmographia* (1st ed. 1532)
Lib. III				Lib. III
– Rising and setting of signs – Chronic, solar risings – Right ascensions – Inequality of days – Oblique ascensions – Living at equator – Arctic circle – Seven climes	7. The climes	11. The climes 12. Technique for measuring the earth 13. Breatdth of the earth 14. Longitudes of cities 15. Units of measurement	6. The climes 7. Longitude 8. Latitude 9. Calculating solar altitude 10. Longitudes of places 11. Units of measurement …	– Rising and setting of stars – Right and oblique ascensions
Lib. IV				Lib. IV
– Movement of the sun – Equant, deferent, epicycle – Lunar eclipse	8. Winds 9. Division of the earth, the ends of the sea, and distances of various islands [10.] A quadrant useful to the cosmographer	16. Winds 17. Ptolemy's method for depicting the world 18. The image of the world 19. Putting paper on a globe (i.e. gores) 20. Geographical names 21. Division of the earth 22–40. Europe… Africa…Asia…and regions beyond Ptolemy		– Inequality of days – Calculating equal/inequal times – Calculating solar altitude Lib. V – Parallels – Longitude and latitude – Calculating distances – Winds – Sines and chords …

References

Manuscripts

Glarean, Heinrich. ca. 1513. *De gaeometriae principiis ad sphaerae astronomicae noticiam necessariis*. Brown University, John Carter Brown Library, MS Cod.Lat.1.

Lefèvre d'Étaples, Jacques. ca. 1495. *De magia naturali.* Olomouc, University Library, MS M.I.119.

Sacrobosco, Joannes. 1493. *Tractatus de Sphera*. Paris: Georg Mittelhus. Munich, Universitätsbibliothek der Ludwig-Maximilians-Universität, 4 Inc.lat.310 no. 6. *A copy heavily annotated by Heinrich Glarean.*

Primary Sources

Apian, Peter. 1529. In *Cosmographia...Additis de eadem re ipsius Gemmae Phyr. libellis*, ed. Gemma Frisius. Antwerp: Arnold Berckmann.

Ciruelo, Pedro. 1498. *Uberrimum sphere mundi commentum intersertis etiam questionibus domini Petri de Aliaco*. Paris: Jehan Petit. http://hdl.handle.net/21.11103/sphaera.100038.

de Bovelles, Charles. 1511. *Que hoc volumine continentur: Liber de intellectu; Liber de sensu; Liber de nichilo; Ars oppositorum; Liber de generatione; Liber de sapiente; Liber de duodecim numeris; Epistole complures. Insuper mathematicum opus quadripartitum: De numeris perfectis; De mathematicis rosis; De geometricis corporibus; De geometricis supplementis*. Paris: Henri Estienne.

de Budweys, Wenceslaus Faber. 1508. *Opus sphericum Ioannis de Sacro Busto figuris et perutili commento illustratum*. Cologne: Heinrich Quentell. http://hdl.handle.net/21.11103/sphaera.100183.

de Lattes, Bonetus. 1493. *Anuli per eum compositi super astrologiam utilitates*. Rome: Per A. Fritag de Argentina.

de Sacrobosco, Johannes. 1472a. *Spaera mondi*. Ferrara. Gallus: Andreas Belfortis. http://hdl.handle.net/21.11103/sphaera.101122.

———. 1472b. *Tractatum de spera*. Venice: Florentius de Argentina. http://hdl.handle.net/21.11103/sphaera.100685.

———. 1488. *Sphaera mundi*. Venice: J. L. Santritter and H. de Sanctis. http://hdl.handle.net/21.11103/sphaera.100822.

———. 1490. *Spaerae mundi compendium foeliciter inchoat*. Venice: Ottaviano Scotto. http://hdl.handle.net/21.11103/sphaera.100885.

———. 1531. *Liber de sphaera*. Wittenberg: Joseph Clug. http://hdl.handle.net/21.11103/sphaera.100138.

———. 1538. *Libellus de sphaera*. Wittenberg: Joseph Klug. http://hdl.handle.net/21.11103/sphaera.101106.

Ficino, Marsilio. 1491. *Opera omnia Platonis*. Venice: Andrea Toresani de Asula.

———. 1496. *Commentaria in Platonem*. Florence: Lorenzo di Alopa.

Fine, Oronce (ed.). 1515. *Theoricarum novarum Textus Georgii Purbachii, cum expositione Capuani, Item de Prierio; Insuper Jacobi Fabri Stapulensis astronomicon: Omnia nuper summa diligentia emendata cum figuris ac commodatisssimis longe castigatius insculptis quam prius suis in locis adiectis*. Paris: Jean Petit.

———. 1532. *De mundi sphaera, sive Cosmographia*, part 3 of *Protomathesis*. Paris: Morhii. http://hdl.handle.net/21.11103/sphaera.101190.

Lefèvre d'Étaples, Jacques. 1492. *Totius Aristotelis philosophiae naturalis paraphrases*. Paris: Johann Higman.

———. 1495. *Textus de sphera Johannis de Sacrobosco, cum additione (quantum necessarium est) adiecta: novo commentario nuper edito ad utilitatem studentium philosophice parisiensis academie: illustratus*. Paris: Wolfgang Hopyl. http://hdl.handle.net/21.11103/sphaera.101126.

———. 1496. *Arithmetica elementa; Musica elementa; Epitome in libros arithmeticos divi Severini Boetii; Rithmimachie ludus que et pugna numerorum appellatur*. Paris: Johannes Higman and Wolfgang Hopyl.

——— (ed. and comm.). 1506. *Contenta. Politicorum libro octo. Commentarii. Economicorum duo. Commentarii. Hecatonomiarum septem. Economiarum publ. unus. Explanationis Leonardi in Oeconomica duo* [trans. Leonardo Bruni]. Paris: Henri Estienne.

———. 1516. *Textus de Sphaera Ioannis de Sacrobosco...Introductoria additione...illustratus*. Paris: Henri Estienne. http://hdl.handle.net/21.11103/sphaera.100990.

——— (ed.). 1517. *Euclidis Megarensis mathematici clarissimi Elementorum geometricorum libri xv, Campani Galli transalpini in eosdem commentariorum libri xv, Theonis Alexandrini Bartholamaeo Zamberto Veneto interprete, in tredecim priores, commentariorum libri xiii, Hypsiclis Alexandrini in duos posteriores, eodem Bartholomaeo Zamberto Veneto interprete, commentariorum libri ii*. Paris: Henri Estienne.

———. 1538. *Textus de Sphaera Ioannis de Sacrobosco...Introductoria additione...illustratus. Cum compositione Annuli astronomici Boneti Latensis et geometria Euclidis Megarensis (a latinum a Boetio translata)*. Paris: Simon de Colines. http://hdl.handle.net/21.11103/sphaera.101011.

Lefèvre d'Étaples, Jacques and Josse Clichtove, comm. 1517. *Introductorium astronomicum theorias corporum coelestium duobus libris complectens, adiecto commentario declaratum*. Paris: Henri Estienne.

Lefèvre d'Étaples, Jacques, Josse Clichtove, and Charles Bovelles. 1503. *Epitome compendiosaque introductio in libros arithmeticos divi Severini Boetii, adiecto familiari [Clichtovei] commentario dilucidata. Praxis numerandi certis quibusdam regulis (auctore Clichtoveo). Introductio in geometriam Caroli Bovilli. Astronomicon Stapulensis*. Paris: Wolfgang Hopyl and Henri Esienne.

Maestlin, Michael. 1597. *Epitome astronomiae: qua brevi explicatione omnia, tam ad sphaericam quàm theoricam eius partem pertinentia, ex ipsius scientiae fontibus deducta*. Tübingen: Gruppenbachius.

Quiccheberg. 1565 [2013]. *The first treatise on museums: Samuel Quiccheberg's Inscriptiones, 1565*. Trans. Mark A. Meadow, and B. Robertson. Los Angeles: Getty Publications.

Ringmann, Matthias. 1507. *Cosmographiae introductio cum quisbusdam geometriae ac astronomiae principiis ad eam rem necessariis insuper quatuor Americi Vespucii navigationes universalis Chosmographiae descriptio*. St. Dié: Vautrin Lud, Nicolas Lud.

———. 1509. *Grammatica figurata*. Strassburg: Gauthier Lud and the Author.

Secondary Literature

Allen, Michael J.B. 1994. *Nuptial arithmetic: Marsilio Ficino's commentary on the fatal number in book VIII of Plato's Republic*. Berkeley: University of California Press.

———. 1999. Ficino, daemonic mathematics, and the spirit. In *Natural particulars: Nature and the disciplines in Renaissance Europe*, ed. Anthony Grafton and Nancy G. Siraisi, 121–137. Cambridge, MA: MIT Press.

Armstrong Tyler, Elizabeth. 1952. Jacques Lefèvre d'Etaples and Henri Estienne the Elder, 1502–1520. In *The French mind: Studies in honour of Gustave Rudler*, ed. W. Grayburn Moore, 17–33. Oxford: Sutherland and Starkis.

Axworthy, Angela. 2016. *Le Mathématicien renaissant et son savoir. Le statut des mathématiques selon Oronce Fine*. Paris: Classiques Garnier.

———. 2017. La notion géométrique de flux du point à la Renaissance et dans le commentaire des Éléments de Jacques Peletier du Mans. In *Miroir de l'amitié. Mélanges offerts à Joël Biard à l'occasion de ses 65 ans*, ed. Christophe Grellard, 17–33. Paris: Vrin.

———. 2018. The debate between Peletier and Clavius on superposition. *Historia Mathematica* 45: 1–38.

Baldasso, Renzo. 2007. *Illustrating the Book of Nature in the Renaissance: Drawing, painting, and printing geometric diagrams and scientific figures*. Ph.D. diss., Columbia University.

Barker, Peter. 2011. The reality of Peurbach's orbs: Cosmological continuity in fifteenth and sixteenth century astronomy. In *Change and continuity in early modern cosmology*, ed. Patrick J. Boner, 7–32. Berlin/New York: Springer.

Bedouelle, Guy. 1976. *Lefèvre d'Étaples at l'intelligence des Écritures*. Geneva: Droz.

Besse, Jean-Marc. 2009. Cosmography and geography in the sixteenth century, the position of Oronce Fine between mathematics and history. In *The worlds of Oronce Fine. Mathematics, instruments and print in Renaissance France*, ed. Alexander Marr, 100–113. Donington: Shaun Tyas.

Bietenholz, Peter G. 1971. *Basle and France in the sixteenth century*. Geneva: Droz.

Bredekamp, Horst. 1995. *The lure of Antiquity and the cult of the machine. The* Kunstkammer *and the evolution of nature, art and technology*. Trans. A. Brown. Princeton: Markus Wiener.

Buisseret, D. 2003. *The mapmaker's quest: Depicting new worlds in Renaissance Europe*. Oxford: Oxford University Press.

Cassirer, Ernst. 1927. *Individuum und Kosmos in der Philosophie der Renaissance*. Leipzig/Berlin: Springer.

Chabás, José, and Bernard R. Goldstein. 2014. *Essays on medieval computational astronomy*. Leiden: Brill.

Clark, Stuart. 2007. *Vanities of the eye: Vision in early modern European culture*. New York: Oxford University Press.

Connors, Joseph. 1990. *Ars Tornandi*: Baroque architecture and the lathe. *Journal of the Warburg and Courtauld Institutes* 53: 217–236.

Crowther, Kathleen M., and Peter Barker. 2013. Training the intelligent eye: Understanding illustrations in early modern astronomy texts. *Isis* 104 (3): 429–470.

Dalché, Patrick Gautier. 2009. *La géographie de Ptolémée en Occident (IVe-XVIe siècle)*. Turnhout: Brepols.

Dupré, Sven, and Michael Korey. 2009. Inside the *Kunstkammer*: The circulation of optical knowledge and instruments at the Dresden court. *Studies in History and Philosophy of Science* 40: 405–520.

Faye, Emmanuel. 1998. *Philosophie et perfection de l'homme: De la Renaissance à Descartes*. Paris: Vrin.

Fenlon, Iain, and Inga Mai Groote. 2013. Heinrich Glarean's books. In *Heinrich Glarean's books: The intellectual world of a sixteenth-century musical humanist*, ed. Iain Fenlon and Inga Mai Groote, 303–383. Cambridge: Cambridge University Press.

Gal, Ofer, and Raz Chen-Morris. 2013. *Baroque science*. Chicago: University of Chicago Press.

Gingerich, Owen. 1988. Sacrobosco as a textbook. *Journal for the History of Astronomy* 19: 269–273.

———. 1999. Sacrobosco illustrated. In *Between demonstration and imagination: Essays in the history of science and philosophy presented to John D. North*, ed. Lodi Nauta and Arie Johan Vanderjagt, 211–224. Leiden: Brill.

Hamel, Jürgen. 2004. Johannes Sacrobosco: Handbuch der Astronomie, Kommentierte Bibliographie der Drucke der *Sphaera* 1472 bis 1656. In *Wege der Erkenntnis: Festschrift für Dieter B. Herrmann zum 65. Geburtstag*, ed. Dietmar Fürst, Dieter B. Herrmann, and Eckehard Rothenberg, 115–170. Frankfurt am Main: Harri Deutsch Verlag.

———. 2006. Johannes de Sacroboscos *Sphaera*. *Gutenberg-Jahrbuch* 81: 113–136.

———. 2014. *Studien zur Sphaera des Johannes de Sacrobosco*. Leipzig: Akademische Verlagsanstalt.

Hankins, James. 2003. Ptolemy's geography in the Renaissance. In *Humanism and Platonism in the Italian Renaissance*, 2 vols., 1, 457–468. Rome: Edizioni di storia e letteratura.

Heawood, Edward. 1905. Glareanus: His geography and maps. *The Geographical Journal* 25 (6): 647–654.

Johnson, Francis R. 1953. Astronomical textbooks in the sixteenth century. In *Science, medicine, and history: Essays on the evolution of scientific thought and medical practice written in honour of Charles Singer*, ed. E. Ashworth Underwood, vol. 1, 285–302. Oxford: Oxford University Press.

Johnson, Christine R. 2006. Renaissance German cosmographers and the naming of America. *Past & Present* 191 (1): 3–43.

———. 2013. Between the human and the divine: Glarean's *De Geographia* and the span of Renaissance geography. In *Heinrich Glarean's books: The intellectual world of a sixteenth-century musical humanist*, ed. Iain Fenlon and Inga Mai Groote, 139–158. Cambridge: Cambridge University Press.

Klinger-Dollé, Anne-Hélène. 2016. *Le* De sensu *de Charles de Bovelles (1511): conception philosophique des sens et figuration de la pensée*. Geneva: Droz.

Korey, Michael. 2007. *The geometry of power: Mathematical instruments and princely mechanics around 1600*. Berlin: Deutscher Kunstverlag.

Kremer, Richard L., and Matthieu Husson. 2012. Introduction: Astronomical tables in late Medieval Europe. *Journal for the History of Astronomy* 43: 267–268.

Laubenberger, Franz. 1982. The naming of America. Trans. Steven Rowan. *The Sixteenth Century Journal* 13 (4):91–113.

Lefèvre, Wolfgang, ed. 2004. *Picturing machines 1400–1700*. Boston: MIT Press.

Leitão, Henrique. 2009. Pedro Nunes against Oronce Fine: Content and context of a refutation. In *The worlds of Oronce Fine. Mathematics, instruments and print in Renaissance France*, ed. Alexander Marr, 156–171. Donington: Shaun Tyas.

Mandosio, Jean-Marc. 2013. Le *De magia naturali* de Jacques Lefèvre d'Étaples: Magie, alchimie et cabale. In *Les Muses secrètes: kabbale, alchimie et littérature à la Renaissance*, ed. Rosanna Camos Gorris, 37–79. Geneva: Droz.

———, ed. 2018. *La Magie naturelle/De magia naturali de Jacques Lefèvre d'Étaples: I. L'Influence des astres*. Paris: Les Belles Lettres.

Margolin, Jean-Claude. 1972. Mathias Ringmann's *Grammatica Figurata*, or, grammar as a card game. *Yale French Studies* 47: 33–46.

Massaut, Jean-Pierre. 1968. *Josse Clichtove, l'humanisme et la réforme du clergé*, 2 vols. Paris: Société d'Edition "Les Belles Lettres."

Maurice, Klaus. 1985. *Der drechselnde Souverän: Materialien zu einer fürstlichen Maschinenkunst*. Zürich: Ineichen.

———. 2004. The princely art of turning: On the aesthetic significance of the natural sciences and technology. In *Turned treasuries*, ed. George Laue, 17–24. Munich.

Monfasani, John. 1976. *George of Trebizond: A biography and a study of his rhetoric and logic*. Leiden: Brill.

Mosley, Adam. 2006a. Objects of knowledge: Mathematics and models in sixteenth-century cosmology and astronomy. In *Transmitting knowledge: Words, images, and instruments in early modern Europe*, ed. Sachiko Kusukawa and Ian Maclean, 193–216. Oxford: Oxford University Press.

———. 2006b. Spheres and texts on spheres: The book-instrument relationship and an armillary sphere in the Whipple Museum of the History of Science. In *The Whipple Museum of the history of science: Instruments and interpretations*, ed. Liba Taub and Frances Willmoth, 301–318. Cambridge: Whipple Museum of the History of Science.

———. 2009. The cosmographer's role in the sixteenth century: A preliminary study. *Archives Internationales d'Histoire des Sciences* 59 (2): 423–439.

Oosterhoff, Richard J. 2014. The fabrist origins of Erasmian science: Mathematical erudition in Erasmus' Basle. *Journal of Interdisciplinary History of Ideas* 3 (6): 3–37.

———. 2015. A Book, a Pen, and the Sphere: Reading Sacrobosco in the Renaissance. *History of Universities* 28 (2): 1–54.

———. 2016. Lovers in paratexts: Oronce Fine's republic of mathematics. *Nuncius* 31 (3): 549–583.

———. 2018. *Making mathematical culture: University and print in the circle of Lefèvre d'Étaples*. Oxford: Oxford University Press.

———. 2019a. Jacques Lefèvre d'Étaples and Charles de Bovelles on platonism, theurgy, and intellectual difficulty. In *Plotinus' legacy: The transformation of Platonism from the Renaissance to the Modern Era*, ed. Stephen Gersh, 73–95. Cambridge: Cambridge University Press.

———. 2019b. Apprenticeship in the Renaissance university: Student authorship and craft knowledge. *Science in Context* 32 (2): 119–136.

———. 2020. Visualising observation: Data management in astronomy before Galileo. In *Coping with Copia: Epistemological excess in early modern Europe*, ed. Fabian Krämer and Itay Sapir. Amsterdam: Amsterdam University Press.

Pantin, Isabelle. 2001. L'illustration des livres d'astronomie à la renaissance: l'évolution d'une discipline à travers ses images. In *Immagini per conoscere: Dal Rinascimento alla Rivoluzione scientifica*, ed. Fabrizio Meroi and Claudio Pogliano, 3–42. Firenze: Leo S. Olschki.

———. 2009a. Altior incubuit animus sub imagine mundi. L'inspiration du cosmographe d'après un gravure d'Oronce Finé. In *Les méditations cosmographiques à la Renaissance*, 69–90. Paris: Presses de l'Université Paris-Sorbonne.

———. 2009b. Oronce Fine's role as royal lecturer. In *The worlds of Oronce Fine. Mathematics, instruments and print in Renaissance France*, ed. Alexander Marr, 13–30. Donington: Shaun Tyas.

———. 2012. The first phases of the *Theoricae planetarum* printed tradition (1474–1535): The evolution of a genre observed through its images. *Journal for the History of Astronomy* 43: 3–26.

Popplow, Marcus. 2007. Setting the world machine in motion: The meaning of *Machina Mundi* in the Middle Ages and the early modern period. In *Mechanics and cosmology in the Medieval and early modern period*, ed. Massimo Bucciantini, Michele Camerota, and Sophie Roux, 45–70. Florence: Leo S. Olschki.

Reedijk, Cornelis. 1989. The story of a fallacy: Erasmus's share in the first printed edition in Greek of Ptolemy's geography (Basel, 1533). In *Theatrum orbis librorum: Liber amicorum presented to Nico Israel on the occasion of his seventieth birthday*, ed. A.R.A. Croiset van Uchelen, K. van der Horst, and G. Schilder, 250–276. Utrecht: HES Publishers.

Rice, Eugene F. 1972. *The prefatory epistles of Jacques Lefèvre d'Étaples and related texts*. New York: Columbia University Press.

Schmidt, Charles. 1879. *Histoire littéraire de l'Alsace à la fin du XVe et au commencement du XVIe siècle*. Vol. 2. Paris: Sandoz et Fischbacher.

Shank, Michael H. 2002. Regiomontanus on Ptolemy, physical orbs, and astronomical fictionalism: Goldsteinian themes in the "defense of Theon against George of Trebizond". *Perspectives on Science* 10 (2): 179–207.

———. 2007a. Regiomontanus as a physical astronomer: Samplings from the defence of Theon against George of Trebizond. *Journal for the History of Astronomy* 38: 325–349.

———. 2007b. Mechanical thinking in European astronomy (13th–15th centuries). In *Mechanics and cosmology in the Medieval and early modern period*, ed. Massimo Bucciantini, Michele Camerota, and Sophie Roux, 3–27. Florence: Leo S. Olschki.

———. 2009. Setting up Copernicus? Astronomy and natural philosophy in Giambattista Capuano da Manfredonia's *Expositio on the Sphere*. *Early Science and Medicine* 14: 290–315.

Tessicini, Dario. 2011. Definitions of "cosmography" and "geography" in the wake of fifteenth- and sixteenth-century translations and editions of Ptolemy's geography. In *Ptolemy's geography in the Renaissance*, ed. Zur Shalev and Charles Burnett, 31–49. London: Warburg Institute.

Thorndike, Lynn, ed. 1949. *The sphere of Sacrobosco and its commentators*. Chicago: University of Chicago Press.

Valleriani, Matteo. 2017. The tracts on the *Sphere*: Knowledge restructured over a network. In *The structures of practical knowledge*, ed. Matteo Valleriani, 421–473. Dordrecht: Springer.

Veyrin-Forrer, Jeanne. 1995. Simon de Colines, imprimeur de Lefèvre d'Etaples. In *Jacques Lefèvre d'Etaples (1450?–1536)*, ed. Jean-François Pernot, 97–117. Paris: Honoré Champion Éditeur.

Wallace, William A. 1981. *Prelude to Galileo: Essays on Medieval and sixteenth-century sources of Galileo's thought*. Dordrecht: Reidel.

Weiss, Benjamin. 2011. The geography in print: 1475–1530. In *Ptolemy's geography in the Renaissance*, ed. Zur Shalev and Charles Burnett, 91–120. London: Warburg Institute.

Westman, Robert S. 2011. *The Copernican question: Prognostication, skepticism, and celestial order*. Berkeley: University of California Press.

7

Fashioning Cosmology: Franco Burgersdijk as the Author of the Dutch *Tractatus de sphaera*

Marius Buning

Abstract In 1625, the highest authorities in the province of Holland prescribed the use of *De sphaera* in all the regional Latin schools. For the occasion, they decided that a new edition of the famous text by Johannes de Sacrobosco was needed, primarily to purify the textbook of its "barbaric" Latin. The task of editing the text was assigned to Leiden professor Franco Burgersdijk. In what follows, I situate Burgersdijk's 1626 edition of *De sphaera* in the broader context of Leiden Aristotelianism, discussing the relationship between religion, educational practices, and astronomy in the first decades of Dutch independence. Teasing out some of the differences with the other contemporary manuals circulating on the market, I argue that Burgersdijk's version of Sacrobosco's textbook must be seen as an attempt to make the Dutch less dependent on Catholic literature and modes of knowledge production.

1 Life and Work

Burgersdijk was a farmer's son, born in 1590 in De Lier, Delfland, not far from The Hague, where the Dutch authorities had declared their independence from landlord Philip II (1527–1598) and the Spanish Habsburg Empire in 1581.[1] According to his first biographer, the young Burgersdijk attended the Latin school at Amersfoort (1604–1606) and the Delft Gymnasium (1606–1610), where he first became familiar with the art of logic and rhetoric. Like many farmer's sons, he then attended the University of Leiden to study theology. He was thoroughly immersed in the art of disputation and undertook private lessons with the famous scholastic adversary of

[1] I am referring to the *Plakkaat van Verlatinghe* (1581), the unofficial declaration of independence. On the history of the Dutch Republic, see for instance (Israel 1995). On Burgersdijk, see (Dibon 1954, 90–119; van Meurs 1625, 340–42; Bos and Krop 1993; Molhuysen et al. 1927, 229–31).

René Descartes (1596–1650), Gisbertus Voetius (1589–1676), who was at the time still the vice-principal of the *Staten Collegie* (a boarding school for students of theology established by the States of Holland, the highest provincial authorities). By the time he finished his education, around 1614, Burgersdijk was conversant in Latin and Greek, in Roman History, and in the salient details of the Protestant faith. The next step in his education was undertaking a grand tour, in the course of which he found himself for some time in the Academy of Saumur. There, he worked with the Scottish philosopher Mark Duncan (ca. 1570–1640) and the strict Calvinist theologian Franciscus Gomarus (1563–1641), who had fled to the Huguenot University in 1611 to escape his Dutch Arminian opponents. And it was in Saumur, in 1614, that Burgersdijk would first attain to a professorship in philosophy.

Burgersdijk's academic career cannot be understood without some degree of insight into the religious troubles that flared up in the period right after the young Dutch Republic had agreed on an armistice with the Spanish enemy. The ink on the treaty, concluded in 1609, was not yet dry when the followers of the Leiden theology professor Jacobus Arminius (1560–1609) requested the States of Holland to call for a church synod in order to decide on a number of sensitive religious issues in the series of heated debates between Arminius and Gomarus (contingent on their respective leaders, followers of both theologians were also called Gomarists or Arminians). The issue at stake in the debates, which later took a violent turn, was the correct interpretation of the Holy Scriptures; whereas the Arminians reckoned with the possibility of a very moderate form of free will, the Gomarists insisted on a strict interpretation of Calvin's notion of double predestination and denied man any form of influence on his future salvation.[2]

The ecclesiastical conflict had been deeply intermingled with Dutch internal power politics from the start. Whereas the Gomarists favored a form of state religion—having the Calvinist Basel in mind as an ideal—the Arminians were generally more liberal in their views on the relationship between the church and the state.[3] The disagreements lingered on for years, suddenly escalating in 1618, when the central authorities of the federal Dutch state, the States-General, agreed to summon an international synod despite the objections of the mighty province of Holland, run mostly by regents who supported the Arminian faction. The States of Holland reacted by raising a military force to restore order, which brought them in direct conflict with Stadtholder Prince Maurice of Orange (1567–1625). In what has been identified as nothing less than "a coup," Maurice set off on a military campaign and quickly took power of the entire country.[4] The political skirmishes concluded dramatically with the beheading of the Land's Advocate of Holland, Johan van

---

[2] Arminians were willing to pretend a certain ignorance on the issue, and to speak of God's 'stationary' or 'permitting' will (*voluntas remissa*, pace Perkins and others). The Arminians were strongly influenced by the Spanish scholastics on this point, and notably by Suarez and Molina (Dekker 1993, 115, 188).

[3] Speaking in general terms, one could say that the Arminians consisted of people from the regent class, whereas the Gomarists were found in greater numbers among the common people (Prak 2005, 31).

[4] For more details, see (Israel 1995, 426–57).

Oldenbarnevelt (1547–1619), on the Inner Court in The Hague. The religious bickering was settled with the international Synod of Dort, which had its last meeting on the 9th of May 1619.[5] The Synod of Dort finally ruled in favor of the Gomarists and some two hundred Arminian preachers were dismissed from office (de Nijs and Beukers 2002, 311).[6]

Arminianism was rooted out at all levels in the Republic, including at the universities. With numerous positions opening up in the "purified" University of Leiden, Burgersdijk immediately saw his chance to return to Holland. The university initially offered him an *extraordinarius* professorship in logic, which he accepted on the 25th of March 1620, obtaining the title of *Magister philosophiae* on the 31st of March 1620, during a promotion ceremony led by the famous astronomer and mathematician Willebrord Snellius (1580–1626). In November 1620, ethics was added to Burgersdijk's teaching duties, and on the 9th of February 1621, Burgersdijk was promoted to *ordinarius* professor. In 1623, he finally completed his theological education by defending a disputation entitled "The Clarity of the Bible and Its Interpretation" (Krop 2014, 2).

Meanwhile, Burgersdijk had maintained an ongoing interest in natural philosophy. In two series of disputations, the first held in 1623 and the second in 1627, his students had dealt with the Aristotelian knowledge corpus on natural knowledge. It brought him into direct conflict with the Leiden Professor Jacchaeus, his former teacher, who bitterly complained at the Leiden Senate that Burgersdijk was overstepping the boundaries of his discipline (Dibon 1954, 97). Jacchaeus had held the chair of natural philosophy since 1617 and remained active in that capacity until his death in 1628.[7] Soon after the chair had become vacant, on the 9th of May 1628, Burgersdijk exchanged it with his chair in ethics. It was under these circumstances that

> …Burgersdijk devoted two manuals [to physics]. The *Idea* is conceived as a "guide" to be used in disputations, presenting the subject matter in definitions and short theses. By referring only to "new" authorities such as Jacopo Zabarella, Benedict Pereira, Franciscus Toletus, and the commentators of Coimbra, Burgersdijk stated as his goal to open the debate on the "text of Aristotle." In the more extensive *Collegium*, the same doctrinal tradition is elaborated in a synthetic order. (Krop 2014, 3)[8]

[5] The first meeting of the Synod of Dort had been on the 13th of November 1618.

[6] The Synod of Dort confirmed the authority of the Belgic Confession and the Heidelberg Catechism, and drew up ninety-three new canonical rules. Furthermore, the Synod decided upon several other issues, such as the production of a Bible translation into the vernacular funded by the States-General (which eventually resulted in the 1637 *Statenbijbel*).

[7] The Scot Gilbert Jack (Lat. Jacchaeus) had first held the chairs for logic and ethics at the University of Leiden (1605, 1607), before publishing his *Institutiones physicae* in 1614, which has been identified as "the first work of any note on physics to issue from Leiden's faculty" (Ruestow 1973, 14). Jacchaeus was made a full professor of natural philosophy in 1617. On Jack, see (Smolenaars 2004). He "taught a reformed Aristotelianism which also made use of recent scholarship, including Zabarella, Suarez, and the Coimbra commentators" (Mijers 2017, 91). Burgersdijk would do the same. He had also taken over the teaching duties of Jacchaeus, when the latter was suspended for 3 months on the suspicion of Arminian sympathies in 1619, in the wake of the Synod of Dort.

[8] On Burgersdijk's *Physics*, see also (Dibon 1954, 90–119; Ruestow 1973, 5–33). An overview of Burgersdijk's publications can be found in (Bos and Krop 1993, 167).

Although Burgersdijk did recognize the birth of new stars (Ruestow 1973, 30) and that heavens act on the terrestrial world by occult qualities, not just by light and motion (Vermij 2016, 293; Vermij 2002, 132), the ensemble of his natural philosophy has nevertheless been characterized as "the Aristotelian principles in an extreme and simplistic (and distorted) rendering…presented largely in the guise of first matter and substantial form" (Ruestow 1973, 22). Indeed, it was not for his natural philosophy that he would be remembered when he died on the 19th of February 1635, eleven days after the end of his last term as Rector Magnificus of Leiden University.[9]

Instead, the Leiden Professor became known for his compendium on logic, the *Institvtionvm logicarvm libri duo* (Leiden, Commelius, 1626), which, "[a]lthough abhorred by students[,]…lingered in the curriculum of the British universities well into the eighteenth century" (Krop 2014, 1).[10] Like his other manuals, the *Institutionum* was characterized by a strong inclination towards order and method.[11] Although an Aristotelian in heart and soul, Burgersdijk deplored the messy presentation in the work of the Stagerite. On the other hand, he believed that in the work of humanists such as Valla, Agricola, and Ramus, where method did prevail, content had fallen behind. It was thus best to follow the example of Bartholomaeus Keckermann (1672–1609) by adopting the method of Ramus and the material of Aristotle. Burgersdijk claimed that, by doing so, he followed in the footsteps of his former mentor in Saumur, Mark Duncan.[12]

Aside from his works on logic and natural philosophy, Burgersdijk wrote a number of *compendia* on moral and political thought, again heavily inspired by Aristotle. If one was to summarize his academic career in one sentence, one could say that Burgersdijk was perhaps not a very original thinker, but a dedicated educator who brought stability to Leiden University in the wake of the Synod of Dort. It is indeed in this light that we must also read Burgersdijk's 1626 edition of Sacrobosco's textbook on the *Sphaera*—a booklet produced under the direct order of the Dutch political authorities.

[9] Burgersdijk was the rector of Leiden University in 1629, 1630, and 1634.

[10] On the influence of Burgersdijk in England, see also (Feingold 1993). On the status of Philosophy course at Leiden University (Sassen 1959; Dibon 1954, 71–76). Like *De sphaera*, the *Institutionum* was commissioned by the States of Holland in the framework of the 1625 School Order, which was intended to provide unity in the Dutch educational program.

[11] On "order" in the *Philosphiae naturalis*, see (Dibon 1954, 95). On the general emphasis on "method" at the time, see (Gilbert 1963).

[12] On the influence of Duncan on Burgersdijk, see (Prost 1907, 42–44). On the Scottish influence on Dutch universities, see (Mijers 2017). On Leiden University more generally, see (Otterspeer 2000; Ruestow 1973; Wiesenfeldt 2002).

2 The *Schoolorde* (1625)

In the hive-off of the Spanish Habsburg Empire, the Northern Netherlands became a kind of federal republic, in which seven semi-independent provinces were each represented with one vote in the Assembly of the States-General seated in The Hague. The provinces, of which Holland was by far the most powerful, were in turn managed by a provincial assembly led by an individual Stadtholder (whose office was a strange remnant from the Habsburg period, resembling a cross between a prince and a president). In reality, the Stadtholder of Holland observed the Stadholdership of the majority of other provinces.

In addition, there were many other interest groups in the political domain, including the different trading companies and various congregations within the Dutch Reformed Church, which met at irregular intervals in provincial synods. It was at one of these provincial synods that the call for a reform of the Dutch educational system could first be heard. During the South Holland Synod held in Rotterdam in 1581, members of the congregation expressed their explicit concern about the lack of properly schooled ministers. The radical break with the Catholic Habsburg landlord in their eyes called for immediate joint action by the Republican authorities to ensure the establishment of a single Dutch reformed identity. Though similar sentiments could be heard at various other provincial synods between 1583 and 1618, the political leadership did not take any action.[13] The issue of religious education was taken up again at the international Synod of Dort, where various ecclesiastical representatives insisted that the States-General should be urged to bring "unity in the education of the Youth, in particular in the rules of Grammar, Dialectics and Rhetoric."[14] An official request to the States-General was indeed submitted on the 18th of May 1619; yet again it did not prompt any action. The congregation continued to complain about the matter at various synods, and in 1622, a commission was even appointed "to study the school order made in the Pfaltz, in Gelderland, in Groningen and elsewhere" in order to come up with a proposal to reform the educational system.[15] Eventually, it all had little effect.

Simultaneously, complaints about the inadequate level of secondary education could be heard at Leiden University, where the professors claimed that students were not well prepared to follow even the most basic university courses in Latin. They attributed this shortcoming to the fragmentation of pre-university school programs and argued that school pupils would benefit from having a more standardized

[13] Kuiper provides a list with no less than sixteen instances of complaint (Kuiper 1958, 39–40).

[14] "...eenparigheyt gehouden werde in de onderwijsinghe van de Jeught, voor al in de regelen van de Grammatica, Dialectica en Retorica." Kerkelijk Placaetboek van N. Wilkens (1722), I, p. 127 sq. As cited in (Kuiper 1958, 40). All translations of the source material are by the author unless otherwise stated.

[15] (Kuiper 1958, 40): "...visiteren ende confereren de schoolordeninghe, die gemaect syn in de CHURFORSTELICHE PFALTZ, in GELDERLAND, in GROENINGEN ende elders om daeruyt een project te maecken." The commission was formed during the South Holland synod of Gorinchem in July 1622.

regimen of studies.[16] After various failed efforts, the Senate of the University, in 1624, impressed upon the Assembly of the States of Holland the necessity of concrete measures. Clearly aware of the earlier efforts in the various synods, a plan was handed in that would (1) guarantee the quality of the secondary educational program and (2) ensure that the youth could move from city to city without having to use new books at every location (Kuiper 1958, 48–51). Should new measures not be implemented, the Senate warned, it was "to be feared that the young adolescents will degenerate again *ad barbariam.*"[17]

The States of Holland gave a full and sympathetic consideration to the complaints of the University, asking the Curators to draft a program that could be implemented in the Latin Schools in the province. On the 4th of February 1625, the Curators organized a meeting with some deputies of the University Senate and of the Holland Synodes in order to come up with a proposal. The deputies of the University Senate were Antonius Walaeus (1573–1638), Antonius Thysius the Elder (1565–1640), Gerhard Johannes Vossius (1577–1649), Petrus Cunaeus (1586–1638), Daniel Heinsius (1580–1655), and Burgersdijk, who was thus closely involved in the making of the order from the very start. Indeed, no less than three members of the senatorial committee managed to put their own books on the obligatory reading list; it seems not to have bothered the commission nominated on the 22nd of July 1625 to study the proposals made by the University.[18] The draft program was approved and on the 1st of October 1625, the States of Holland enacted the School Order (*Schoolorde*) that would remain in force until the early nineteenth century.

3 Print and Privilege

The School Order meticulously prescribed the use of specific books on specific hours, laying out an intense program for school-aged children (ca. twelve to eighteen years old) from Monday morning through to Saturday afternoon. Willem Frijhoff has provided such an eloquent and concise overview of this program that it deserves to be quoted here in full:

> When the child had completed the writing classes, he entered the sixth (lowest) class, where he learned declensions and conjugations and parts of speech (the *Rudimenta*), and practiced reading Erasmus's dialogues. Subject matter for the fifth class included the letters of Cicero and the Disticha of Cato, as well as Erasmus's book of etiquette. The fourth class brought more Cicero, some Terence, and—as first exercises in prosody—Ovid's *Tristia* and Virgil's

[16] Kuiper convincingly identifies examination methods based on memorization as one of the main "culprits" in this entire problematic (Kuiper 1958, 45).

[17] (Kuiper 1958, 49): "…staet te vreesen, dat de jonge jeugt wederom soude mogen ad barbariem vervallen."

[18] The committee appointed by the States of Holland consisted of representatives from Dordrecht, Haarlem, Alkmaar, Amsterdam, and Hoorn (all cities with famed Latin Schools).

> *Bucolica*. From that point on the pupil also delved seriously into Greek. The focus of the second class was on rhetoric, explained with the aid of Vossius's handbook, and more Cicero, who was now studied in great detail. The program of the highest class included the logic of the Leiden professor Franco Burgersdijk, as well as Horace and Homer, and again the inescapable Cicero. The top group of the highest class was initiated into the world of the quadrivium (music, arithmetic, geometry, and astronomy), with lessons in ethics, the arithmetic of Gemma Frisius, and the Sphaera of Johannes de Sacrobosco. (Frijhoff and Spies 2004, 245)[19]

With the existence of a fixed catalogue with books that had to be printed on a regular basis, there was a group of printers in the Republic who were attracted by the prospect of making financial gains. One way to increase their profit margin was by securing a so-called "printing privilege" from the local authorities. A printing privilege provided the exclusive rights for a limited time (generally around 8 years) to print a book for the Dutch market.[20] Regularly issued for the printing of catechisms and schoolbooks written in Dutch, the Republican authorities did not issue many printing privileges for the production of works in Latin. In 1581, for example, the schoolmaster Cornelius Schonaeus (1540–1611) had tried in vain to obtain a privilege for a Latin grammar book, which would prevent instruction "from any other grammar book in the province of Holland."[21] His request was denied, and the States of Holland opted to give a reward instead. The same year, the famous printer Christophe Plantin (ca. 1520–1589) made another attempt to obtain a printing privilege for a grammar book by Cornelius Valerius (1512–1578), a former professor of Latin at the famed Collegium Trilingue in Leuven (Molhuysen 1913, 18 (27th of April 1581). Although supported by the Leiden University Senate, Plantin's attempt to claim the exclusive rights to the Latin grammar book were similarly unsuccessful. The book by Valerius, first published in 1567, would indeed never be printed in the Republic, but in Antwerp, in 1583, under the earlier privilege that had been issued to Plantin by the Spanish Habsburg authorities in 1566 for a set of books once written by the Leuven Professor for the instruction of the youth:

> Grammaticarum institutionum libri quatuor.
> Tabula summam artis Rhetorica complectens etc.

---

[19] The short overview by Frijhoff is obviously far from complete; a full transcription of the Order can be found in (Kuiper 1958, 5–25).

[20] Printing privileges were mainly issued by the (inter)provincial authorities in the Dutch Republic, not by the cities. In addition to sanctions against copying, they often contained a clause that prohibited the import of printed materials from abroad. The first printing privileges issued independently by the Dutch authorities date from the 1580s; they were issued for books as well as for other printed materials, such as globes and copper engravings. On the history of printing privileges in the Dutch Republic, see (Hoftijzer 1993; Schriks 2015; Buning 2019; Orenstein 2006).

[21] (Kuiper 1958, 39–40): "van dat geene andere grammatica binnen Holland geleert soude mogen worden,"18th of March 1581. Schonaeus was the rector of the Latin School in Haarlem; the book was commissioned by the States of Holland, being part of the first political interference in the Dutch educational program by Prince William of Orange, who consulted a commission that included high-level scholars such as Douza, Lipsius, and Donellus. Schonaeus eventually obtained fifty guilders instead of a printing privilege.

Tabulae totius Dialectices etc.
De sphaera & primis astronomia rudimentis libellus.
Physicae, sue de natua philosophia institutio etc.[22]

In the wake of the Council of Trent (1545–1563), the Spanish Habsburg authorities had clearly chosen to privilege a strong and uniform teaching program in preparation for the Faculty of the Arts. The School Order, as we shall see, was in many ways an attempt to keep up with the educational program as it was implemented in Leuven and elsewhere.

In contrast to what happened in Leuven, however, the Dutch authorities chose not to issue any printing privilege for the books produced in the framework of the School Order. The magistrate of Leiden had initially pushed in the States of Holland for a privilege on "the first Edition of the Books of the School Order, revised and addressed by the Deputies of the Senate of the University, in Leyden, so that the first Copies may be printed perfectly."[23] Yet most members of the Assembly abhorred the idea of a Leiden privilege and it was soon decided that printing the books in the School Order "should remain free" to all.[24] Nonetheless, Burgersdijk initially managed somehow to obtain a printing privilege for his compendium on logic (the *Institvtionvm logicarvm*), which inevitably led to problems when he presented the finished book in the Assembly on the 23rd of September 1625. At that point, the authorities decided "that the privilege will not have force, and [instead] to honor him [=Burgersdijk] with a hundred and fifty guilders for his trouble and Dedication."[25] After strong objections by the author, the States reconsidered its decision one week

---

[22] The example comes from (Valerius 1583, A-IV) but can be found in any of the books listed here. See also (Imhof 2014, 702–03). *De sphaera* by Valerius, for that matter, again followed closely the order used by Sacrobosco.

[23] (Koninklijke Bibliotheek, KW 402 A 31, 607): "…om den eersten Druk van de Boeken van de Schoolordre, by de Gedeputeerden van den Senaat van de Universiteyt gerevideert ende geaddresseert, te moogen doen doen binnen Leyden, op dat de eerste Exemplaren perfect moogen werden gedrukt…," 20th of December 1625. The source material as well as translations of the "Resolutions concerning the Dutch School Order" can be found online at (Bently and Kretschmer 2017). Note that a couple of months later, the Leiden authorities again probed "whether one had to grant the Professors (who corrected the Books) a privilege for some years to have the Books printed at Leyden, so that these would be printed well and correctly for the first journey" (*of men de Professoren (de Boeken gecorrigeert hebbende) soude moogen toestaan Octroy voor eenige jaaren, om de Boeken tot Leyden te doen drukken, op dat se wel en correct souden moogen worden gedrukt voor de eerste reyse*). Koninklijke Bibliotheek, KW 402 A 31, 663 (8th of April 1626). The Provincial Assembly again decided against it.

[24] (Koninklijke Bibliotheek, KW 402 A 31, 607): "…het selve moeten aan alle de Leeden vry gelaaten werden," 20th of December 1625. Translation and transcription available online at (Bently and Kretschmer 2017). It is remarkable that all of the first editions were nevertheless printed in Leiden, and that also later on, Leiden printers kept a privileged position in practice.

[25] (Koninklijke Bibliotheek, KW 402 A 31, 751): "…dat het Octroy van geen kragt en sal weesen, ende dat men hem voor de moeyte en Dedicatie vereeren sal met honderd vyftig guldens," 23rd–24th of September 1626. Translation and transcription available online at (Bently and Kretschmer 2017).

later, agreeing "that the aforesaid Privilege will serve him from now on for another year in order to get rid of his Copies at a price deemed appropriate by the Delegates States."[26] Burgersdijk was forbidden, however, to print any additional copies mentioning the existence of a printing privilege.[27] It was, indeed, without any privilege that Bonaventura Elzevir (1583–1652) and Abraham Elzevir (1592–1652), the printers to the University of Leiden,[28] one year later marketed the octavo edition of:

> Sphaera Johannis de Sacrobosco, Decreto Illustr. & Potent. DD. Ordinum Hollandiae & West-Frisiae, in usum scholarum ejusdem provinciae, sic recensita, ut & Latinitas, & methodus emendata sit, multaque addita, quae ad huius doctrinae illustrationem requirebantur. Operam & studio Franconis Burgersdicii. Lugduni Batavorum, Ex Officina Bonaventurae & Abrahami Elzevir. Academiae Typogr. 1626 (Burgersdijk and de Sacrobosco 1626).

The professors involved in the editing and production of schoolbooks that had become mandatory under the School Order would eventually obtain a total of 2000 guilders for their collective efforts.[29] Alas, we do not know the exact share of this that Burgersdijk received. What we do know, based on a declaration signed by Burgersdijk upon finishing his project, is that the 1626 edition of *De sphaera* had a first print run of 600 exemplars, which were sent to twenty different cities in the province of Holland; the biggest school had to purchase ninety-six exemplars and the smallest school six (Kuiper 1958, 86).[30] Apparently, the Hoorn school master Jacobus Hovius managed somehow to get the schoolbooks for his students for free (Abels 2002, 409), but the rest of books were sold for exactly one *stuiver* and thirteen *penningen* per book.[31] Considering the large numbers of books that were sent to the schools, we may assume that the books were used by students individually in class and passed on from year to year (i.e. they remained in the school).

---

[26] (Koninklijke Bibliotheek, KW 402 A 31, 758): "...om middelertydt syn exemplaren quyt te maaken, ten pryse als de HH. Gecomm. Raden sullen goedvinden ende geen meer moogen doen drucken," 1st of October 1626. Translation and transcription available online at (Bently and Kretschmer 2017).

[27] That printers did not always adhere to the state regulations can be deduced from the fact that, in 1645, an abridged version of Burgersdijk's Logic was still printed with an indication of the 1626 privilege issued by the States General (Burgersdijk 1645). The original printing privilege had by then long lost its legal validity. On the value and meaning of these expired printing privileges, see (Buning, 2019).

[28] On the Elzeviers, see (Willems 1880; Dongelmans et al. 2001).

[29] (Koninklijke Bibliotheek, KW 402 A 31, 780–81), 19th of November 1626. Available online at (Bently and Kretschmer 2017).

[30] The total printing cost of the project was 4604 guilders and fourteen stuivers. Although we lack the exact numbers, the size of the Latin school student body was considerable in the seventeenth century Dutch Republic. It has been reported that, at some point, 2200 students attended a Latin school in Amsterdam (Fortgens 1958, 31). In that light, 600 copies of *De sphaera* was not necessarily such a large number.

[31] A guilder was worth twenty stuivers; one stuiver equaled sixteen penningen.

4 Situating *De sphaera*

To fully appreciate the status of *De sphaera* in the educational program of the School Order, it is necessary to say a few words more about the educational system in the Dutch Republic. That system was divided into several phases. The youth first learned to read and write, as well as the fundamentals of the Heidelberger catechism, in so-called "Dutch Schools" (*Nederduytsche scholen*), or occasionally in "French Schools" (*Franse scholen*), where they were taught a few words of French and where usually some more attention was given to mathematics and geography.[32] Youngsters attended these schools (with a lot of variation) between the age of five and twelve, after which most of them dropped out to take up a trade or a craft. Only some continued to follow courses on specific topics in private or evening schools, and only a select group of boys would continue to attend a Latin school that prepared them for university education.[33] Only at the Latin schools was education carried out in groups, in contrast to the other school types, where teaching was done on an individual basis.

The Latin schools were subdivided into six classes, of which the highest class (class 1) was sometimes split into two divisions. For most pupils, the lowest division was the normal *terminus* of their education. But sometimes a select group of pupils formed a second division in which more attention was given to topics that were part of the curriculum at the University Faculty of the Arts. It was exclusively in these upper divisions that fragments of *De sphaera* were read.[34] In fact, the School Order had been quite precise about when to read the textbook by Sacrobosco, namely on any day except Wednesday and Saturday:

> In the highest division, when such a division exists in accordance with local circumstances and tradition, the beginnings of higher philosophy will be taught. And to this end will be taught: The *Physics* by Magirus, the *Ethics* by Walleus, the *Arithmetic* of Gemma Frisius, and the *Sphaera* by Sacobosco, with as much improvement of the Latin as possible. Furthermore, for the General History of Rome and the Outside World, Plorus and Justinus. Moreover, from Mela or Dodonaeus, and from 6 or 7 of the most important maps of Ortelius, the situation of the known parts of the world.[35]

---

[32] There were many variations and even crossovers possible between French and Dutch Schools. Some major cities also had *Grote Scholen*, a remainder of the Middle Ages, which combined primary and secondary education. The *Grote Scholen* were paid for by the urban governments, however, and thus the influence of the Church was limited. This paragraph is based on (Booy 1980, 23–50).

[33] For the history of the Latin schools see (Bot 1965; Fortgens 1958; Bastiaanse and Evers 1986; Bastiaanse et al. 1985; Emmius 1951; Witsenburg and Van der Noordaa 1988).

[34] Such classes seem to have flourished mainly in cities where at some point either an Athenaeum Illustre (Deventer 1630, Amsterdam 1632, Breda 1646, Rotterdam 1681) or a university would be founded (Leiden 1575, Franeker 1585, Groningen 1614, Utrecht 1636, Hardewijk 1648, Nijmegen 1655).

[35] (Kuiper 1958, 20): "In superiori ordine, ubi is pro loci conditione admitti solet, initia solidioris Philosophiae traduntur. Atque hic docebitur Physica Magiri, Ethica Wallaei, Arithmetica Gemmae Frisii, Sphaerica Sacrobosci, Latinitate ubi poterit emendata. Item propter Universatem Historiam Romanam et Externam Florus et Justinus. Item ex Mela aut Dodonaeo, et sex septem tabulis praecipuis Ortelii, docebitur situs nobilium partium orbis terrarium."

The authors of the Order envisioned a comprehensive program, in which the pupils in the highest class would gain a broad overview of geography, astronomy, ethics, and natural philosophy. The next section will explore in which ways the Dutch were influenced by earlier initiatives of the Jesuits in this respect.

5 Burgersdijk's Edition of *De sphaera*

The 1625 School Order had not been the first attempt to organize the educational system in the independent Northern Netherlands. The province of Friesland had introduced a first School Order in 1588, and similar directives are known in Zeeland (1590) (Cau et al. 1664, 2:column 2207–08) and Gouda (1593) (Abels 2002, 409). Moreover, there were many urban regulations regarding the Latin Schools (an inventory of all Latin schools of the Republic in 1650 shows that ninety-two towns had a Latin school (Frijhoff and Spies 2004, 245)). These policy measures were a local affair, but also played an important role within an international framework of continuous ideological warfare. Accordingly, the Holland School Order of 1625 should be seen as a concrete response to Jesuit efforts to streamline the educational program, including the 1566 Leuven initiative mentioned earlier, or the *Ratio Studiorum* of 1599.[36] The Curators of Leiden University emphasized this aspect in their 1624 guidance letter to the Senate, where they unambiguously stated that the School Order was needed:

> Just as similar orders unifying the class schedules have been implemented with enormous success by the Enemies of our Religion in the schools of the Jesuits, who, much to our disadvantage and prejudice, have managed that way to direct and prepare their youth much better for the further understanding of the higher disciplines, which are afterwards studied in the universities.[37]

The Holland School Order was thus part of a broader process to gain control over the educational system in an attempt to establish the Dutch Republic as a Calvinist state. Other elements in that process included the appointment of school masters who, from 1619 onwards, were obliged to sign a confession that they supported the

---

[36] I am not elaborating here on the various efforts made in the Catholic world to gain control over the educational system. For an excellent analysis on the impact of the *Ratio Studiorum*, with a focus on mathematics and France, see (Romano 1999).

[37] (Kuiper 1958, 49–50): "Gelyck deselffde eenparige ordre van lessen met sonderlinge goet succes werdt gepractiseert by de Vyanden onser Religie in alle scholen der Jesuyten, die, tot onse groote affbreucke ende naerdeel, hare jeucht inde so eerste beginselen met dusdanige eenparige ordre veel beter aenleyden ende bequaem maecken tot het vordre begryp van hoogere disciplinen, die daernaer in de Universiteyten werden geleert." In 1624, the Curators had still pushed for a "national" Order, but due to practical considerations the Senate later decided to first realize a provincial order in the hope that other provinces would follow the example set by Holland.

Canons of Dort (Kuiper 1958, 51) as well as the ban on Catholic teaching to counter the spread of "popery superstitions" (States of Holland 1589).[38]

Amidst these various power games to claim hegemony in the religious sphere, one also finds a reference to *De sphaera* by Sacrobosco—a teaching book, originally written around 1230, that had served for several centuries to instruct the youth of the Western Latin world in the first principles of astronomy.[39] Especially with the breaking apart of the Christian world in sixteenth-century Europe, several editions of *De sphaera* had come to compete fiercely with one another. The Lutheran reformer Philipp Melanchthon (1497–1560), for instance, had used the text of Sacrobosco to formulate a panegyric on astronomy that "taught about Providence" (Kusukawa 1995, 129) (Chap. 10), whereas the Jesuit mathematician Christopher Clavius (1538–1612) used *De sphaera* from 1570 onwards as book of *loci* containing the *Summa* of his cosmological insights. It was particularly in the paratext and the commentaries that *De sphaera* concealed particular worldviews. The 1626 edition by Burgersdijk was no exception in that respect.

Burgersdijk's version of *De sphaera* was preceded by a short preface (see Appendix). There, the author explained that the representatives in the States of Holland had made it his assignment to purify the text by Sacrobosco from its "barbaric" (*barbariem*) Latin, making the textbook perspicuous and easy to understand. The Leiden professor clearly stated that his edition of *De sphaera* had been chiefly inspired by the "highly learned" (*doctissimus*) work of Christophorus Clavius, to which he had added some of Peuerbach's *Theorica novae planetarum* (actually, Clavius had done the same). In contrast to his Jesuit counterpart, however, Burgersdijk did not include any excursions on mathematical demonstrations, since this would go beyond the understanding of his readers. His aim was primarily to provide an effective textbook "from which adolescents could learn the first principles of Astronomy, later to be studied more fully and accurately in the Academy."[40]

The unconcealed reference to a famous Jesuit mathematician in a book by a Calvinist Aristotelian in the Dutch Republic is, at first sight, remarkable. First published in 1570, Clavius's work had been privileged by Pope Pius V (1504–1572) "to the common benefit of all the scholarly community as well as his Jesuit comrades (*sociorum*)," and it had become a standard work in Jesuit defense against the Copernican worldview (Clavius and de Sacrobosco 1570, A1v).[41] Clavius had com-

[38] The placard against printing seditious religious literature, first issued by the States of Holland on 20th of December 1581, was later renewed in various forms and variations. As an introduction to censorship in the Republic, see (Weekhout 1998).

[39] For a first introduction to (the commentary tradition on) *De sphaera*, see (Hamel 2014; Pedersen 1985; Thorndike 1949; Valleriani 2017).

[40] (Burgersdijk and de Sacrobosco 1626, sig. a1–a2v): "ut ex eam adolescentes prima Astronomiae principia deliberarent, quam in Academiam deinde discerent plenius accuratiusque." See Appendix.

[41] "ad communem omnium studiosorum vtilitatem suis sociorumque…." Note that the printer Victorius Helianus (Vittorio Eliano) obtained a highly interesting privilege for a set of books to promote the Jesuit production of knowledge. Eliano was a converted Jew, who specialized as a censor of Hebrew texts (Casetti 1993).

plemented each fragment of the text by Sacrobosco with an extensive commentary, which had resulted in a hefty book with an index on chapters, persons, and topics (the first edition had consisted of no less than 506 folio pages). The Jesuit Father then continued to update his book in the seventeen subsequent editions that followed in the years 1581–1611, which represented something of a highpoint in the early modern commentary tradition on *De sphaera*.[42] Yet another important strand in that tradition were the annotations by French humanist Élie Vinet (1509–1587), first published in 1556, and later reprinted on almost a yearly basis.[43] Over time, Vinet's commentary was complemented by other commentaries, and it was often printed together with additional *compendia*; yet it remained the basis for the proper understanding of Sacrobosco's text in the sixteenth and seventeenth centuries. As an illustration, a 1573 edition published by the Antwerp printer Jean Bellère (1526–1595) included, in addition to a version of *De sphaera* annotated by Vinet and Franciscus Junctinus (Francesco Giuntini, 1523–1590), Sacrobosco's *Computus ecclesiaticus* as well as his *Alogarismus*, a compendium to *De sphaera* by Ioannes Pierius Valerianus (Pierio Valeriano, 1477–1558), and the commentary by Pedro Nuñes (1502–1578) on the climate zones (Sacrobosco et al. 1573). The idea to print *De sphaera* together with the *Computus ecclesiasticus* had first come from Melanchton and Georg Joachim Rheticus (1514–1574), whose so-called Wittenberg editions might be identified as a third strand in the early modern commentary tradition on Sacrobosco's text (Rosen 1974, 245; Westman 1975) (Chap. 9).[44] Indeed, just as interesting as Burgersdijk's open reference to Clavius in the Leiden edition is his omission of any references to other editions of *De sphaera* that circulated on the market at the time.

In comparison to those other editions, the version put out by the printers of Leiden University was rather short. After 120 pages of text on the sphere followed a simple table of contents without any index on name or topic. The aim was to deliver a book without curlicues. To that end, Sacrobosco's non-humanist Latin had to be "improved" (*emendata*) as well.[45] Burgersdijk faithfully followed the original order, yet rewrote *De sphaera* in such a rigorous manner that some scholars have come to argue that it essentially resulted in a new text (Hamel 2014, 44). It suffices to compare the opening words in the original with the retranslation to illustrate the radicalism of Burgersdijk's editing methods:

---

[42] Carefully revised reprints of the work appeared in Rome (1581 and 1585), Venice (1591), Lyon (1592, 1593 and 1594), Venice (1596 and 1601), Rome (1606), San Crispino (1608), and finally in Mains (1611–1612).

[43] The commentary by Vinet was mainly printed in Paris but also in other cities, including Lyon, Koln, Venice, and Antwerp (Valleriani 2017). Database accessible through: https://sphaera.mpiwg-berlin.mpg.de/. Accessed June 2019.

[44] I would like to thank Matteo Valleriani for pointing out that there was a long tradition of combining the *Computus* and *De sphaera* together in manuscript form; the idea to do so in print marked essentially a return to an older tradition.

[45] Citation taken from the School Order, see (Kuiper 1958).

> Sphaera igitur ab Euclide sic describitur: Spaera est transitus circunferentiae dimidii circuli quae (fixa diametra) eousque circunducitur, qousque ad locum suum redeat. id est: Shaera est tale rotundum & solidum, quod describitur ab arcu semicirculi circunducto. (de Sacrobosco et al. 1574, 6)

Except for some minor issues of orthography, all of the earlier editions had made use of exactly the same wording. Instead, Burgersdijk rendered this little fragment into:

> Sphaera, defieniente Euclide, est figura, quae, converso circa diametrum quiescentem semicirulo, & in eundem rursus locum restituto, à quo coeperat moveri, continetur. lib. II. d. 14 hoc est, Spheara est rotundum, quod semicirculo circumducto describitur. (Burgersdijk and de Sacrobosco 1626, 6)[46]

The brief example gives an impression of how Burgersdijk went about rewriting the original Latin without, however, essentially changing the original design of the text.[47]

This is not to say that Burgersdijk did not make numerous subtle changes. In doing so, he was—as he had indicated in the preface—primarily inspired by Clavius's extensive commentaries, merging them integrally, albeit in a highly abbreviated form, with the original text by Sacrobosco. His "method" (*methodo*) becomes immediately apparent when comparing his text to the earlier versions of his Jesuit counterpart.[48] To illustrate, Clavius had made an extensive twenty-two-page "commentary" (*commentarius*) in his 1570 edition on Sacrobosco's chapter on the size of the earth, listing various alternative opinions and calculations, beginning from Aristotle (384–322 BCE) and Alfarganus (al-Farghani) (ca. 805–870) and extending to more recent authors like Jean François Fernel (1497–1558) (Clavius and de Sacrobosco 1570, 265–67).[49] In other commentaries, such as those by Vinet and Giuntini, these numbers could not be found. Yet Burgersdijk listed the exact same numbers as Clavius in a throwaway sentence (Burgersdijk and de Sacrobosco 1626, 27–28), omitting a direct reference to his source, or to any tables and mathematical demonstrations that could be found in the work of his Jesuit predecessor.

In many ways, Burgersdijk's text reads as a highly condensed abridgement of Clavius's masterpiece. The Leiden Aristotelian proceeded freely in summarizing his predecessor's insights, occasionally merging together different comments that had been presented separately (such as Clavius's expositions on the composition and on centrality of the earth).[50] Every now and then, Burgersdijk incorporated information

[46] The reference to book 11 in Euclid could be found in the editions annotated by Vinet, but not in the works of Clavius; although Burgersdijk could have taken this reference from anywhere, it seems likely that he was aware of the reference in Vinet's works.

[47] It should be noted that Burgersdijk did divide the four sections of the original text into four actual parts (*partes*), each with individual chapters, instead of using four chapters with unnumbered subheadings, as had been customary. Yet he did not change the order of the text.

[48] The use of *methodo* comes from Burgersdijk's letter to the reader (Burgersdijk and de Sacrobosco 1626, sig. a2). See Appendix.

[49] The entire chapter *De ambitu terrae* by Clavius on the size of the earth runs from page 250 to 272. Like many other commentaries, it was copied into later editions.

[50] Chapter VII in Burgersdijk (Burgersdijk and de Sacrobosco 1626, 23–25) provided a summary of (Clavius and de Sacrobosco 1570, 151–74) and (Clavius and de Sacrobosco 1570, 175–244).

from other textbooks as well.[51] But at all times he meticulously followed the argumentation and sequential layout of his Jesuit example. Clavius, for example, had divided the original chapter in Sacrobosco on the meridian and the horizon into two different chapters, each with its own commentary.[52] Burgersdijk did exactly the same, whereas such a division cannot be found in any edition based on the comments by Vinet and Giuntini.[53] Following the Jesuit's lead, Burgersdijk also wrote about the various movements in the eighth, ninth, and tenth spheres. There was no doubt that this motion "had been unknown to Aristotle" and his predecessors but confirmed by recent astronomical observations (Chaps. 5 and 8).[54] And yet the "earth was immobile" at the center of the universe (Burgersdijk and de Sacrobosco 1626, 26).[55] In this respect too, Burgersdijk closely followed Clavius, defending geocentrism at a time when Galileo Galilei's (1564–1642) discoveries were commonly known and Copernican cosmology widely discussed.[56]

The question then, perhaps, is why the Dutch authorities considered the text by Clavius particularly appropriate to shape the minds of young Dutch youth.[57] James Lattis has listed in this connection a number of points characterizing what he called Clavius's "Orthodox Cosmology" (Lattis 1994, 65–85). These points include, among others, (1) the reconciliation of the Ptolemaic planetary system with Aristotelian physics, (2) the consensus on the overall size of the cosmos, and

---

[51] Burgersdijk listed, for example, exact numbers for the cycles of the planets (Burgersdijk and de Sacrobosco 1626, 13), whereas these had only been indicated by approximation in Clavius (Clavius and de Sacrobosco 1570, 97; Clavius and de Sacrobosco 1611, 30). One possible source is (Simi 1555, 70), yet the same numbers can be found in other handbooks as well.

[52] Clavius's separate treatment of the meridians had become necessary to include the enormous amount of new information that had become available in the wake of the European voyages of discovery. The chapters on the meridian and the horizon start at pp. 338 and 361, respectively. The influence of new voyages returned in various contexts. Vinet, too, had argued for example that the disagreement on the inhabitability of the tropics was "held by experience" (*habetur ab experientia*), identifying Columbus's 1491 journey as the major transition point (Bellère 1573, 43). Burgersdijk did exactly the same, highlighting how recent experience had shown that various parts of the world were indeed inhabitable (Burgersdijk and de Sacrobosco 1626, 51).

[53] Burgersdijk's division of Sacrobosco's original chapter on the meridian and the horizon conveys by comparing Burgersdijk's chapter V (Burgersdijk and de Sacrobosco 1626, 41–46) and chapter VI (Burgersdijk and de Sacrobosco 1626, 42–47) with, for instance, (Sacrobosco, Vinet, and Giuntini 1573, 35).

[54] (Burgersdijk and de Sacrobosco 1626, 12): "Hic motus Aristoteli, aliisque ante Aristotelem Philosophis & Astrologis, fuit incognitus." Burgersdijk also used the example of Timocharis here, just as Clavius had done (Clavius and de Sacrobosco 1570, 81).

[55] "*Terram est immobilem.*" Compare with (Clavius and de Sacrobosco 1611, 106).

[56] On the comets of 1577 and 1618 and the so-called "decline of the Aristotelian worldview" see also (Van Nouhuys 1998). It may be worth noting that, after visiting the astronomer in the Tuscan countryside, the Elzeviers also published Galileo's *Discorsi* in 1638; a year before they reprinted Burgersdijk's *Sphaera* (see below in text).

[57] It remains an open question where Burgersdijk got the idea to integrate the text of Clavius with the original text of Sacrobosco. In 1601 and in 1610, a mixture of Vinet and Clavius had been printed in Cologne by the Jesuit Hofbuchdrucker Peter Cholinus (J. de Sacrobosco et al. 1610). Yet Clavius's remarks in that edition had been limited to only to a few short remarks, primarily at the beginning of each of the four main parts. Burgersdijk's notes were much more precise.

(3) the acceptance of the order of planetary spheres according to Ptolemy (ca. 100–ca. 160).[58] On all of these issues, Burgersdijk clearly followed his Jesuit paragon, using the same examples and the same style of argument, also when speaking for instance of the existence of an immobile empyrean heaven beyond the moving spheres as "a seat of the blessed" (Burgersdijk and de Sacrobosco 1626, 14). As Lattis has argued, this quantitative world picture, where the afterlife had a distinct place, was vital for maintaining a Christian vision of the cosmos (Lattis 1994, 82).[59] It must have been precisely this aspect that made Clavius's orthodox views so appealing to the Leiden Neo-Aristotelians and their allies.

Yet, in some places, Burgersdijk remained selective in his rendering of Clavius's comments. He copied only the bits and pieces that he considered relevant and strategically eliminated certain specifics. Whereas Clavius, for example, had spoken of the empyrean heaven as a "a happy seat and home of the angels and the blessed,"[60] the Leiden professor simply spoke of a "a seat of the blessed."[61] Hence, in line with his Protestant faith, he placed less emphasis on the existence of angels. And Burgersdijk omitted other, perhaps more important aspects too. He did not mention, for instance, Clavius's ideas of the existence of an eleventh sphere, introduced by the Jesuit author in his reworks from 1593 onwards, to accommodate Copernican precession theory (Lattis 1994, 171; Grant 1994, 318).[62] The reason for clinging on to earlier ideas on the topic of ten moving spheres was probably not that Burgersdijk lacked access to Clavius's later editions,[63] but his unwillingness to reject Alfonsine trepidation theory in favor of a model based on Copernican mathematics.[64]

[58] The order of planetary spheres is the moon, Mercurius, Venus; *then*, the Sun, Mars, Jupiter, Saturn, the eight and the ninth sphere, and the *primum mobile*, beyond which there was only an empyrean heaven. On rival cosmologies, see (Lattis 1994, 73–77; 86–106) (Chap. 5).

[59] On the various opinions regarding the existence of an empyrean heaven, see (Grant 1994, 270–89).

[60] (Clavius and de Sacrobosco 1570, 62; Clavius and de Sacrobosco 1611, 24): "felicem angelorum beatorum sedem ac patriam."

[61] (Burgersdijk and de Sacrobosco 1626, 14): "sedes beatorum."

[62] Clavius added a moving sphere, following the ideas of Giovani Antonio Magini. Adopting "the adapted Copernican scheme for the motions of the fixed stars" probably resulted from his work on the calendar reform (Lattis 1994, 173). Burgersdijk sticks to ten moving spheres; see note 58.

[63] It is extremely difficult to pin down with certainty which edition of Clavius's work Burgersdijk might have used, yet a number of editions can be reasonably excluded; Burgersdijk mentioned Pope Gregory XIII's calendar reform of 1582 (Burgersdijk and de Sacrobosco 1626, 40) using the same terminology and order as in (Clavius and de Sacrobosco 1611, 147). Obviously, the remark on the calendar reform had been absent in editions earlier than 1582. See also (Clavius and de Sacrobosco 1570, 331). Perhaps, a future comparative analysis of the imagery in the various editions can provide an answer to the question of which version Burgersdijk used as a model (the 1585 Rome edition of Clavius was, for instance, very different in terms of imagery and it is hard to imagine that this edition served as an example).

[64] Although he did not mention it specifically, Burgersdijk may have objected to the existence of an eleventh sphere on the basis of theological grounds. It is impossible to track down a unique source for Burgersdijk's ideas on this matter. Calvin had rejected the notion of a tenth sphere and remained silent on the existence of a ninth sphere (Kaiser 1988, 83). Perhaps, Burgersdijk was rather inspired in this regard by Coimbra Jesuits (Grant 1994, 321) (Chap. 10), whom he cites approvingly in

For Burgersdijk, scriptural revelation remained the basis for acquiring knowledge of heavenly motions. It was perhaps for similar reasons that Burgersdijk refrained from mentioning the existence of the telescope, whereas Clavius in his 1611 edition (the final edition completed just before his death) had indicated that the recent discoveries made with an "oblong tube" (*tubi...oblongi*) required astronomers to rethink how the celestial orbs could be arranged in order to save the phenomena (Clavius and de Sacrobosco 1611, 75). Omitting such particulars, which had the potential to shake up worldviews, allowed Burgersdijk to embrace Clavius's cosmology without thereby endorsing Clavius's "ontologically loaded" physical mathematics (Chap. 6).[65]

Whereas Burgersdijk's edition of *De sphaera* was primarily inspired by Clavius in terms of content, it included elements from the 1573 Antwerp edition published by Bellère in terms of imagery (which in turn had been based on the so-called 1538 Wittenberg edition) (Chap. 8). The illustrations to the text were interlaced at similar locations.[66] The volvelle in Burgersdijk, for instance, came at the end of chapter one of the third part (on the rise of zodiacal signs), just as in the Antwerp edition; the tool was not to be found in Clavius (Sacrobosco et al. 1573, 48; Burgersdijk and de Sacrobosco 1626, 56).[67] The design for the illustration of the five climate zones (Burgersdijk and de Sacrobosco 1626, 51) was clearly imitative of the Antwerp edition (Sacrobosco, Vinet, and Giuntini 1573, 43), whereas the layout was entirely different in the edition prepared by Clavius (Clavius and de Sacrobosco 1611, 187).[68] A set of images in the chapter on "natural and civil day" (Burgersdijk and de Sacrobosco 1626, 62–89) was again inspired by Bellère, and absent in Clavius. Yet elsewhere, Burgersdijk freely copied visual elements from Clavius's work as well. For instance, an illustration supporting the notion of the roundness of the earth (Burgersdijk and de Sacrobosco 1626, 19) was bluntly plagiarized from Clavius (Clavius and de Sacrobosco 1611, 20; Clavius and de Sacrobosco 1570, 143) and absent in Bellère. The table on different climates around the world (Burgersdijk and de Sacrobosco 1626, 88–95), the only table in Burgersdijk's work, was directly copied from Clavius (Clavius and de Sacrobosco 1570, 480–90; Clavius and de Sacrobosco 1611, 288–89) and was not to be found in that form in Bellère either.

---

other works as well, or by the work of Jean Bodin, who was widely read in the Republic, and who argued explicitly against the eleventh sphere in his *Universae Naturae Theatrum*, making extensive references to theological interpretations of the crystalline sphere (Bodin 1596 in (Stimson 1917, 211–12).

[65] I would like to thank Peter Barker for his comments and suggestions on the issue of Burgersdijk's approach concerning Clavius's cosmology. On Burgersdijk's gradual change of view over time with regard to new cosmologies, primarily reflected in his *Collegium physicum* (1632), see (Vermij 2002, 132–33). The change did not reflect in his booklet on *De sphaera*.

[66] On illustrations of *De sphaera*, see (Crowther and Barker 2013; Gingerich 1999; Hamel 2014, 23–49).

[67] Moving parts were first introduced by Joseph Klug in the Wittenberg edition of 1538 (Gingerich 1999, 215). Alas, we do not know who made the engravings, or who was responsible for the editorial decisions concerning illustrations in Leiden.

[68] The exact same illustration can be found in other Antwerp editions published by Bellère as well. For example, in (Appianus 1584, 11).

In short, Burgersdijk freely proceeded to mix and match illustrations as it best suited him to support and clarify his ideas.

All the same, Burgersdijk's edition made use of considerably less imagery than any of the previous editions. There were, for instance, no additional illustrations for geometrical explanations, like in Clavius's editions, nor was there any place for ornamental imagery. Instead, the thirty-six images in Burgersdijk's edition were newly engraved, in the course of which well-known images got a touch of the Dutch. Two warriors depicted under a sphere were, for example, decked out with a hat typical of Dutch traditional clothing (Figs. 11.1, 11.2 and 11.3).[69] Elsewhere, a building was made to resemble a Dutch church (Figs. 11.4 and 11.5).[70] Thus, the work of Sacrobosco was not just adapted to the Dutch market but fashioned to strengthen the sense of Dutch distinctness as well.

The text of *De sphaera* was never translated into Dutch (i.e. nor Burgersdijk's version, nor any other version). This is somewhat remarkable for a Latin school book, and for the fact that other works by Burgersdijk did appear in translation.[71] Yet for those who were interested in learning something about astronomy, there were numerous alternatives on the market. Of particular importance was, for instance, the work of Adriaan Metius (1581–1635), who taught in Dutch at the University of Franeker on various topics that were of interest to navigators, engineers, and surveyors.[72] His 1614 bestseller *Institutiones astronomicae et Geographicae* (Metius 1614)—despite its title entirely written in Dutch—was a practical textbook on cosmography connecting the theory of the sphere with geography and cartography; it clearly included elements of Sacrobosco, but went much further than his predecessor by including elements of arithmetic and geometry.[73] And there were other booklets available too, such as the practical guide to the use of globes issued by the

---

[69] On this illustration, see also (Crowther and Barker 2013, 453).

[70] On this illustration, see also (Gingerich 1999, 217–19; Hamel 2014, 39). Notice also the topgallant sail, only present in the Dutch edition.

[71] The other works by Burgersdijk (*Institutio physica*, 1648; *Idea philosophiae naturalis*, 1648; *Logica practica*, 1648, *Institutio metaphysica*, 1649) were translated by Allardus Lodewijk Kok, who translated Vossius's textbook *Elementa rhetorica* (Amsterdam, 1648) as well.

[72] The Frisian example was followed in Leiden with the establishment of the *Duytsche Mathematique*, a mathematical school associated to the University where teaching was carried out in Dutch. For more background, see (Dijksterhuis 2017; Dijkstra 2012).

[73] In a way, one could compare the work of the Franeker mathematician with that of Clavius, except that Metius followed a somewhat different order than Clavius had done in his Sacrobosco. He also introduced new information that had become available due to the voyages of discovery (Chap. 8). On the mixing of the *Sphaera* with cosmography, see also (Chap. 9) and (Valleriani 2017, 438). An interesting detail in this connection is that Metius used a slogan from Ovid, prominently present in Clavius 1570 panegyric on mathematics, as "his" inscription in various *alba amicora* (Balck 1618; van Wijckel 1635). *Felices animae, quibus haec cognoscere primum, Inque domos superas scandere, cura fuit* ("Ah happy souls, who first took thought to know | these things and scale the heavenly mansions!" (Ovid 1989, 23). Metius therewith aligned himself with the conviction that astronomy is extremely useful and even necessary to all aspects of life. On this issue, see also (Omodeo 2017).

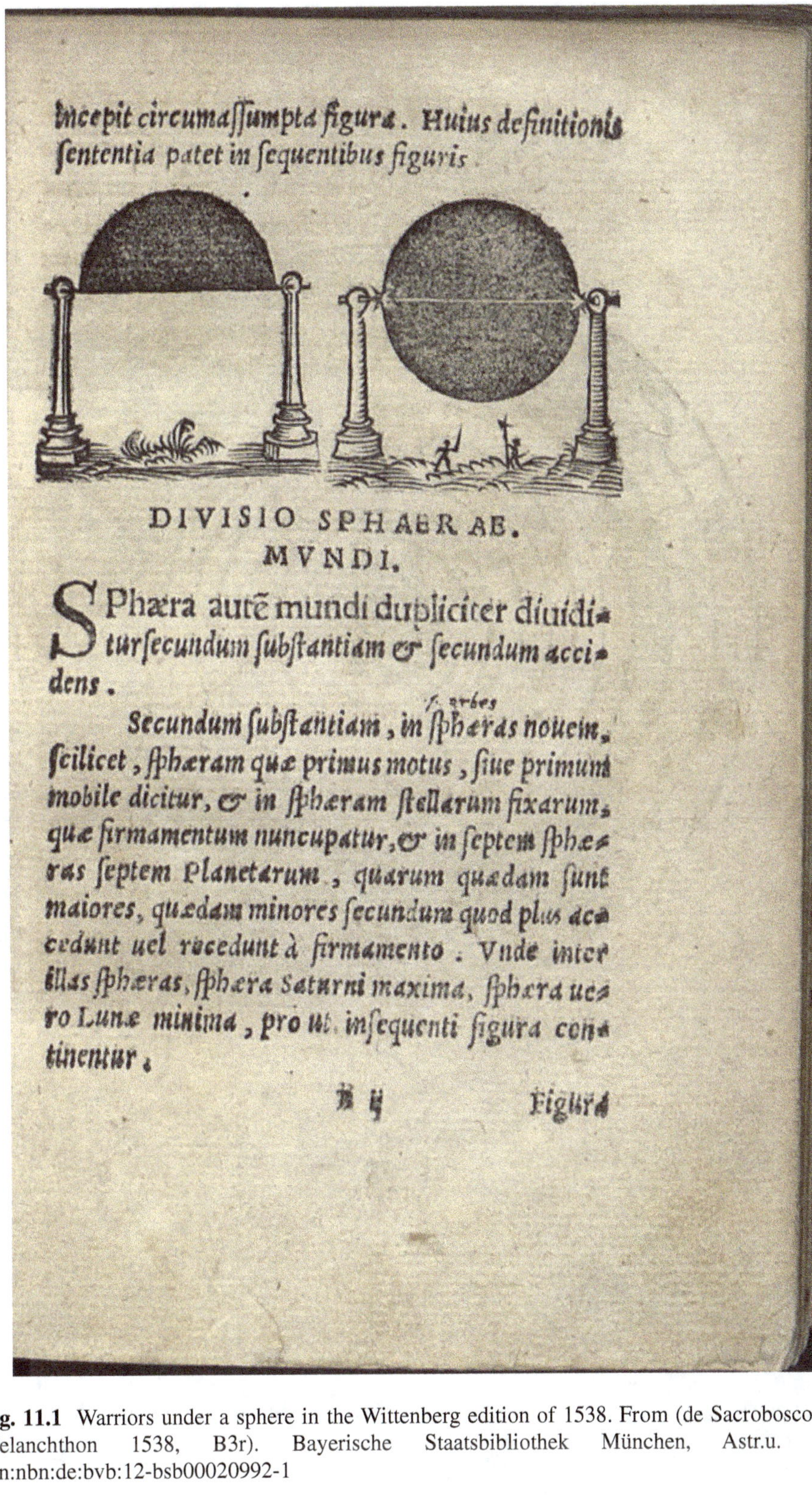

incepit circumassumpta figura. Huius definitionis sententia patet in sequentibus figuris

DIVISIO SPHAERAE MVNDI.

Sphæra autē mundi dupliciter diuiditur secundum substantiam & secundum accidens.

Secundum substantiam, in sphæras nouem, scilicet, sphæram quæ primus motus, siue primum mobile dicitur, & in sphæram stellarum fixarum, quæ firmamentum nuncupatur, & in septem sphæras septem Planetarum, quarum quædam sunt maiores, quædam minores secundum quod plus accedunt uel recedunt à firmamento. Vnde inter illas sphæras, sphæra Saturni maxima, sphæra uero Lunæ minima, pro ut in sequenti figura continentur.

B ij Figura

Fig. 11.1 Warriors under a sphere in the Wittenberg edition of 1538. From (de Sacrobosco and Melanchthon 1538, B3r). Bayerische Staatsbibliothek München, Astr.u. 154, urn:nbn:de:bvb:12-bsb00020992-1

DE SAC. BOS. CAP. I. 3

circunductus semicirculus in idem rursus redit, vnde cœpit circunscribi figura illa. Huius definitionis sententia ex subiectis picturis patere potest. Quam autem posuit auctor, corruperant Græcorum Euclidis interpretes. Si enim linea in latus fluens describit tantùm superficiem, circunferentia semicirculi circumacta, describet superficiem globi solùm, non globum solidum.

DIVISIO SPHAERAE MVNDI.

Sphæra autem mundi dupliciter diuiditur, secundum substantiam, & secundum accidens. Secundum substantiam in sphæras nouem. *Mundi partes insignióres duæ.*

SCHOLION VINETI.

In sphæras nouem.) Recentiores Astrologi decimam spheram addiderunt, propter tertium in octaua sphera animaduersum motum, quem dixerunt motum trepidationis, & motum accessus & recessus: de quo Purbachius in Theoricis.

Scilicet sphæram nonam, quæ primus motus, siue primũ mobile dicitur: & in sphæram stellarum fixarum, quæ firmamentum nuncupatur: & in septem sphæras septem planetarum, *Quòd nouẽ sint orbes cælestes. Firmamẽtum.*

a 2

Fig. 11.2 The same warriors in the *Sphaera emendata* issued by the Heirs of Birckmann (Antwerp 1566). From (de Sacrobosco et al. 1566, a2r)

6 SPHÆRA

PARS PRIMA.

CAP. I.

Definitio Sphæræ, Centri, Axis & Poli.

SPhæra, definiente Euclide, eſt figura, quæ, converſo circa diametrum quieſcentem ſemicirculo, & in eundem rurſus locum reſtituto, à quo cœperat moveri, continetur, lib. 11. d. 14. hoc eſt, Sphæra eſt rotundum, quod ſemicirculo circumducto deſcribitur.

A Theodoſio ſphæra definitur ſolidum unâ ſuperficie contentum, in cujus medio punctum eſt, à quo undequaque lineæ rectæ ad ſuperficiem ductæ, æquales ſunt. def. 1.

Prior harum definitionum, modum fabricundæ ſphæræ oſtendit; poſterior explicat ſphæræ fabricatæ naturam: unde illam deſcriptionem; hanc, definitionem vocare liceat.

Theodoſii definitio ex Platonis Timæo ſumta videtur:

Fig. 11.3 The re-engraved image in the Leiden edition of 1626. From (Burgersdijk and de Sacrobosco 1626, 6). Courtesy of the History of Science Collections, University of Oklahoma Libraries

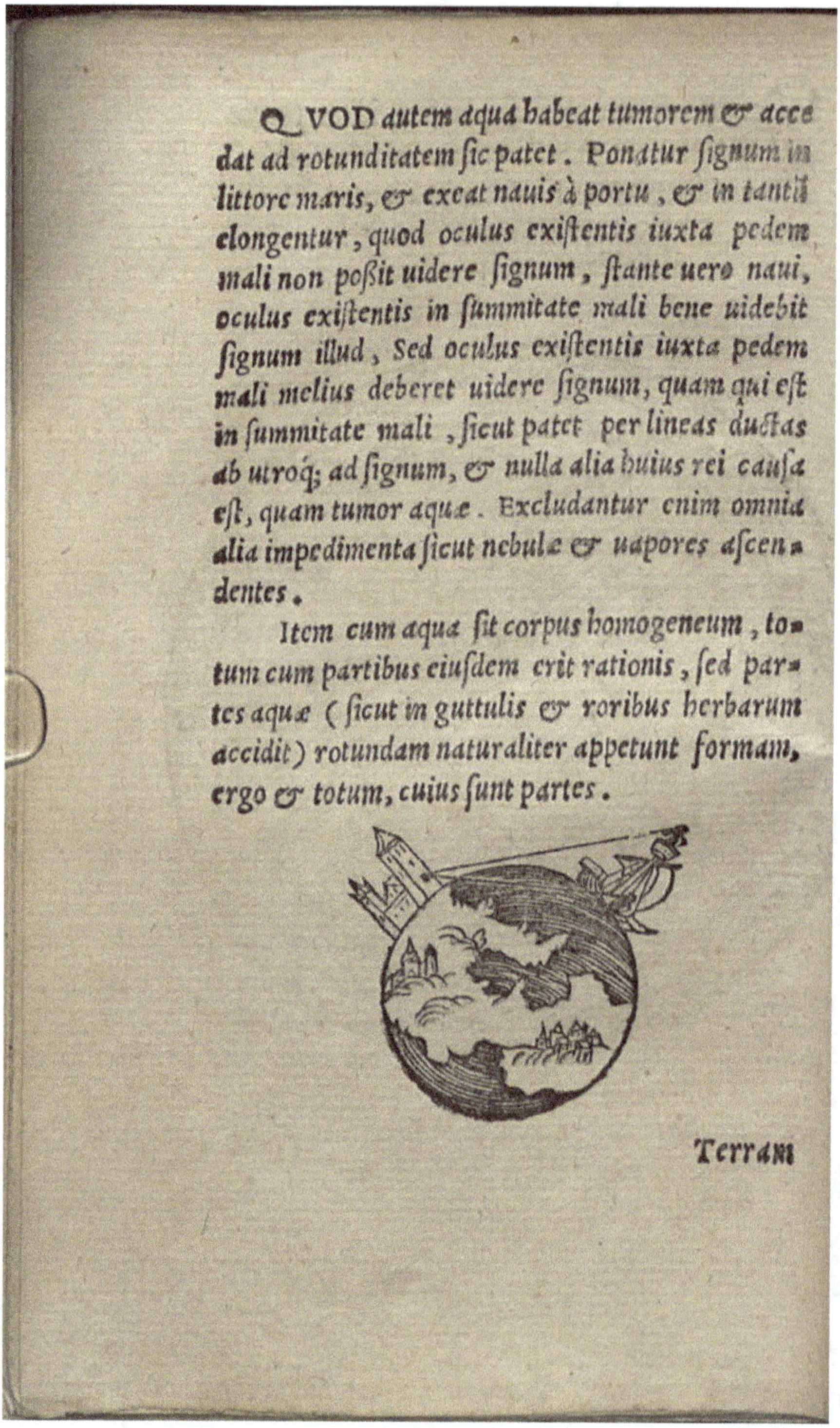

QVOD autem aqua habeat tumorem & accedat ad rotunditatem ſic patet. Ponatur ſignum in littore maris, & exeat nauis à portu, & in tantũ elongentur, quod oculus exiſtentis iuxta pedem mali non poßit uidere ſignum, ſtante uero naui, oculus exiſtentis in ſummitate mali bene uidebit ſignum illud, Sed oculus exiſtentis iuxta pedem mali melius deberet uidere ſignum, quam qui eſt in ſummitate mali, ſicut patet per lineas ductas ab utroq; ad ſignum, & nulla alia huius rei cauſa eſt, quam tumor aquæ. Excludantur enim omnia alia impedimenta ſicut nebulæ & uapores aſcendentes.

Item cum aqua ſit corpus homogeneum, totum cum partibus eiuſdem erit rationis, ſed partes aquæ (ſicut in guttulis & roribus herbarum accidit) rotundam naturaliter appetunt formam, ergo & totum, cuius ſunt partes.

Terram

Fig. 11.4 Curvature of the ocean in the Wittenberg edition of 1538. From (de Sacrobosco and Melanchthon 1538, Cv). Bayerische Staatsbibliothek München, Astr.u. 154, urn:nbn:de:bvb:12-bsb00020992-1

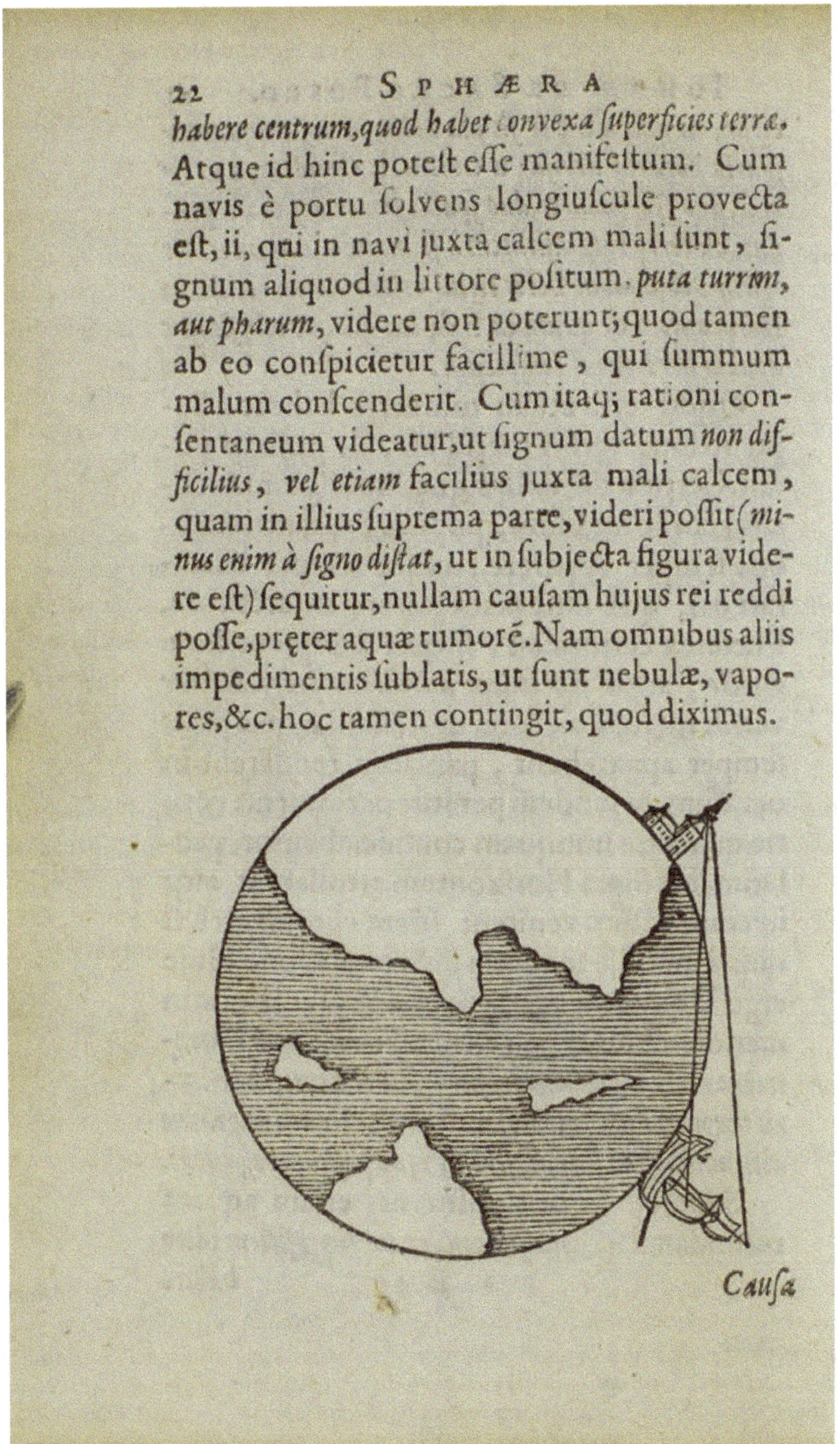

22 SPHÆRA

habere centrum,quod habet convexa ſuperficies terræ. Atque id hinc poteſt eſſe manifeſtum. Cum navis è portu ſolvens longiuſcule provecta eſt, ii, qui in navi juxta calcem mali ſunt, ſignum aliquod in littore poſitum, *puta turrim, aut pharum*, videre non poterunt; quod tamen ab eo conſpicietur facillime, qui ſummum malum conſcenderit. Cum itaq; rationi conſentaneum videatur,ut ſignum datum *non difficilius, vel etiam* facilius juxta mali calcem, quam in illius ſuprema parte,videri poſſit(*minus enim à ſigno diſtat*, ut in ſubjecta figura videre eſt) ſequitur,nullam cauſam hujus rei reddi poſſe,prȩter aquæ tumoré.Nam omnibus aliis impedimentis ſublatis, ut ſunt nebulæ, vapores,&c. hoc tamen contingit, quod diximus.

Cauſa

Fig. 11.5 Curvature of the ocean as re-engraved in the 1626 Leiden edition. From (Burgersdijk and de Sacrobosco 1626, 2). Courtesy of the History of Science Collections, University of Oklahoma Libraries

famed publisher Willem Janszoon Blaeu (1571–1638), who placed the astronomy of Ptolemy and Nicolaus Copernicus (1473–1543) on an equal footing (*Tweevovdigh onderwiis van de hemelsche en aerdsche globen* was printed in Dutch in 1634, 1638, 1647, 1666, and in Latin in 1634, 1652, 1657, 1668, and 1690, under the title *Institvtio astronomica*).

Despite the existence of alternatives, Sacrobosco's *Sphaera* could still to be found on Dutch school desks long after Copernican astronomy had changed into an established worldview. In conformity with the School Order, the medieval textbook was, for example, still prescribed in the 1692 statutes of the municipal school in The Hague, where the highest class was to be schooled in "the *Logic* by Burgersdijk, the *Sphaera* by Sacrobosco, Florus, The Greek History by Heridianus, Vergilius, Horatius, Juvenalis, etc."[74] Although the statutes did not specify exactly which version of *De sphaera* the youngsters were supposed to use, it could have very well still have been Burgersdijk's edition; reprints of his 1626 *Sphaera* followed in 1639, 1647 and 1656 (Burgersdijk and de Sacrobosco 1639, 1647, 1656). We still grope in the dark as regards the timing of these reprints. Perhaps the establishment of new schools in the Generality Lands played a role (compare in this connection the publication dates with the program for the Latin School in 's-Hertogenbosch below). Perhaps the Latin schools merely needed new books every 10 years because of wear and tear. In any case, an initial comparison between the 1626 and the 1647 edition has not brought to light any significant changes.

Even if the 1625 School Order was legally valid in the various towns of the province of Holland, schoolmasters apparently had considerable freedom when implementing the specifics of the program. Thus, we do not know exactly how intensively *De sphaera* by Sacrobosco was studied in the different Latin schools, just as we are in the dark on whether the textbook was still used at university. The Leiden Orientalist and mathematician Jacob Golius (1596–1667), for example, taught a course *elementa Astronomica exponit* in September 1662 and in February 1663, but it remains unclear what books he used to this end (Molhuysen 1918, 174∗, 176∗). Other evidence on the use of *De sphaera* by students in the Dutch Republic has not yet turned up.[75]

The School Order would eventually not be implemented outside of the province of Holland, except partly in the city of 's-Hertogenbosch in Brabant, where, after the takeover by the Republican army, the authorities wanted to erect a new Latin

[74] (Kuiper 1958, 199): "Hyer sal geleert werden *Logica Burgersdijck, Sphaera Johannis de Sacrobosco, Florus, Historia Graeca Herodiani, Virgilius, Horatius, Juvenalis* etc."

[75] All the same, it is clear that the book was widely read. Christophorus Wittichius, for instance, literally copied a section on the vacuum in a dissertation on outer limits of the universe; (Wittichius 1653, 168–69): "Si caelum non esset…non essent figura orbiculari." Compare (Burgersdijk and de Sacrobosco 1626, 16–17). The dissertation by Wittichius was compiled in a book that would make him famous for bringing together the philosophy of René Descartes and the theology of the Gisbertus Voetius, a dogmatist neo-scholastic from the Gomarist faction.

school to replace a Jesuit school with an excellent reputation.[76] In the program for the new school, drafted in 1639, the original segment on Sacrobosco in the Holland School Order was replaced by some other mandatory measures. Nonetheless, the students had to follow a course over two or three years, reading *De sphaera* by Sacrobosco and the *Logic* by Burgersdijk, the *Physics* (*Physiologiae Peripateticae libri sex*) by Johannes Magirus (ca. 1560–1596), the *Enchiridion Metaphysicum* (1621) by Caspar Bartholin the Elder (1585–1629), *The Compendivm Ethicae Aristotelicæ ad normam veritatis christianæ revocatum* by Antonius Walaeus, Justinus, Florus, and Sulpicius complemented with Johannes Sleidanus (1506–1556).[77] Except for the classics, these were all well-known and determinedly Protestant writers who also appeared on the Holland School Order with the intent to counterbalance the Jesuit influence on the educational system.

6 Conclusions

Burgersdijk's neo-Aristotelianism was closely related to the rise of the Contra-Remonstrant movement within the Dutch Reformed Church. Against that background, reflected in and articulated during the Synod of Dort and in Holland's School Order of 1625, the agenda of the last editor of *De sphaera* becomes clear: to produce a "clean" copy of a textbook that had proved its use for instruction, adapted to the needs of the young Protestant youth. In staunch competition with the Jesuits, Burgersdijk made an attempt to make the Dutch less dependent on Catholic literature and modes of knowledge production. Yet in reality, his neo-Latin retranslation of the *Sphaera* effortlessly integrated many of the insights of the work by the Jesuit mathematician Christophorus Clavius. Indeed, Leiden neo-Aristotelianism was not substantially incompatible with Jesuit principles on all fronts. The outcome was that Dutch school pupils were still instilled with a geocentric worldview at a time when the heliocentric model had already become widely available.

[76] Even if it was not directly copied, the Holland Order did have an influence on the other provinces (Kuiper 1958, 161–73). I did not find any additional evidence for the use of *De sphaera* in (Bastiaanse, Bots, and Evers 1985; Bastiaanse and Evers 1986; Booy 1980).

[77] On the implementation of the school program in 's-Hertogenbosch, see (Kuiper 1958, 161–63). The direct influence of Holland School Order was clearly present (Kuiper 1958, 163): "Cum hos authores Scholarum Hollandicarum publicus ordo, quem hic noster, quam maxime potest, proxime imitatur, etiam praescripserit." Note that the original segment on geographical knowledge had disappeared.

Appendix

Letter to the reader from (Burgersdijk and de Sacrobosco 1626, sig. a2–a2v).

Benevolo Lectori.

Cum illust. & Potentes D D. Ordines Holandiae & VVest-Frisiae decrevissent, ut iidem linguarum & artium autores in omnibus scholis Hollandicis praelegerentur, inter caeteros libros Sphaeram Ion. de Sacro-Bosco censuerunt juventuti proponendam, ut ex eam adolescentes prima Astronomiae principia deliberarent, quam in Academiam deinde discerent plenius accuratiusque. Verum enim vero, cum hic autor no tam bonus esset, quam celebris, iidem Ordines nobis mandavarunt, ut eum inter multos alios libros scholis destinatos recenseremus, Latinatem que emendaremus. Cuirei, cum mihi obtigisset, ut admoverem manum, deprehendi statim, abstergeri non posse barbariem ab hoc autore, nisi totus, non in aliam linguam, sed in alium sermonem converteretur. Itaque sententiam Iohannis, quantum fieri potuit, meis verbis extuli, methodo etiam leviter emendata. Neque tamen hoc egi, ut sermone uterer elaborato, & comto sed simplici & facili, ac tantum non barbaro. Addidi praeterea multa ex aliis autoribus, praecipue ex Christophoro Clavio, qui [a2v] doctissimis commentariis hunc nostrum illustravit, eaq; passim suis locis inserui, & quo facilius à praeceptis autoris distinguerentur, alio charactere edenda curavi. Ad finem pauca quaedam adjeci de Theoricis Planetarum ex Georgio Purbachio, ut universae Astonomiae compendium, hoc libello contineretur. Neque tamen necessarium puto, haec omnia pueris inculcare (quaedam enim talia sunt, ut puerorum captum superent) sed praeceptorum judicio permittemdum censui, ut hinc petant, quae discipulorum suorum captui judicabunt esse accommodata. Unum etiam aut alterum Lucani testimonium, à Ioanne citatum, praeterii; quia Jospehus Scaliger in Prolegomenis in Manilium, Lucanum ipsum in iis locis erravisse, accuratè demonstravit. Nihil in toto hoc opere, benigne Lector, mihi vendico, praeter laborem & studium; aliorum enim vestigiis institisse me, libenter profiteor. Neque laudem hinc capto aut gloriam. Hoc unum mihi laboris erit satis amplum praemium, si intellexero, studium meum tibi non improbari. Vale.

References

Abels, P.H.A.M. 2002. *Duizend jaar Gouda: een stadsgeschiedenis*. Hilversum: Verloren.

Appianus, Petrus. 1584. *Cosmographia, siue descriptio vniuersi Orbis, Petri Apiani & Gemmae Frisij, Mathematicorum insignium, iam demùm integritati suae restituta. adiecti sunt alij, tum Gemmae Frisij, tum aliorum auctorum eius argumenti tractatus ac libelli varij*. Antwerp: Jean Bellère.

Balck, Everhardus. 1618. *Album amicorum*. Steinfurt/Franeker/Leiden [e.a.]: Koninklijke Bibliotheek: 75 A 2 / 1.

Bastiaanse, R.M., and Meindert Evers. 1986. Ontwikkelingen binnen het onderwijs aan de Gelderse Latijnse Scholen in de 17e en 18e eeuw. *Bijdragen en mededelingen Vereniging Gelre* 77: 68–91.

Bastiaanse, René, Hans Bots, and Meindert Evers. 1985. *Tot meesten nut ende dienst van de jeught: een onderzoek naar zeventien Gelderse Latijnse scholen, ca. 1580–1815*. Zutphen: Walburg.

Bently, Lionel, and Martin Kretschmer. 2017. *Resolutions concerning the Dutch school order, The Hague (1625–1626). Primary sources on copyright (1450–1900)*. www.copyrighthistory.org. Accessed June 2019.

Bos, Egbert P., and H.A. Krop, eds. 1993. *Franco Burgersdijk (1590–1635): Neo-Aristotelianism in Leiden*. Amsterdam: Rodopi.

Bot, Petrus Nicolaas Maria. 1965. *Humanisme en onderwijs in Nederland*. Utrecht: Spectrum.

Buning, Marius. 2019. Privileging the common good. The moral economy of printing privileges in the seventeenth-century Dutch Republic. In *Buying and selling: The business of books in early modern Europe*, ed. Shanti Grahelli, 88–108. Leiden: Brill.

Burgersdijk, Franco. 1645. *Franconis BurgersdicI Institutionum logicarum synopsis, sive Rudimenta logica. In quibus præcipuæ definitiones, divisiones, & regulæ, ad artem logicam pertinentes, per quæstiones & responsiones, breviter & dilucide proponuntur; in usum scholarum Hollandicarum*. Leiden: Abraham Commelinus.

Burgersdijk, Franco, and Johannes de Sacrobosco. 1626. *Sphaera Iohannis de Sacro-Bosco, decreto illustr. Et potent. D.D. Ordinum Hollandiae et West-Frisiae, in usum scholarum ejusdem provinciae sic recensita, ut & Latinitas & methodus emendata sit, multaque addita, quae ad huius doctrinae illustrationem requirebantur*. Leiden: Bonaventura & Abraham Elzevir. http://hdl.handle.net/21.11103/sphaera.101041.

———. 1639. *Sphaera Iohannis de Sacro-Bosco, decreto illustr. et potent. D.D. Ordinum Hollandiae et West-Frisiae, in usum scholarum ejusdem provinciae sic recensita, ut & Latinitas & methodus emendata sit, multaque addita, quae ad huius doctrinae illustrationem requirebantur*. Leiden: Bonaventura & Abraham Elzevir. http://hdl.handle.net/21.11103/sphaera.100589.

———. 1647. *Sphaera Iohannis de Sacro-Bosco, decreto illustr. et potent. D.D. Ordinum Hollandiae et West-Frisiae, in usum scholarum ejusdem provinciae sic recensita, ut & Latinitas & methodus emendata sit, multaque addita, quae ad huius doctrinae illustrationem requirebantur*. Leiden: Bonaventura & Abraham Elzevir. http://hdl.handle.net/21.11103/sphaera.100590.

———. 1656. *Sphaera Iohannis de Sacro-Bosco, decreto illustr. et potent. D.D. Ordinum Hollandiae et West-Frisiae, in usum scholarum ejusdem provinciae sic recensita, ut & Latinitas & methodus emendata sit, multaque addita, quae ad huius doctrinae illustrationem requirebantur*. Leiden: Johannes Elzevir.

Casetti, Carla. 1993. Eliano, Vittorio. In *Dizionario biografico degli italiani*. Roma: Istituto della Enciclopedia italiana. http://www.treccani.it/enciclopedia/vittorio-eliano_(Dizionario-Biografico)/. Accessed June 2019.

Cau, Cornelis, Simon van Leeuwen, Jacobus Scheltus, Paulus Scheltus, Isaac Scheltus, Didericus Lulius, and Joannes van der Linden (eds.). 1664. *Groot placaet-boeck, vervattende de placaten, ordonnantien ende edicten van de doorluchtige, hoogh mog. heeren Staten Generael der Vereenighde Nederlanden ende vande ed. groot-mog. heeren Staten van Hollandt en West-Vrieslandt, mitsgaders vande ed. mog. heeren Staten van Zeelandt: waer by noch ghevoeght zijn eenige placaten vande voorgaende graven ende princen de selver landen, voor soo veel de selve als noch in gebruyck zijn*. Vol. 2. 9 vols. The Hague: weduwe ende erfgenamen van wylen Hillebrandt Jacobsz van Wouw [etc.].

Clavius, Christoph, and Joannes de Sacrobosco. 1570. *Sphaeram Ioannis de Sacro Bosco commentarivs*. Rome: Vittorio Eliano. http://hdl.handle.net/21.11103/sphaera.100365.

Clavius, Christoph, and Johannes de Sacrobosco. 1611. *Christophori Clavii Bambergensis e Societate Iesu Operum mathematicorum V tomis distributa*. Moguntiae [Mainz]: Hierat, Anton I. http://hdl.handle.net/21.11103/sphaera.100383.

Crowther, Kathleen M., and Peter Barker. 2013. Training the intelligent eye: Understanding illustrations in early modern astronomy texts. *Isis* 104 (3): 429–470. http://hdl.handle.net/10.1086/673269.

de Booy, Engelina Petronella. 1980. *Kweekhoven der wijsheid: basis- en vervolgonderwijs in de steden van de provincie Utrecht van 1580 tot het begin der 19e eeuw*. Zutphen: Walburg.

Nijs, Thimo de, and Eelco Beukers. 2002. *Geschiedenis van Holland*. Vol. 2. 3 vols. Hilversum: Verloren.

Sacrobosco, Johannes de and Philipp Melanchthon. 1538. *Ioannis de Sacro Busto libellus, De sphaera: Eiusdem autoris libellus, cuius titulus est Computus, eruditissimam anni & mensium descriptionem continens. Cum praefatione Philippi Melanth. & novis quibusdam typis, qui ortus indicant*. Wittenberg: Joseph Klug. http://hdl.handle.net/21.11103/sphaera.101106.

Sacrobosco, Johannes de, Élie Vinet, Francesco Giuntini, and Pedro Nuñes. 1566. *Sphaera Ioannis de Sacro Bosco emendata. Cum additionibus in margine, et Indice rerum et locorum memorabilium, et familiarissimis scholijs, nunc recenter compertis, et collectis à Francesco Iunctino Florentino sacrae Theologia Doctore. Interserta etiam sunt Eliae Vineti Santonis egregia Scholia in eandem Sphaeram. Adiunximus eiusdem auctoris Computum Ecclesiasticum. Et Petri Nonii Salaciensis demonstrationem eorum, quae in extremo capite de climatibus Sacroboscius scribit de inaequali climatum latitudine, eodem Vineto interprete*. Antwerp: Heirs of Arnold Birckmann. http://hdl.handle.net/21.11103/sphaera.101115.

de Sacrobosco, Joannes, Elie Vinet, Pedro Nunes, and Pierio Valeriano. 1574. *Sphaera Ioannis de Sacro Bosco emendata*. Venice: Heirs of Girolamo Scotto. http://hdl.handle.net/21.11103/sphaera.100309.

Sacrobosco, Joannes de, Christophorus Clavius, Élie Vinet, Albertus Hero, Francesco Giuntini, Pedro Nunes, and Pierio Valeriano. 1610. *Sphaera Ioannis a Sacrobosco emendata, aucta et illustrata. Eliae Vineti Santonis, & Francisci Iunctini Florentini scholia in eandem Sphaeram restituta. Quibus praeter Alberti Heronis scholia nunc accessere R. P. Christoph. Clavij breves commentarij. Adiunximus huic libro compendium in Sphaeram per Pierium Valerianum Bellunensem: et Petri Nonii Salaciensis Demonstrationem eorum, quae in extremo capite de climatibus Sacroboscus scribit de inaequali climatum latitudine, eodem Vineto interprete*. Cologne: Peter Cholinus. http://hdl.handle.net/21.11103/sphaera.100357.

Dekker, E. 1993. *Rijker dan Midas: vrijheid, genade en predestinatie in de theologie van Jacobus Arminius (1559–1609)*. Zoetermeer: Boekencentrum.

Dibon, Paul Auguste Georges. 1954. *La philosophie Néerlandaise au siècle d'or*. Amsterdam: Elsevier.

Dijksterhuis, Fokko. Jan. 2017. Duytsche Mathematique and the building of a new society: Pursuits of mathematics in the seventeenth-century Dutch Republic. In *Mathematical practitioners and the transformation of natural knowledge in early modern Europe*, ed. Lesley B. Cormack, Steven A. Walton, and John A. Schuster, 167–181. Cham: Springer. https://doi.org/10.1007/978-3-319-49430-2_9.

Dijkstra, Arjen. 2012. *Between academics and idiots: A cultural history of mathematics in the Dutch province of Friesland (1600–1700)*. Ph.D. Leeuwarden: University of Twente.

Dongelmans, Berry P.M., Paul Gerardus Hoftijzer, and Otto Stephanus Lankhorst. 2001. *Boekverkopers van Europa: het 17de-eeuwse Nederlandse uitgevershuis Elsevier*. Zutphen: Walburg.

Emmius, Ubbo. 1951. *Programma bij de aanvaarding van het rectoraat der Latijnse School (thans Praedinius-Gymnasium) te Groningen in 1594*. Trans. Antoon Gerard Roos. Groningen: Wolters.

Feingold, Mordechai. 1993. The ultimate pedagogue: Franco petri Burgersdijk and the English speaking academic learning. In *Franco Burgersdijk (1590–1635). Neo-Aristotelian in Leiden*, ed. Egbert P. Bos and H.A. Krop, 151–165. Amsterdam: Rodopi.

Fortgens, H.W. 1958. *Schola latina, uit het verleden van ons voorbereidend hoger onderwijs, door Dr. H.W. Fortgens*. Zwolle: Tjeenk Willink.

Frijhoff, Willem, and Marijke Spies. 2004. *Dutch culture in a European perspective: 1650, Hard-Won Unity*. Assen: Van Gorcum.

Gilbert, Neal Ward. 1963. *Renaissance concepts of method*. New York: Columbia University Press.

Gingerich, Owen. 1999. Sacrobosco illustrated. In *Between demonstration and imagination*, ed. Lodi Nauta and A.J. Vanderjagt, 211–224. Leiden: Brill.

Grant, Edward. 1994. *Planets, stars, and orbs: The medieval cosmos, 1200–1687*. Cambridge: Cambridge University Press.

Hamel, Jürgen. 2014. *Studien zur Sphaera des Johannes de Sacrobosco*. Leipzig: AVA.

Hoftijzer, Paul. 1993. Nederlandse boekverkopersprivileges in de 17e en 18e eeuw. *Jaarboek Nederlands Genootschap Van Bibliofielen* 1: 49–62.

Imhof, Dirk. 2014. *Jan Moretus and the continuation of the Plantin press: A bibliography of the works published and printed by Jan Moretus I in Antwerp (1589–1610)*. Vol. 2. 2 vols. Leiden: Brill.

Israel, Jonathan Irvine. 1995. *The Dutch Republic. Its rise, greatness, and fall, 1477–1806*. Oxford: Clarendon.

Kaiser, Christopher B. 1988. Calvin's understanding of Aristotelian natural philosophy: Its extent and possible origin. In *Calviniana: Ideas and influence of Jean Calvin*, Sixteenth century essays & studies Vol. 10, ed. Robert V. Schnucker, 77–93. Kirksville: Sixteenth Century Journal Publishers.

Krop, Henri. 2014. Burgersdijk, Franco. In *Encyclopedia of renaissance philosophy*, ed. Marco Sgarbi, 1–6. Cham: Springer. https://doi.org/10.1007/978-3-319-02848-4_271-1.

Kuiper, E.J. 1958. *De Hollandse 'Schoolordre' van 1625: een studie over het onderwijs op de Latijnse scholen in Nederland in de 17de en 18de eeuw*. Groningen: Wolters.

Kusukawa, Sachiko. 1995. *The transformation of natural philosophy: The case of Philip Melanchthon*. Cambridge: Cambridge University Press.

Lattis, James M. 1994. *Between Copernicus and Galileo: Christoph Clavius and the collapse of Ptolemaic cosmology*. Chicago: University of Chicago Press.

Metius, Adriaan Adriaansz. 1614. *Institutiones Astronomicae et Geographicae. Fondamentale ende grondelijcke onderwysinge van de Sterrekonst, ende beschryvinghe der Aerden, door het ghebruyck van de Hemelsche ende Aerdtsche Globen*. Amsterdam: Willem Jansz. Blaeu.

Mijers, Esther. 2017. *Addicted to Puritanism*: Philosophical and theological relations between Scotland and the United Provinces in the first half of the seventeenth century. *History of Universities* 29 (2): 69–96.

Molhuysen, Philip Christiaan, (ed.). 1913–1924. *Bronnen tot de geschiedenis der Leidsche Universiteit. 1574–[1811]*. Vol. 1. 7 vols. Rijks Geschiedkundige Publicatiën 20. The Hague: M. Nijhoff.

———, (ed.). 1918. *Bronnen tot de geschiedenis der Leidsche Universiteit. 1574–[1811]*. Vol. 3. 7 vols. Rijks Geschiedkundige Publicatiën 20. The Hague: M. Nijhoff.

Molhuysen, Philip Christiaan, Petrus Johannes Blok, Laurentius Knappert, and Friedrich Karl Heinrich Kossmann. 1927. *Nieuw Nederlandsch biografisch woordenboek*. Vol. 7, 10. Leiden: Sijthoff.

Omodeo, Pietro D. 2017. *Utilitas astronomiae* in the renaissance: The rhetoric and epistemology of astronomy. In *The structures of practical knowledge*, ed. Matteo Valleriani, 307–331. Dordrecht: Springer. https://doi.org/10.1007/978-3-319-45671-3_12.

Orenstein, Nadine. 2006. Sleeping caps, city views and state funerals: Privileges for prints in the Dutch Republic 1580–1650. In *In his milieu: Essays on netherlandish art in memory of John Michael Montias*, ed. Amy Golahny, Mia M. Mochizuki, and Lisa Vergara, 313–346. Amsterdam: Amsterdam University Press.

Otterspeer, Willem. 2000. *Het bolwerk van de vrijheid: De Leidse Universiteit, 1575–1672*. Amsterdam: Bakker.

Ovid. 1989. *Fasti*. Cambridge, MA: Harvard University Press.

Pedersen, Olaf. 1985. In quest of Sacrobosco. *Journal for the History of Astronomy* 16 (3): 175–221. https://doi.org/10.1177/002182868501600302.

Prak, Maarten Roy. 2005. *The Dutch Republic in the seventeenth century: The golden age*. Trans. Diane L. Webb. Cambridge: Cambridge University Press

Prost, Joseph. 1907. *La philosophie à l'Académie protestante de Saumur (1606–1685)*. Paris: H. Paulin. http://gallica.bnf.fr/ark:/12148/bpt6k55160737. Accessed June 2019.

Romano, Antonella. 1999. *La contre-réforme mathématique. Constitution et diffusion d'un culture mathématique jésuite à la Renaissance (1540-1640)*. Rome: Bibliothèque des Ecoles françaises d'Athènes et Rome.

Rosen, Edward. 1974. Rheticus as editor of the Sacrobosco. In *For Dirk Struik: Scientific, historical and political essays in honour of Dirk J. Struik*, ed. Robert S. Cohen, J.J. Stachel, and Marx W. Wartofsky, 245–248. Berlin: Springer.

Ruestow, Edward Grant. 1973. *Physics at seventeenth and eighteenth-century Leiden: Philosophy and the new science in the university*. The Hague: Nijhoff.

Sacrobosco, Joannes, Élie Vinet, and Francesco Giuntini. 1573. *Sphaera Ioannis de Sacro Bosco emendata. Cum additionibus in margine, et Indice rerum et locorum memorabilium, et familiarissimis scholijs, nunc recenter compertis, et collectis à Francesco Iunctino Florentino sacrae Theologia Doctore. Interserta etiam sunt Eliae Vineti Santonis egregia Scholia in eandem Sphaeram. Adiunximus eiusdem auctoris Computum Ecclesiasticum. Et Petri Nonii Salaciensis demonstrationem eorum, quae in extremo capite de climatibus Sacroboscius scribit de inaequali climatum latitudine, eodem Vineto interprete*. Antwerp: Heirs of Arnold Birckmann. http://hdl.handle.net/21.11103/sphaera.101115.

Sassen, Ferd. 1959. *Geschiedenis van de wijsbegeerte in Nederland tot het einde der negentiende eeuw*. http://www.dbnl.org/tekst/sass011gesc01_01/. Accessed June 2019.

Schriks, C. F. J. 2015. *The history of copyright in the Netherlands in the 16th–19th century: The book as legal entity and provincial and national legislation; state's copyright in the Netherlands in the 17th–21st century; public access in the Trias Politica and the right of copy and copyright*. Zutphen: Walburg.

Simi, Nicolò. 1555. *Theoricae planetarum in compendium redactae, et pluribus figuris auctae*. Basel: Johannes Oporinus. https://doi.org/10.3931/e-rara-16422.

Smolenaars, Marja. 2004. Jack [Jacchaeus], Gilbert (Bap. 1577, d. 1628), philosopher and physician. In *Oxford dictionary of national biography*. http://www.oxforddnb.com/view/10.1093/ref:odnb/9780198614128.001.0001/odnb-9780198614128-e-14512. Accessed June 2019

States of Holland. 1589. *Placcaet inhoudende Verbot van te Drucken, ofte te stroyen, eenighe seditieuse boucken*. Delft: Aelbrecht Hendricksz.

Stimson, Dorothy. 1917. *The gradual acceptance of the Copernican theory of the universe*. New Hampshire: Hanover. http://archive.org/details/gradualacceptanc00stimrich. Accessed June 2019.

Thorndike, Lynn. 1949. *The Sphere of Sacrobosco and its commentators*. Vol. 2. Chicago: University of Chicago Press.

Valerius, Cornelius. 1583. *Grammaticarvm Institvtionvm liber IIII*. Antverp: Plantin.

Valleriani, Matteo. 2017. The tracts on the *Sphere*: Knowledge restructured over a network. In *The structures of practical knowledge*, ed. Matteo Valleriani, 421–473. Dordrecht: Springer. https://doi.org/10.1007/978-3-319-45671-3_16.

van Meurs, Johannes. 1625. *Athenae Batavae. Sive, de vrbe Leidensi et academià, virisque claris; qui utramque ingenio suo, atque scriptis illustrarunt: libri dvo*. Leiden: Andreas Cloucquius and Elsevirs.

Van Nouhuys, Tabitta. 1998. *The ages of two-faced Janus: The comets of 1577 and 1618 and the decline of the Aristotelian world view in the Netherlands*. Leiden: Brill.

van Wijckel, Hans. 1635. *Album Amicorum*. Franeker/Groningen/Leiden: Koninklijke Bibliotheek: 79 J 52.

Vermij, Rienk. 2002. *The Calvinist Copernicans: The reception of the new astronomy in the Dutch Republic, 1575–1750*. Amsterdam: KNAW.

———. 2016. Seventeenth-century Dutch natural philosophers on celestial influence. In *Unifying heaven and earth essays in the history of early modern cosmology*, ed. Miguel á Granada, Dario Tessicini, and Patrick J. Boner, 291–316. Universitat de Barcelona: Barcelona.

Weekhout, Ingrid Marlies. 1998. *Boekencensuur in de Noordelijke Nederlanden: De vrijheid van drukpers in de zeventiende eeuw*. Den Haag: SDU.

Westman, Robert S. 1975. The Melanchthon circle, Rheticus, and the Wittenberg interpretation of the Copernican theory. *Isis* 66 (2): 165–193.

Wiesenfeldt, Gerhard. 2002. *Leerer Raum in Minervas Haus: Experimentelle Naturlehre an der Universität Leiden, 1675–1715*. Amsterdam/ Berlijn: Edita KNAW/ Verlag für Geschichte der Naturwissenschaften und der Technik.

Willems, Alphonse. 1880. *Les Elzevier: histoire et annales typographiques*. Bruxelles: G.-A. van Trigt.

Witsenburg, K., and H.M. Van der Noordaa, eds. 1988. *Catalogus van 125 Nederlandse Latijnse scholen*. Almelo: Stichting Stadsherstel.

Wittichius, Christophorus. 1653. *Dissertationes duae*. Amsterdam: Abraham Elzevir.

Permissions

All chapters in this book were first published by Springer; hereby published with permission under the Creative Commons Attribution License or equivalent. Every chapter published in this book has been scrutinized by our experts. Their significance has been extensively debated. The topics covered herein carry significant findings which will fuel the growth of the discipline. They may even be implemented as practical applications or may be referred to as a beginning point for another development.

The contributors of this book come from diverse backgrounds, making this book a truly international effort. This book will bring forth new frontiers with its revolutionizing research information and detailed analysis of the nascent developments around the world.

We would like to thank all the contributing authors for lending their expertise to make the book truly unique. They have played a crucial role in the development of this book. Without their invaluable contributions this book wouldn't have been possible. They have made vital efforts to compile up to date information on the varied aspects of this subject to make this book a valuable addition to the collection of many professionals and students.

This book was conceptualized with the vision of imparting up-to-date information and advanced data in this field. To ensure the same, a matchless editorial board was set up. Every individual on the board went through rigorous rounds of assessment to prove their worth. After which they invested a large part of their time researching and compiling the most relevant data for our readers.

The editorial board has been involved in producing this book since its inception. They have spent rigorous hours researching and exploring the diverse topics which have resulted in the successful publishing of this book. They have passed on their knowledge of decades through this book. To expedite this challenging task, the publisher supported the team at every step. A small team of assistant editors was also appointed to further simplify the editing procedure and attain best results for the readers.

Apart from the editorial board, the designing team has also invested a significant amount of their time in understanding the subject and creating the most relevant covers. They scrutinized every image to scout for the most suitable representation of the subject and create an appropriate cover for the book.

The publishing team has been an ardent support to the editorial, designing and production team. Their endless efforts to recruit the best for this project, has resulted in the accomplishment of this book. They are a veteran in the field of academics and their pool of knowledge is as vast as their experience in printing. Their expertise and guidance has proved useful at every step. Their uncompromising quality standards have made this book an exceptional effort. Their encouragement from time to time has been an inspiration for everyone.

The publisher and the editorial board hope that this book will prove to be a valuable piece of knowledge for researchers, students, practitioners and scholars across the globe.

List of Contributors

Tayra M. C. Lanuza Navarro
El Instituto de Historia de la Medicina y de la Ciencia "López Piñero," Universitat de València

Richard J. Oosterhoff
University of Edinburgh

Kathleen M. Crowther
Department of the History of Science, Medicine, and Technology at Johns Hopkins University

Matteo Valleriani
Department I, Max Planck Institute for the History of Science, Berlin
Technische Universität of Berlin
University of Tel Aviv

Isabelle Pantin
École Normale Supérieure in Paris (PSL University)

Roberto de Andrade Martins
Brazilian National Council for Scientific and Technological Development (CNPq)
São Paulo Federal University (UNIFESP)
Municipal Higher Education Bragança Paulista Foundation (FESB)
University of São Paulo (GHTC-USP)

Marius Buning
European University Institute
Dahlem Humanities Center of the Freie Universität Berlin

Index

www.ingramcontent.com/pod-product-compliance
Lightning Source LLC
LaVergne TN
LVHW080630210926
848485LV00003B/240